Advances in Plant Defense Mechanisms

Edited by Josphert Ngui Kimatu

Published in London, United Kingdom

Advances in Plant Defense Mechanisms
http://dx.doi.org/10.5772/intechopen.98128
Edited by Josphert Ngui Kimatu

Contributors
Khursheed Hussain, Sameena Lone, Sumati Narayan, Faheema Mushtaq, Ajaz Malik, Majid Rashid, Gazala Nazir, Anjali Tiwari, Kapil Kesarwani, Arushi Sharma, Tapan Ghosh, Nisha Bisht, Shailja Punetha, Shahzad Ali, Muhammad Usman, Asmat Ali, Waqar Ali, Majid Khan, Muhammad Aasim, Nadir Zaman, Ananya Baidya, Kousik Atta, Mohammed Anwar Ali, Partha S. Basu, Sushil Kumar Chaturvedi, Pooran Mall Gaur, Biswajit Mondal, Surendra Kumar Meena, Krishnashis Das, Vaibhav Kumar, Kalpana Tewari, Kusum Sharma, Ananya Prova, Md. Saeed Sultan, Aydın Unay, Volkan Mehmet Cinar, Serife Balci, Najimu Adetoro, Sikirou Mouritala, Talaat El Sebai, Maha Abdallah, Ayan Raichaudhuri, Shubham Dey, Maduraimuthu Djanaguiraman, Theivasigamani Parthasarathi, Saiyyeda Firdous, Einstein Mariya David, Kuppan Lesharadevi, Golden Gokhale, Guru Dutt Sharma, Maryam Dahajipour Heidarabadi, Fatima Batool, Batcho Anicet Agossa, Zainab Y. Sandhu, Muhammad Bilal Sarwar, Sameera Hassan, Bushra Rashid, Flávio José Rodrigues Cruz, Raphael Leone da Cruz Ferreira, Susana Silva Conceição, Edson Ugulino Lima, Cândido Ferreira de Oliveira Neto, Jessivaldo Rodrigues Galvão, Sebastião da Cunha Lopes, Ismael de Jesus Matos Viegas, Olena Nedukha, Asmat Ara, Mahroofa Jan, P.A Sofi, Munezeh Rashid, M.A. Rather, Musharib Gull, Ajaz Ahmad Lone

First published in London, United Kingdom, 2022 by IntechOpen
IntechOpen is the global imprint of INTECHOPEN LIMITED, registered in England and Wales, registration number: 11086078, 5 Princes Gate Court, London, SW7 2QJ, United Kingdom

British Library Cataloguing-in-Publication Data
A catalogue record for this book is available from the British Library

Additional hard and PDF copies can be obtained from orders@intechopen.com

Advances in Plant Defense Mechanisms
Edited by Josphert Ngui Kimatu
p. cm.
Print ISBN 978-1-80355-801-1
Online ISBN 978-1-80355-802-8
eBook (PDF) ISBN 978-1-80355-803-5

For EU product safety concerns:
IN TECH d.o.o., Prolaz Marije Krucifikse Kozulić 3, 51000 Rijeka, Croatia,
info@intechopen.com or visit our website at intechopen.com.

Meet the editor

Prof. Josphert Ngui Kimatu is an associate professor in the School of Sciences and Computing, Department of Life Science, South Eastern Kenya University (SEKU). He has a Ph.D. in Plant Molecular Epigenetics from Northeast Normal University (NENU), China; an MPhil in Forestry Pathology from Moi University, Kenya; a bachelor's degree in Botany and Zoology from Moi University, Kenya; and a PGDE from Maseno University, Kenya. He trained in post-harvest management in maize, legumes, and rice in a four-module certificate course at Stellenbosch University, South Africa; Sydney University, Australia; and Kwame Nkrumah University of Science and Technology, Ghana. He has a certificate in Molecular Diagnostics from the University of Nairobi, Kenya. Dr. Kimatu received a training certificate in integrated pest management and pesticide safety (IPM) from the Soybean Innovation Lab, and a training certificate in COVID-19 from the London School of Hygiene and Tropical Medicine.

Contents

Preface

Studies involving *Arabidopsis thaliana* and other crop plants have continued to advance our knowledge of the mechanisms used by plants to survive in dynamic biotic and abiotic environments. Plants have adapted to diverse ecologies for ages; however, recent human migrations, agricultural activities, and climate change-induced phenomenon have forced plants to increasingly adapt to new environments. Furthermore, agricultural production is becoming increasingly vulnerable to climate variability and change characterized by altered frequency, timing, and magnitude of precipitation, temperature, and wind.

This book discusses morphological, anatomical, physiological, molecular, and genomic advances in plant defenses. It examines advances involving plant stress signaling mechanisms and molecular defense responses involving recently discovered biomolecules. It also discusses advances in gene expression changes, which can be inherited as described in epigenetic mechanisms and linked to observed phenotypic plant plasticity. The book also looks at emerging areas like plant growth-promoting microorganisms (PGPMs), which consist of a vast group of microorganisms found in the rhizosphere, on the root surface, or associated with it. It examines the mechanisms of action of PGPMs, including their ability to produce different plant growth regulators or phytohormones. Studies indicate that researchers have found intriguing plant interactions with their physical, microbial, and biotic environments. These interactions have been highlighted and linked to novel mechanisms, allowing us to understand deeper ways by which sessile plants have adapted and thriven in challenging environments.

The study examined issues on hailstorms and reports that to control secondary damage due to insect pests and diseases, the removal of fallen fruit to reduce the spread of disease and pests during their decay, large wounds on trunks and branches should be covered with a water-based paint to avoid desiccation and disease infection, then fruit thinning by removal of hail-damaged fruits are shown to improve the yield and quality of the remaining fruit. However, bud-breaking chemicals and growth/bio regulators may be applied to induce vegetative growth in the orchard crop along with fertilizers. In general, the major and chronic threats to ecosystem stability and sustainable agriculture are phytopathogenic microorganisms, the unlimited and regular use of chemical pesticides and fungicides could destroy soil ecology, degrade the fertility of the soil, and disrupt environment resulting in show harmful effects on human health and contaminating groundwater.

However, our studies found that PGPMs are environmentally friendly and offer a promising sustainable way of obtaining sustainable soil fertility and plant growth, although indirectly. Nevertheless, antibiotics production is considered one of the most powerful and studied biocontrol mechanisms of PGPMs against phytopathogens.

Josphert Ngui Kimatu, BSc., Mphil., PGDE, Ph.D.
Professor,
South Eastern Kenya University,
School of Sciences and Computing,
Department of Life Sciences,
Kitui County, Kenya

Section 1

Mechanisms of Plant Responses to Abiotic Stress

Chapter 1

Impact of Abiotic Stress on Phytoplankton and Zooplankton with Special Reference to Food Web

Golden Gokhale and Guru Dutt Sharma

Abstract

In aquatic ecosystems, bacterial colonies constitute an important aspect of biological diversity and biogeochemical cycling. Phytoplankton is the primary producer of the food web and zooplanktons are an important part of freshwater food webs and biogeochemical cycles, as they serve as the main trophic connection between primary producers (phytoplankton) and fish. This chapter conducts abiotic stress effects on phytoplankton and zooplankton along with the impact of abiotic stress on their energy succession. Abiotic stress shows the decreasing supply of essential vitamins due to abiotic stress can have huge consequences for the aquatic food web. Abiotic factors had a significant impact on the biomass of phytoplankton and zooplankton communities exposed including increased temperature, acidification, nutrient enrichment and increasing ultraviolet (UV) environment of the aquatic ecosystem that significantly affect their survival, behaviour, nutritional procurement, reproduction and their overall population dynamic. Oxygen stress also is a widespread occurrence in freshwater environments, with the depletion of DO in the water layers under the epilimnion becoming increasingly common. At moderately high salinities, a decreased top-down control by zooplankton on phytoplankton may be an indirect result, leading to a worsening of eutrophication symptoms.

Keywords: abiotic stress, phytoplankton, zooplankton, antioxidants, UV radiation

1. Introduction

In aquatic ecosystems, planktons form the base of the food web. Planktons are microscopic organisms that flow with streamlined water. The term "Plankton" is used to describe all non-motile, water-current-resistant organisms that are present in both freshwater and marine environments. Water current carries them along. Its range in size is from 0.2 mm to more than 20 cm ranging from tiny microorganisms to enormous creatures like jellyfish. Their distribution changes depending on the amount of light and nutrients available. Planktons are a source of food for large aquatic species. It shows the dynamic nature of the aquatic ecosystem due to which they act as the most dominating group among all water bodies. Planktons survival depends upon the structure of the aquatic ecosystem and the availability of nutrients. They adapt themselves into strategic adaptations. Planktons are acts as an important biological indicator of water quality and also determine the trophic status of an aquatic ecosystem. They are categorised into Phytoplankton and Zooplankton.

Phytoplankton also known as microalgae are chlorophyll-containing organisms that require sunlight to grow and survive. The majority of phytoplankton remains buoyant and floats in the upper layers of the ocean where sunlight can reach them. Additionally, phytoplankton needs inorganic elements like nitrates, phosphates and Sulphur which are essential for making proteins, lipids and carbohydrates. The phytoplankton plays important role in the production of oxygen by the process of photosynthesis in the presence of light, e.g. during the daytime, the water column forms the photic layer. Firstly, oxygen enters in water and then evaporates into the air from the water surface thus; it is contributing oxygen to the atmosphere. A decrease in the oxygen productivity by phytoplankton can disturb the balance between earth and life which further cause extinction in organisms. The phytoplankton forms the base of the food chain in an aquatic ecosystem. They also convert solar energy into chemical energy thus, they act as energy transducers. This converted energy is transferred by zooplankton to the higher trophic levels while providing the link between producers and consumers. The growth and morphological properties of phytoplankton are examined by the assessments of biological and physico-chemical parameters of an aquatic ecosystem.

Zooplankton is microscopic organisms that are present in the water column of almost all water bodies, such as lakes, ponds, and seas. However, they are largely unable to survive in rivers and streams. They may contain the larval stages of larger organisms like mussels and fish and range in size from a few millimetres to a few microns (1 μm is equal to 1/1000 of a mm).

Zooplanktons are microscopic organisms that are present in the water column of almost all water bodies, such as lakes, ponds and seas. However, they are unable to survive in rivers and streams. They may contain the larval stages of larger organisms like mussels and fish which range in size from a few millimetres to a few microns (1 μm = 1/1000 of mm). They also act as a food source for invertebrates and fishes. They play important role in the transmission of energy to higher trophic levels. The seasonal variations in an aquatic ecosystem can cause fluctuation in environmental characteristics to result from abiotic stress. This abiotic stress causes a greater impact on patterns of energy succession.

In an aquatic ecosystem, extrinsic and intrinsic interaction is observed. The intrinsic interaction is defined as competition, parasitism, predation and mutualism and the extrinsic is known as an interaction between aquatic organisms and their environment. They both cause an impact on the dynamic pattern of taxa because seasonal variations could vary the communities. The energy flow of the biotic factors ecosystem starts with photosynthetic organisms which use solar energy to convert inorganic compounds to organic compounds. The diatoms and dinoflagellates are more dominant species in the marine ecosystem, which are primary producers in the classical food web. The primary predators are Zooplankton for phytoplankton which is get affected by variations in environmental factors [1–7].

In this chapter, the impact of abiotic stress on phytoplankton and zooplankton energy succession has been discussed. Analysis, the regulation of zooplankton can be examined by phytoplankton composition [8]. In Mesotrophic Lake, Predation can cause the breakdown of blooms of bacteria-plankton and also varies the phytoplankton structure [9]. The interaction between phytoplankton and zooplankton is depending upon the trophic status of an aquatic ecosystem [10, 11] (**Figure 1**).

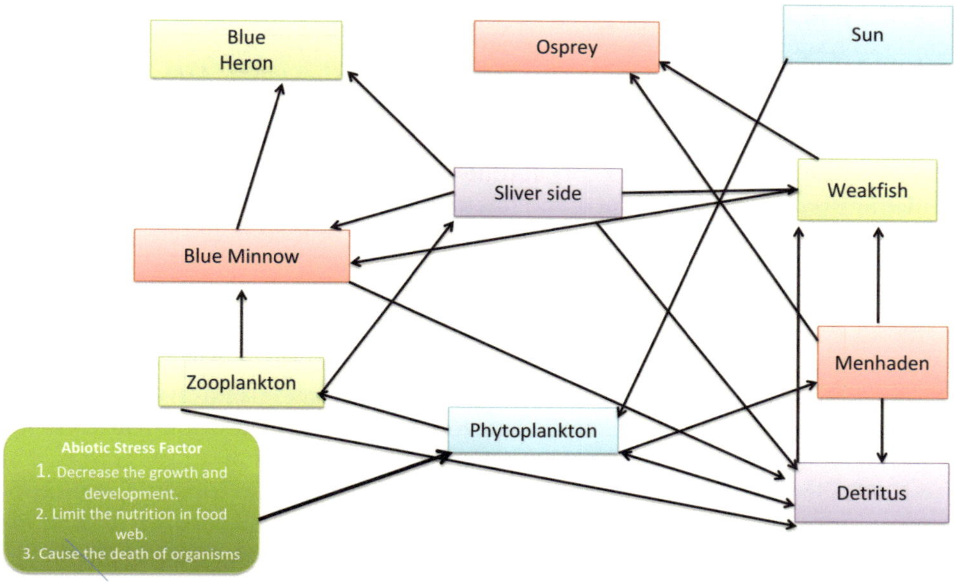

Figure 1.
Abiotic stress impact on food web.

2. Abiotic stress impacts α-tocopherol and β-tocopherol antioxidants

In aquatic ecosystems, the phytoplankton is a major source of non-enzymatic antioxidants and their precursors for primary consumers and other organisms higher up in the food web [12–14]. Antioxidants can be categorised into two categories: (1) enzymatic antioxidants such as superoxide and (2) non-enzymatic antioxidants such as glutathione, α-tocopherol (vitamin- E) and β- carotene. In an aquatic ecosystem, the phytoplankton the antioxidants α-tocopherol (vitamin E) and β-carotene in phytoplankton depend on changes in abiotic factors.

The α-tocopherol plays important role in the protection and prevention of membrane lipids whereas β-carotene act as antennae of complex photosystem II. The maximum variations in antioxidants in the aquatic ecosystem are because of temperature, density and salinity fluctuations. The antioxidant value in phytoplankton directly depends upon the composition of species, physiological characteristics and their strategies to deal with oxygen species. The α-tocopherol and β-tocopherol are non-enzymatic antioxidants which are produced by photosynthetic organisms at higher trophic levels in the form of dietary intake due to system shifts toward the phytoplankton can cause environmental variations which cause a larger impact on aquatic food webs. The composition of species decreased due to unbalancing of tocopherols.

3. Impact on vitamin B1 (thiamine)

Vitamins play important roles for plants and animals. Thiamine (vitamin B1) compound is soluble in water. The structure of pyrophosphate is consisting of pyrophosphate ester and thiamine diphosphate (TDP) which act as a cofactor in many reactions of metabolic. It also plays role in acetyl-coenzyme A and it also has a role in the Kerb cycle. Thiamine monophosphate (TMP) is intermediate in thiamine metabolism. The ethology of thiamine deficiencies is not well known but some researchers illustrate that transmission of thiamine gets decreases from phytoplankton to other higher trophic levels in an aquatic ecosystem. This result that variation in thiamine causing an impact on the composition of species with lower thiamine content and also decreasing the thiamine cellular concentration. In a marine food web, the synthesis of thiamine is done by prokaryotes and phytoplankton which are further transmitted to higher trophic levels through energy succession. But only some species of phytoplankton can secrete thiamine but not all species synthesise thiamine. It has been observed that both thiamine autotrophs and thiamine producers can uptake external thiamine.

The salinity in the water caused a larger impact which was further followed by temperature and photon flux density were examined in phytoplankton species. This effect can cause a decrease in thiamine concentration availability for the aquatic food chain and food webs. The decrease in thiamine concentration shows a larger impact on the aquatic food webs. It is an unclear statement that vitamin variations are more efficient to cause an impact on grazers in climatic conditions. The mechanism for example selective grazing in zooplanktons also causes an impact on their nutritional habits. Vitamin B1 in diatoms increases at high temperature whereas it lowers at salinity.

The decreasing fatty acid level shows an adverse impact on growth and also affects the transmission of carbon in the food web [7, 15, 16]. The zooplankton fatty acid depends upon the fatty acid of phytoplankton and their composition but zooplankton causes greater inter-specific differences [15, 17]. The thiamin content of copepod is closely related to the thiamin concentration of micro, nano and picoplankton while there is some difference in thiamine content of copepod species. Hence, aquatic food web processes depend upon the concentration of thiamin at the higher trophic level.

4. Eutrophication

The assessment of physico-chemical parameters is necessary for examining the water quality and energy succession of an aquatic ecosystem. The abundance and diversity of phytoplankton and zooplankton depend upon the physico-chemical parameter of an aquatic ecosystem [18]. Eutrophication is defined as the process by which all nutrients are accumulated in all aquatic ecosystems. Eutrophication is a major problem for the aquatic food web. This is caused by the increase in the concentration of phosphorus and nitrate. The formation of algal blooms limits the intake of oxygen for phytoplankton and another organism which can cause the death of an organism. The water of algal bloom smells foul and it also causes the death of fish. The negative effects of eutrophication on water bodies include a decrease in biodiversity can also toxicity in water bodies and a change in species abundance (**Figure 2**).

4.1 Classification of eutrophication

4.1.1 Anthropogenic eutrophication

Anthropogenic eutrophication is occurring through human activity such as golf courses, lawns etc. in the form of fertilisers which get a transfer into water bodies.

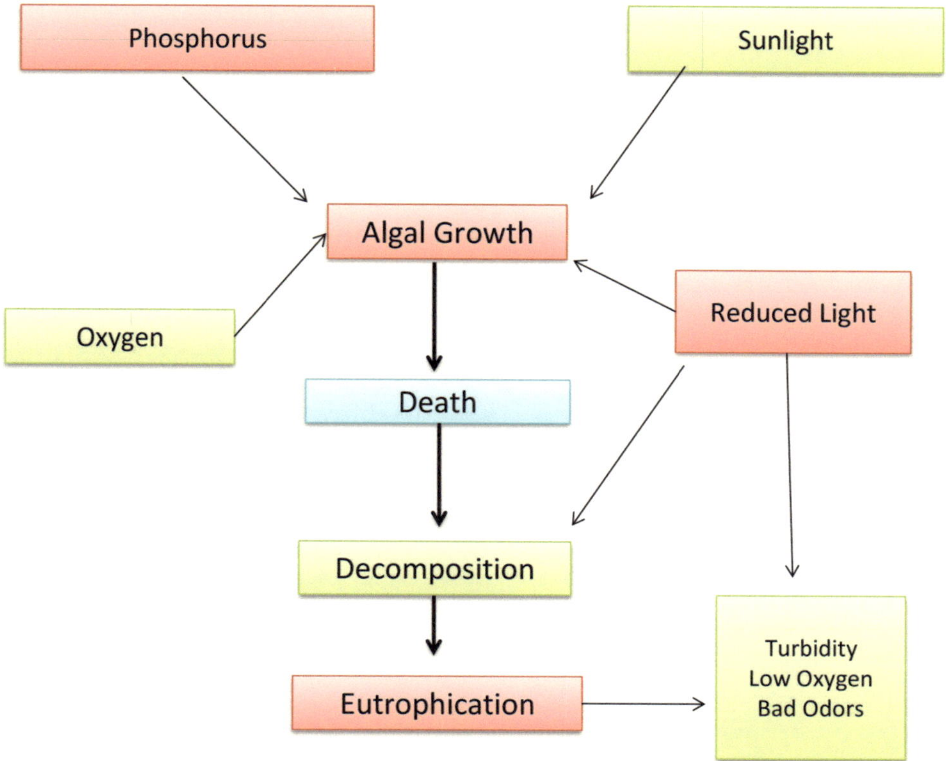

Figure 2.
Eutrophication process.

When it comes in contact with algae and plankton it causes eutrophication. The variation of nutrient which enters water bodies by human intervention increase eutrophication rapidly. This eutrophication process does not require decades. Cultural eutrophication begins with phosphorus which is present in fertilisers along with partially sewage treatment. Phosphorus is considered the strongest indicator of the growth of algae. Deforestation and soil erosion are the major cause of eutrophication.

4.1.2 Natural eutrophication

Natural eutrophication is defined as the enrichment of water bodies with help of natural calamities. This process takes a slow time as compared to anthropogenic eutrophication. It depends upon the environmental temperature. Natural eutrophication duration is up to 100 years because in this process organic matter takes a long time for the deposition. Natural eutrophication occurs in natural conditions like landslides and floods with environmental characteristics such as temperature, carbon dioxide and light also play important role in natural eutrophication. The natural eutrophication primary stage starts in oligotrophic water in which accumulation of nutrient take place. The accumulation of nutrients and their utilisation get continues which further causes eutrophication. The duration of eutrophication depends upon the variation of water quality (**Figure 3**).

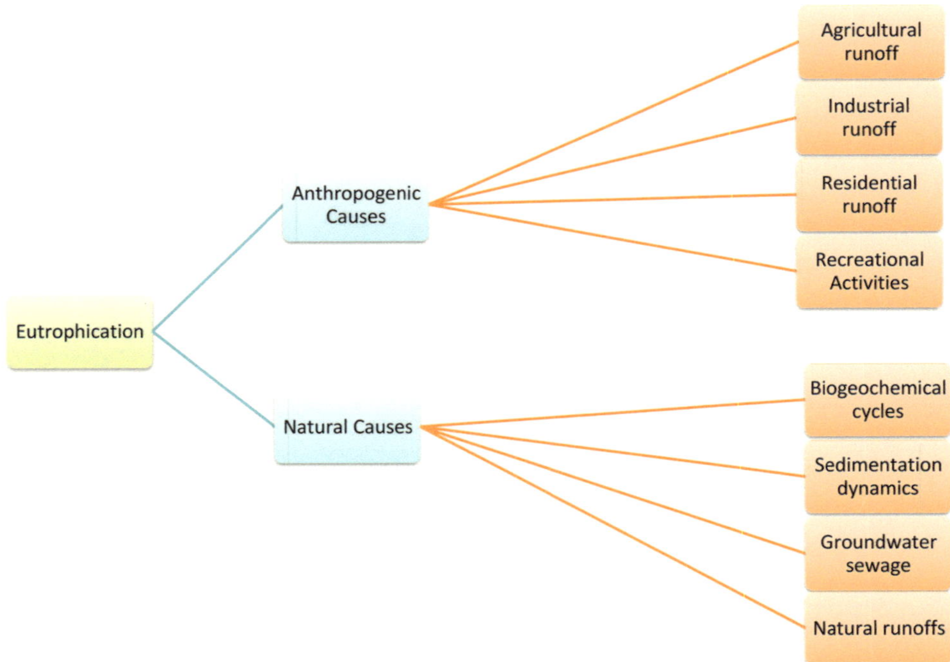

Figure 3.
Types of eutrophication and their causes.

4.1.3 Effects of eutrophication

During eutrophication, phytoplankton grows faster. These phytoplankton species are toxic. In eutrophic water, the gelatinous zooplankton bloom grows faster in this water. Biomass of algae increases in eutrophic water. The loss of transparency in water and water smell and colour. The shellfish and harvestable fish population get decreased. The concentration of dissolved oxygen decreases which causes the death of fish.

5. Ocean acidification

Ocean acidification is a steady decrease of the pH concentrations as a result of the release of carbon dioxide (CO_2) from the atmosphere. Human use of fossil fuels is the barrier to ocean acidification. The CO_2 concentration is obtained from the atmosphere which depends upon the interactions between biological activities and physico chemical parameters such as temperature, salinity, intensity and surface chemistry. The decrease in the pH concentration because of anthropogenic CO_2 concentration causes ocean acidification [19, 20]. Ocean acidification is controversial that how it causes an impact on the carbon fixation by photosynthetic organism. The process of calcification in phytoplankton gets affected by the influence of ocean acidification. Ocean acidification caused due to global warming as well as it causes environmental factors on productivity and composition. Ocean acidification serious problem which affects the growth and development along with the nutritional quality of primary producers which further affects the higher trophic levels. The phytoplankton biochemical composition gets disturbed during acidification causing an impact on the taxonomic group. The sensitivity to pCO2 causes low food quality for higher trophic levels. Zooplankton is less affected by acidification but the acclimation process leads to an increase in the rate of respiration and also an increase in the rate of grazing. Because their shell is mainly composed of chitin which prevents them from acidification. At higher trophic levels organisms are more affected because the outer skeleton is made up of calcium carbonate.

6. UV radiation

During summer, growth and development decrease growth of phytoplankton when UV radiation are high this shows that phytoplankton species show complexity toward UV radiation and temperature. Moreover, they are dependent on UV radiation and temperature [21–23]. UV radiation also affects the growth and development of zooplankton and also changes their pattern of species composition. The cell size variation due to UV radiation causes an impact on zooplankton feeding habits. The phytoplankton gets more affected by high UV radiation compare to zooplankton. UV Radiation causes a moulting process for a few zooplankton present in the water bodies. The process of Moulting involves chitinolytic enzyme and apoptosis process done by caspase-3-activity. It has been observed that UV radiation cause an adverse impact on the moulting process as a result it decreases the growth and development of the population dynamic of plankton. The UV-B radiation can strongly cause impairment of photosynthesis whereas inhibition of calcification is done by UV-A radiation. The outer calcified scales form an exoskeleton. At maximum calcium concentration, the zooplankton more resists UV radiation rather than at limited calcium concentration (**Figure 4**).

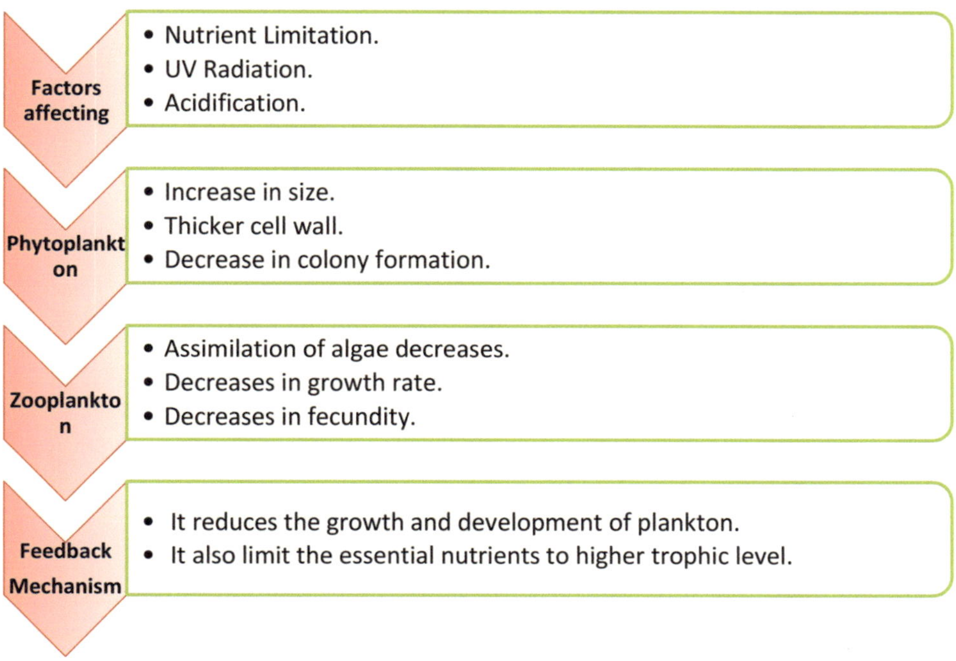

Figure 4.
Factors affecting the phytoplankton and zooplankton.

7. Conclusion

This chapter suggests a well understanding of phytoplankton and zooplankton along with their role in the ecological succession and food web. The seasonal changes show an impact on the physicochemical parameters which changes the patterns of the composition of species. Mostly they depend upon the Temperature. These factors affect the growth and development of higher trophic levels in the food web The abiotic stress such as antioxidants, vitamins, eutrophication, acidification and UV radiation show an impact on phytoplankton and zooplankton. Antioxidants such α-tocopherol and β-tocopherol secretion in the phytoplankton depend upon the seasonal variations. The unbalancing in the α-tocopherol and β-tocopherol also disturbed the composition of species. The thiamine deficiency decreases the growth and development at a higher trophic level. The ethology of thiamine deficiencies is not well known but some researchers illustrate that transmission of thiamine gets decreases from phytoplankton to other higher trophic levels in an aquatic ecosystem. This result that variation in thiamine and also causes an impact on the composition of species with lower thiamine content and decreases the thiamine cellular concentration. Eutrophication is a major problem for the aquatic food web. This is caused by the increase in the concentration of phosphorus and nitrate the negative effects of eutrophication on water bodies include a decrease in biodiversity can also toxicity in water bodies and a change in species abundance. However, acidification affects phytoplankton biochemical composition which gets disturbed during acidification causing an impact on the taxonomic group and UV radiation shows an adverse impact on the phytoplankton which further limits the all type of nutrition for higher trophic levels. These abiotic stress show harmful effects on the food web and the composition of species.

Author details

Golden Gokhale* and Guru Dutt Sharma
Department of Life Sciences, Devi Alhiya University, Indore, Madhya Pradesh, India

*Address all correspondence to: gokhalegolden1432@gmail.com

References

[1] Beaugrand G, Mackas D, Goberville E. Applying the concept of the ecological niche and a macroecological approach to understand how climate influences zooplankton: Advantages, assumptions, limitations and requirements. Progress in Oceanography. 2013;**111**:75-90

[2] Carter JL, Schindler DE. Responses of zooplankton populations to four decades of climate warming in lakes of southwestern Alaska. Ecosystems. 2012;**15**(6):1010-1026

[3] Eisner LB, Napp JM, Mier KL, Pinchuk AI, Andrews AG III. Climate-mediated changes in zooplankton community structure for the eastern Bering Sea. Deep Sea Research Part II: Topical Studies in Oceanography. 2014;**109**:157-171

[4] Guijun Y, Boqiang Q, Xiangming T, Zhijun G, Chunni Z, Hua Z, et al. Contrasting zooplankton communities of two bays of the large, shallow, eutrophic Lake Taihu, China: Their relationship to environmental factors. Journal of Great Lakes Research. 2012;**38**(2):299-308

[5] Mozetič P, Francé J, Kogovšek T, Talaber I, Malej A. Plankton trends and community changes in a coastal sea (northern Adriatic): Bottom-up vs. top-down control in relation to environmental drivers. Estuarine, Coastal and Shelf Science. 2012;**115**:138-148

[6] Shurin JB, Clasen JL, Greig HS, Kratina P, Thompson PL. Warming shifts top-down and bottom-up control of pond food web structure and function. Philosophical Transactions of the Royal Society, B: Biological Sciences. 2012;**367**(1605):3008-3017

[7] Tian W, Zhang H, Zhang J, Zhao L, Miao M, Huang H. Responses of zooplankton community to environmental factors and phytoplankton biomass in Lake Nansihu, China. Pakistan Journal of Zoology. 2017;**49**(2):493-504

[8] Vardi A, Schatz D, Beeri K, Motro U, Sukenik A, Levine A, et al. Dinoflagellate-cyanobacterium communication may determine the composition of phytoplankton assemblage in a mesotrophic lake. Current Biology. 2002;**12**(20):1767-1772

[9] Hennes KP, Simon M. Significance of bacteriophages for controlling bacterioplankton growth in a mesotrophic lake. Applied and Environmental Microbiology. 1995;**61**(1):333-340

[10] Carney HJ, Elser JJ. Strength of zooplankton-phytoplankton coupling in relation to lake trophic state. In: Large Lakes 1990. Berlin, Heidelberg: Springer; 1990. pp. 615-631

[11] Elser JJ, Goldman CR. Zooplankton effects on phytoplankton in lakes of contrasting trophic status. Limnology and Oceanography. 1991;**36**(1):64-90

[12] Häubner N, Sylvander P, Vuori K, Snoeijs P. Abiotic stress modifies the synthesis of alpha-tocopherol and beta-carotene in phytoplankton species. Journal of Phycology. 2014 Aug;**50**(4):753-759

[13] Matsuno T. Aquatic animal carotenoids. Fisheries Science. 2001;**67**(5):771-783

[14] Snoeijs P, Häubner N. Astaxanthin dynamics in Baltic Sea mesozooplankton communities. Journal of Sea Research. 2014;**85**:131-143

[15] Fridolfsson E, Bunse C, Legrand C, Lindehoff E, Majaneva S, Hylander S. Seasonal variation and species-specific concentrations of the essential vitamin B1 (thiamin) in zooplankton and seston. Marine Biology. 2019;**166**(6):1-3

[16] Müller-Navarra DC, Brett MT, Liston AM, Goldman CR. A highly unsaturated fatty acid predicts carbon transfer between primary producers and consumers. Nature. 2000;**403**(6765):74-77

[17] Brett M, Müller-Navarra D. The role of highly unsaturated fatty acids in aquatic foodweb processes. Freshwater Biology. 1997;**38**(3):483-499

[18] Sharma NK, Bhardwaj S. An assessment of seasonal variation in phytoplankton community of Mahi River (India). GEN. 2011;**10**(40):154-164

[19] Caldeira K, Wickett ME. Anthropogenic carbon and ocean pH. Nature. 2003;**425**(6956):365

[20] Gao K, Zhang Y, Häder DP. Individual and interactive effects of ocean acidification, global warming, and UV radiation on phytoplankton. Journal of Applied Phycology. 2018;**30**(2):743-759

[21] Rae R, Vincent WF. Effects of temperature and ultraviolet radiation on microbial foodweb structure: Potential responses to global change. Freshwater Biology. 1998;**40**(4):747-758

[22] Roos JC, Vincent WF. Temperature dependence of UV radiation effects on Antarctic cyanobacteria. Journal of Phycology. 1998;**34**(1):118-125

[23] Williamson CE, Salm C, Cooke S, Saros J. How do UV radiation, temperature, and zooplankton influence the dynamics of alpine phytoplankton communities? Hydrobiologia. 2010;**648**(1):73-81

Chapter 2

Effects of High Temperature on Crops

Theivasigamani Parthasarathi, Saiyyeda Firdous, Einstein Mariya David, Kuppan Lesharadevi and Maduraimuthu Djanaguiraman

Abstract

The effect of high-temperature situations leads to a significant reduction in yield. The elevated temperature on crops is expected to have a widespread negative effect as a consequence of global warming. Meanwhile, the global population is rapidly increasing and is predicted to be 11 billion in 2100. An increase in 70% of global food production is a challenging task to feed the increasing population. Increasing the food crop yield is crucial to meet the global food demand and ensuring food security. An increase in high temperature every year due to global warming and an increase in greenhouse gases leads to a rise in temperature. The rise in temperature significantly affects the yield; so, it is important to understand the mechanism and how to counteract high temperature on food crops. It is also important to neutralize the effect of high temperature on food crops and to increase the yield by minimizing the effect of high temperature and developing heat resistant or tolerant variety. It is essential to develop heat-tolerant crops or transgenic food crops that can assure great yield and food security for future generations. It is essential to examine the metabolic, physiological, and molecular mechanisms of food crops to have an enhanced understanding of high temperature and their effects on crops.

Keywords: heat stress, high temperature, photosynthesis, pollen and root

1. Introduction

Agricultural productivity is prone to change in temperatures. Knowledge of climate change, specifically high temperatures, is essential for agronomists, decision-makers, and crop producers to ensure food security across the globe [1]. Crops thrive at their optimum growth temperatures. However, elevation in temperature level predominantly influences plants' physiological processes, especially photosynthesis, transpiration, respiration, and yield. High temperature causes declining yields in major food crops, which is a major concern for depreciating agricultural productivity [2, 3]. Temperature is classified as minimum, optimum and maximum, and it has been predicted that the temperature will rise 2–5°C in the future climate in 2100 (IPCC, 2014) [4–6]. Agnolucci et al. [7] have demonstrated the significance and impact of climate change with different statistical patterns in 18 crops that contribute

70% of the land and 65% of calorific value. In conclusion, we emphasize the yield disparity in the primary crop is associated with high temperature [8].

The maximum threshold temperature for various crops differs. However, high temperatures above 35°C can cause damage to rice crops. Evident injuries were observed due to high temperatures in different developmental stages. Recent studies exposed that sorghum pistils and pearl millet both are similarly sensitive to high temperatures [9]. Moreover, Zhao et al. [1] have investigated and concluded that the rise in global temperature leads to global yield loss in four significant food crops. In addition, it also raises concern about the increasing temperature, and reduced yield should be neutralized with modern and sustainable modern agricultural techniques to fend off global hunger and to meet the prospective food requirement. Global warming due to greenhouse gas (GHG) emissions, is considered a significant threat to global agriculture productivity [10]. It is evaluated that without the use of CO_2 fertilization, efficient solution, genetic transformation, each 1°C rise in the global

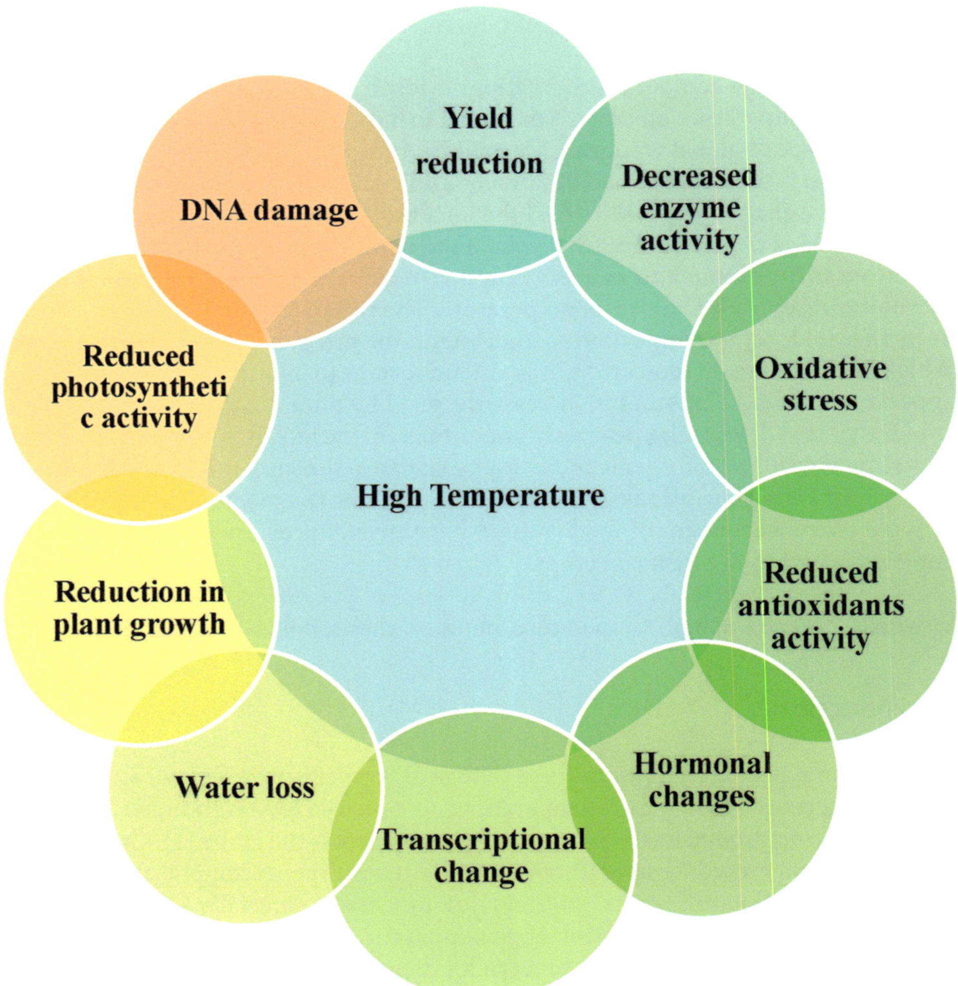

Figure 1.
High-temperature influences on crops.

mean temperature reduces global maize yield by 7.4%, wheat yield by 6.0%, rice yield by 6.2%, overall milled rice by 7.1–8.0%, head rice by 9.0–13.8% and overall milling profit by 8.1–11.0% and soybean yield by 3.1% [1, 11]. According to Food and Agriculture Organization (FAO) data, the relative rates of increase in yield for major cereal crops are reducing. However, crop productivity must be increased as the population is projected to reach 11 billion in 2100 (UN Population Division report) [12] and about a 70% increase in global agricultural productivity is essential; increase in 2°C or increase in the average temperature could lead to 20–40% reductions in cereal grain output, notably in Asia and Africa [13]. The increasing global warming provokes the weather pattern, leading to an increase in global temperature by 2.0–3.5°C in all regions as reported in the fifth assessment report (AR5) by the Intergovernmental Panel of Climatic Change (IPCC, 2018) it will reach 2.5–5.8°C before the 2100 s [6, 14]. High temperature during grain-filling has a significant effect on sunflower seeds and oil constituents [15]. In addition, it also reduces the linoleic acid content in numerous oilseed oils [16]. It also reduces the oil content, seed yield, and speeds up seed maturity as a consequence, erucic acid over seed development was influenced **Figure 1** [17].

2. Stage and intensity

High temperature limits the yield and affects various growth stages in plants. The reproductive stage is the most vulnerable phase of the crop's entire lifespan; this vulnerability during the reproductive stage leads to significant depletion in seed set and crop yield [9, 18]. Numerous food crops such as rice, wheat, soybean, maize, cotton, sorghum, and tomato are tremendously vulnerable to high temperatures [19]. Intense high temperature causes pollen abortion resulting in incomplete pollination. The interactions and stability between pollen and pistil during high temperatures lead to successful reproduction. But pollen and pistil both are extremely susceptible to high-temperature [20]. Crops have a variety of alternatives for resisting, minimizing, and surviving high temperatures during flowering. Plants can survive in high temperatures by keeping a cooler canopy through enhanced transpiration [9]. An elevated temperature could disturb numerous metabolic processes that take place in guard cells, as a result of high temperature; the stomatal response is frequently influenced by transpiration rate, photosynthetic rate, plant water status, and vapor pressure deficit [21]. High temperature considerably influences the crops by affecting several physiological injuries like leaf abscission, leaf scorching, senescence, and root and shoot growth limitation that subsequently leads to a reduction in yield. Moreover, the impact of high temperature affects photosynthetic membranes followed by ion leakage, enlargement of grana stacks, and aberrant stacking. By downregulating particular genes in carbohydrate metabolism, high temperature alters the activities of carbon metabolic enzymes, starch accumulation, and sucrose production. High temperature increases certain essential phytohormones such as abscisic acid, ethylene, and salicylic acid and reduces a few like gibberellic acid, cytokinin, and auxin; it furthermore leads to the fabrication of reactive oxygen species [4]. In addition, the high temperature throughout the day is properly not reported. However, the progressive rise in night temperature needs to be scrutinized considering that it causes prior to time of day of anthesis in cereal grains like rice crop.

3. Changes in mechanism under temperature stress

3.1 Effects of high-temperature stress

Plants endure various factors during growth and development, and high-temperature stress is one of the major abiotic factors that adversely affect crop production. High-temperature stress causes risk at different growth stages and ultimately reduce the yield by affecting the physiological mechanism [22]. The seedling stage of crops is susceptible to high-temperature stress in rice, mungbean, wheat [23, 24] and the reproductive stage in rice, wheat, and other cereals [25, 26]. High temperature modifies cells' morphogenetic structure, leading to a decrease in cell size, enhancing stomatal density and cell membrane permeability, inflating xylem vessels, and impairing mesophyll cells [27].

High-temperature stress causes various physiological changes in crops during the different growth stages and germination is affected at the initial stage. The impact of temperature on seeds has decreased seed germination percentage, plant emergence, poor seedlings vigor, abnormal seedlings, and decrease radicle and plumule growth [4]. Seed germination is inhibited under high temperatures through the stimulation of abscisic acid biosynthesis. Exposure to high temperatures reduces plant height, total biomass, and the number of tillers. In wheat, the germination rate was completely inhibited and caused cell death and the embryo for seedlings establishment rate was additionally decreased. The morphophysiological characteristics such as phenology, plant water relations, dry matter partitioning, and shoot growth were restricted by heat stress in bean plants [28]. In some plants, it reduces total phenological duration, shorter grain filling period, and diminishes the germination period. The loss of cell water content due to high temperature eventually decreases the cell size and growth. In response to high temperature, the net assimilation rate was reduced, and it was directly associated with plant growth. Elevated temperature causes programmed cell death in specific cells or tissues as long as the denaturation of proteins. In addition, the high temperature for a prolonged period might cause gradual death; these injuries may lead to the shedding of leaves, abortion of flowers, or even death of the whole plant [29]. Furthermore, high temperature enhances the evapotranspiration during vegetative and reproductive stages ration which limits plant's water availability and uptake, which influences dehydration that reduces growth at the organ level as well as the whole plant level [30].

3.2 Root physiology

Root size and morphology play a vital role in plant water, and nutrient uptake, whereas the plant root system requires the optimal temperature to grow, if it exceeds the normal temperature it may change the uptake [31]. The temperature of the root system is lower when compared to the shoot. In addition, the optimum root temperature may vary across plant species [32]. Exposure to high temperatures (>29°C) leads to a decrease in primary root length, lateral root density, and root growth in sunflower crops [32, 33]. Even though species share the same environment, their root system architecture (RSA) may differ from species to species by the changes in soil temperatures [34, 35]. In some plants, the increased temperature may produce an expanded root system, whereas, in adult maize plants, it suppresses the lateral root growth by developing the long axial root to take up water from deep soil layers. Similarly, at increased temperature, the initiation and elongation of adventitious and lateral roots were inhibited [36] in potatoes. Swelling of root cap meristem and blends in root tip were found on potatoes [37] with the rise in soil temperature. Root zone temperature in Sorghums causes a decreased rate of cell production and root elongation. Overall, these changes by the high temperature hinder the root growth with a decreased rate of cell division [38]. The impact of elevated temperature on the formation of root growth in lupine species was studied at the initial and lateral stages. Temperature modifies the growth through altering the root architecture [39]. On the other hand, root respiration could vary based on the temperature range. During this root respiration process, cells uptake the oxygen which is surrounded in the spaces among the soil particles at the root zone, increasing with every 10°C on soil temperature, where the solubility of oxygen is in contrast to temperature. This increased requirement for oxygen leads to root hypoxia [32]. Reactive oxygen species (ROS) produced in excess under high-temperature stress results in oxidative stress [40]. The increase in ROS production includes hydrogen peroxide (H_2O_2), superoxide free radicals (O_2^-), and lipid peroxidation, which leads to the enhancement of cell membrane damage [41, 42]. To prevent the effect of ROS, plants produce antioxidant enzymes, such as catalase (CAT), superoxide dismutase (SOD), ascorbate peroxidase (APX), guaiacol peroxidase (GPX), glutathione reductase (GR), dehydroascorbate reductase (DHAR), and glutathione S transferase (GST), and nonenzymatic antioxidants such as anthocyanin, flavonoids, carotenoids, and ascorbic acid (AA) that protect crop growth from various other stresses also. Overall, the high temperature on root growth of plants either be promotive, inhibitory, or first stimulatory and then inhibitive once the optimum temperature is crossed [14].

3.3 Leaf physiology

High-temperature stress showed various morphological symptoms in shoots such as sunburns of leaves and twigs, scorching of leaves and stems, leaf senescence, root and shoot inhibition, fruit damage, and discolorization, which finally caused a decrease in crop productivity. In some cases, high temperature causes drying and rolling of leaves, scorching of leaf tips and margins, and necrosis was found in sugarcane. Leaf photosynthesis is considerably affected by extreme temperatures. The components of photosynthetic metabolism processes are sensitive to temperature [43]. Chloroplast is substantially affected by high temperature during the photosynthesis process including swelling of grana stacks and abnormal stacking [44]. However, chloroplast plays an important role to activate the adaptive process to these extreme conditions [45]. Studies reported that high temperature is associated with the upregulation of two hundred chloroplast-related genes in model rice plants [46]. To evaluate the heat-resistant crops, stay-green physiological traits were used to affirm the mechanism of heat damage interacting with Chl adaptability, antioxidant and photosynthetic capacity. Stay-green genotypes are highly associated with chlorophyll metabolism, these genes encode magnesium dechelatase, which is involved in the degradation of chlorophyll. A chloroplast targeted DnaJ protein (SlCDJ2) which is located in the thylakoids and stroma, protects the Rubisco activity and regulates the CO_2 assimilation in tomato plants to cope with heat stress [47, 48]. In lettuce seedlings, the application of exogenous spermidine regulates the stability of mitochondrial and chloroplast structure. The stomatal opening and density are inhibited by exogenous spermidine, thus leading to an increase in the photosynthetic rate and biomass of lettuce. It significantly alleviates the high-temperature stress and protects the leaves from damage [49].

3.4 Reproductive physiology

The reproductive tissues of plants are more susceptible to elevated temperatures among the other plant tissues, which result in a yield penalty. At the time of reproduction in crops, a short period of extreme temperature could significantly increase flower abortion. Similarly, heat spells at reproductive stages may not produce flowers, or flowers do not produce fruit or seed [29]. Reproductive development in various species is reported to be sensitive under high-temperature stress that disrupts/affects meiosis in male and female gametes, pollen germination and pollen tube growth, pollen/pistil interactions, ovule viability, number of pollen grains, formation of endosperm and embryo development, fertilization, and post-fertilization processes. The effect of high-temperature day/night causes an impact on pollen viability in crops including sorghum, wheat, rice, canola, groundnut, common bean, and soybean [50]. Due to high-temperature stress ROS accumulation, membrane integrity, changes in protein, carbohydrate, and lipid mechanism, and alters in phospholipids profiles in the mature pollen causes loss of pollen viability [41, 51, 52]. High temperature (33°C) subsequently reduces anther dehiscence and pollen fertility rate; it causes a decrease in the number of pollen on stigma which leads to decreased fertilization, spikelet fertility, and sterile seed in rice [53]. The process of anther dehiscence prompts to dispense mature pollen grains from the locules of the anther is responsible for pollination. During this process, the anther wall is opened by degeneration of anther tissues called septum and stomium [54]. The proper release of pollen from the anther needs expanding of endothecial cells and also strengthening and thickening of cell walls. In common bean and rice plants, the anther indehiscence happens due to high temperature, thus disrupting the pollen release. In addition, the inhibition of thickening in the endothecium cell and dissolution of interlocular septa in common bean and tomato were observed, which lead to failure in anther dehiscence and pollination [55, 56]. Heat treatment during panicle initiation significantly reduced the yield through a decreased number of spikelets per panicle and seed setting percent in rice. However, the heat susceptible variety showed poor seed setting percent mainly associated with spikelet fertility which is induced by a decrease in anther dehiscence, reduced pollen viability, and poor shedding of anthers [57]. Exposure to heat stress for a short period alters pollen development during meiosis. Increased spikelet sterility might decrease pollen germination due to high night temperature in rice. Other intrinsic factors of pollen development are carbohydrate metabolism and partitioning [50]. Reduced pollen viability is directly associated with sugar level accumulation in developing anther and pollen walls.

Plant female reproductive organ pistil/carpel includes stigma, style, and ovary. Female gametophyte occurs in the ovary, and the nutritional support and shelter were provided by sporophytic maternal tissue. The development of female gametophytes occurs in two phases: megasporogenesis and megagametogenesis. Generally, it consists of three antipodal cells, two synergid cells, one egg cell, and one central cell. Male reproductive development is susceptible to high-temperature stress when compared to female reproductive development. However, heat stress affects female reproductive development it causes degenerated eggs and synergids, embryo sac malformations, less number of ovules, increase ovule abnormalities and abortion, decrease in the size of transmitting tissue present in the style, and dried and drooping stigma, style and ovary [41, 51]. Heat stress interrupts the pollen-pistil interactions by changing the structural positioning of anthers and stigma. In response to high temperature, the maturation of stigma/style changes it leading to modification in cell division and elongation [58]. Heat stress affects reproductive function and has been reported in many studies. For example, in *Brassica* it decreases the fertility of microgametophytes, in tomato, it impaired meiosis in the male gametophyte which causes pollen germination and pollen tube growth. To counteract this problem, variations in genotypes of various pollen traits were selected to study the heat-tolerant genotypes. In recent study, 12 cultivars of *Brassica napus* L. were used to study heat tolerance based on pollen traits such as pollen germination, pollen viability, and pollen tube growth [59]. Abortion of flowers, young pods, and loss of seed numbers in soybean was observed under high-temperature stress [60]. Even mild heat stress (30°C/25°C) for 5 days, as long as the loss of tapetum differentiation and injuries to the microsporogenesis process was observed in Barley. The tapetum cells consist of enough mitochondria and a high metabolic activity under optimum growing temperature which in turn production of reactive oxygen species (ROS). But the plants exposed to high temperature provokes ROS homeostasis in the tapetum and microspores which cause accumulation of ROS initiation to lead to unplanned programmed cell death in tapetal cells through stimulating the membrane damage, lipid peroxidation, decreased transcription, and translation [61]. Extreme temperatures shortened the plant growing days, which resulted in early maturity, a lower life cycle, and accumulation of lesser biosynthetic products which eventually decreased grain development. The duration of grain filling decides the grain development which is related to grain yield. Crop yield is associated with temperature because even a 1.5°C increase in temperature might cause drastic yield loss in crops. The grain yield was mostly affected via the phenological development process. Many reports show that high temperatures cause a reduction in yield in various crops including pulses, cereals, and oil-yielding crops [29, 62]. Filled seed weight and size were ultimately reduced due to heat stress which directly affects grain yield in sorghum was reported. In addition, heat stress not only affects the yield, but it also correlates to reduce the quality of grain in barley. The concentration of non-structural carbohydrates, starch, raffinose, fructose, and lipids was reduced in barley grain, whereas in okra, the fiber content was decreased [63].

4. Physiological traits associated with high temperature stress

4.1 Cell membrane integrity

In response to heat stress, various physiological modifications might cause cell membrane damage. Exposure to heat stress causes cell membrane damage resulting from denaturing membrane protein, and inactivating enzyme, which leads to cell membrane permeability and integrity causing reduced ion flux, leakage of electrolytes, changes in relative water content, production of toxic compounds, and interruption in homeostasis that result in decreased cell viability. Reduced cell viability ceased plant growth and prompted leaf wilting, leaf abscission, and leaf area reduction [64]. Among the other physiological factors, membrane permeability and relative cell injury were affected at different growth stages under heat stress. Cell membrane stability differs with plant tissue, age, growing season, growth stage, plant species, and severity of heat stress. To measure electrolyte leakage from the cell membrane in an aqueous medium is used to determine the cell membrane malfunction in response to heat stress [65]. Leaf membrane stability index (MSI) was estimated in leaf tissues (100 mg) heated at 40°C for 30 min (C1) and 100°C for 10 min (C2) in water bath.

The conductivity of samples is measured by the using conductivity bridge and MSI was calculated using the following formula (MSI = [1 – (C1/C2)] × 100) [66].

4.2 Leaf gas exchange parameters

Transpiration is the physiological process that occurs in crops where the net radiation energy is converted into heat, underneath physiological control through alters in the stomatal aperture. In plants, photosynthesis and stomatal conductance play a vital role in all aspects. The leaf gas exchange parameters are commonly measured using a portable photosynthesis system including LI-6400 and LI-6800 (LI-COR®, Lincoln, USA), the CIRAS-3 (PP systems, Amesbury, USA), the GFS-3000 (Walz Gmbh, Effeltrich, Germany), and the iFL (Opti-sciences, Hudson, USA) [21]. However, decrease root hydraulic conductance in response to soil drying is an important control mechanism of stomatal closure. Low leaf water potential induced by high transpiration rates results in decreased stomatal conductance. The potential of crops to maintain CO_2 assimilation rate and leaf gas exchange under high temperatures is accompanied by heat tolerance. The water status in the leaf, stomatal conductance, and intercellular CO_2 concentration are affected by high temperatures [67]. Stomatal closure under high temperature is a distinct reason for impaired photosynthesis which disturbs the intercellular CO_2. In many plant species, the stomatal conductance (g_s) and net photosynthetic rate decline due to mild heat stress; it was caused due to reduced activation of Rubisco. High temperature affects vapor pressure density (VPD) but it may change hydraulic conductance and water supply to the leaf area [68]. Higher stomatal conductance accelerates transpirational cooling and canopy temperature depression (CTD). Increased stomatal conductance and related leaf cooling contribute to heat tolerance against high temperature; these conclude a positive interaction between stomatal conductance and the yield of wheat under extreme temperatures [69].

4.3 Chlorophyll pigments

Chlorophyll biosynthesis in plastids plays an important role in light-harvesting was high impact due to high-temperature stress. In addition, the high-temperature stress causes impairment and degradation of chlorophyll pigments in plastids. Chlorophyll biosynthesis was inhibited due to high temperature stress through the eradication of enzymes involved in the mechanism of chlorophyll biosynthesis. The enzymatic activity of 5-aminolevulinate dehydratase (ALAD), the first enzyme of pyrrole biosynthesis was decreased under high-temperature stress [70]. In some cases, the barley seedlings pre-treated with temperature for 4 h or 8 h prohibited the chl biosynthesis which result in reduced protochlorophyllide. Similarly in wheat, the protochlorophyllide (Pchlide) synthesis, Pchlide oxidoreductase, and porphobilinogen deaminase are eventually affected. These conclude that high temperature causes reduction in chlorophyll a, total chlorophyll content, sucrose content, and in contrast, it increases the reducing sugar content and leaf soluble sugar was observed in soybean [29]. However, in celery leaves, chlorophyll biosynthesis was inhibited under extreme temperatures due to the down-regulation of mRNA genes associated with biosynthesis [71]. Chlorophyll content was measured in the leaf using acetone by spectrophotometrically measuring the absorption at 663, 652, and 645 nm [72]. Non-destructively chlorophyll can be measured using the handheld devices such as SPAD chlorophyll meter (SPAD 502 Plus Chlorophyll Meter, Konica Minolta, Japan) [73],

CCM Chlorophyll content meter (CCM-200plus *Chlorophyll Content Meter,* Opti-Sciences, Inc., USA), CL-01 Chlorophyll Meter (Hansatech Instruments Ltd., United Kingdom). The NIR reflectance spectroscopy can be a promising methodology to measure/predict chlorophyll and other pigments under field conditions [74].

4.4 Heat shock proteins (HSPs)

Production of heat shock proteins in plants to protect the cell from various stress factors. These types of proteins were not found in non-stressed plants [75]. These proteins are categorized into high molecular (68kD to 110kD) and low molecular weight proteins (15kD to 27kD). Low molecular weight HSPs are found in higher plants which are plant-specific proteins whereas high molecular weight is found in all types of plants. Some heat shock proteins are known as molecular chaperones but not all HSPs are molecular chaperones [76]. Molecular chaperones are proteins that generate during high-temperature stress. Chaperones bind to denature proteins or unfolded proteins to make them stabilize and protect from thermal aggregation. Heat shock proteins regulate cellular homeostasis by eliminating harmful proteins which rise from aggregation and misfolding [69]. Some classical HSPs show chaperone activity to protect the protein denaturation from thermo-aggregation. The non-classical heat shock proteins consist of plastid protein synthesis elongation factor (EF-Tu) and peptidyl-prolyl cis/trans isomerases that produce low molecular weight proteins and provide heat resistance. Under heat stress, the EF-Tu gene in transgenic wheat exhibits reduced thermo-aggregation and decreased thylakoid membrane damage moreover increasing the photosynthetic system [77]. The chloroplast localized HSPs provide heat resistance to the photosynthetic electron transport chain in isolated chloroplast. Some studies revealed that small heat shock proteins pertain with thylakoid and protect O_2 evolution and oxygen-evolving complex proteins of PSII from heat stress. The heat treatment in tomato leaves protects PSII from temperature-dependent oxidative stress by chloroplast heat shock protein HSP21 present in chloroplast. In addition, the chloroplast HSPs may not repair the stress-induced damage but they avoid damage [78].

4.5 Optimal light and dark reaction

High-temperature stress is directly associated with light intensity which damages the photosystem. Heat stress alters cell respiration and photosynthesis, which shortens the life cycle and reduces crop production [79]. Thermal stress modified the structural changes of chloroplast protein complexes and decreased the activity of enzymes. Light-dependent chemical reactions occur in the thylakoid membrane and carbon metabolism takes place in the stroma, which is an important portion, that is damaged in response to high temperature. Increased temperature of the leaf and photon flux density impacts the thermo-tolerance adjustment of photosystem (PS) II. For measuring the quantum yield of PSII mostly modulated fluorimeters are used such as the Mini-PAM II by Walz, FMS2 by Hansatech instruments (King's Lynn, UK), the OS5+ by Opti-Sciences, or the FluorPen FP 100-MAX of Photo Systems Instruments (Drasov, Czech Republic) [21, 80, 81]. PSII is particularly receptive to temperature and its activity is enormously impacted to some extent and terminated under high temperature stress. Oxygen developing complex is exposed to temperature causing damage; it concludes in an imbalance flow of electrons to the acceptor site of PSII. The synthesis of starch and sucrose was eventually affected being a decrease in the activities of some enzymes such as sucrose phosphate synthase, adenosine diphosphate-glucose pyrophosphorylase, and invertase due to high-temperature stress. Net photosynthesis in many plant species is suppressed because of a decrease in the activation state of the CO_2 binding enzyme, Rubisco. Although these negative impacts of adverse conditions on photosynthesis, the optimum temperature prerequisites for photosynthesis are familiar to rise with increased concentration of CO_2 in the atmosphere [30]. The decrease in photosynthesis was due to damage to the chlorophyll pigments, reduced leaf nitrogen content, a hindrance to PSII reaction and electron flow, reduced quantum efficiency (Fv/Fm), and down-regulation of PSII photochemistry.

Many studies reported PSI is more stable than PSII towards heat stress. PSI activity increased the thylakoid proton conductance and cyclic electron flow under high temperatures. When the activity of PSII is reduced due to adverse conditions, the proton conductance and cyclic electron flow around PSI could be the favorable process that produces ATP. Further, heat stress enhances the dark reduction of plastoquinone and stimulates the thylakoid proton gradient which was elucidated to activate the cyclic electron flow around PSI [76]. There is an increase in ATP under mild temperature stress due to Rubisco activase and active photorespiration. The larger requirement of ATP under mild heat stress as long as a higher in NADPH/ATP ratio is helpful for non-photochemical reduction of the plastoquinone pool from stroma donors which is sequential to activate the NADH-mediated cyclic electron pathway. This process could distribute the energy and produce increased ATP to regulate active CO_2 fixation.

5. Future perspectives

High-temperature stress eventually affects germination, vegetative stage, reproductive stage, and yield. Due to increased global warming, the temperature will increase in upcoming years which drastically affects crop production. Subsequently, the decreased yield may lead to food scarcity in the future. Even though many research projects work on abiotic and biotic stress to protect the plants from adverse conditions. We need to improve the crop from the seedling stage to the harvest stage. At each stage, optimum growth conditions need to be provided for the plants. To counteract the problem, researchers found stress-tolerant genotypes. However, these can act on particular metabolic pathways and regulate the mechanism to improve plant growth. Sometimes, the crops endure both biotic and abiotic stress which can significantly affect the yield. In the future, we need to target the multi-stress-tolerant approach to protect the plants from multiple or unprecedented stress conditions.

6. Conclusion

High temperature is a devastating environmental factor that influences crop growth and yield by affecting numerous crop mechanisms. High temperature is an alarming concern that needs to be considered since it directly affects yield in a situation where have to generate high yield to sort out global hunger and meet the demands of global food hunger and ensure food security. High-temperature tolerance is not only important for the current situation, but also for the future since the simulated predictions show the elevating high temperature that enormously affects yield. New possibilities to resist or enhance high-temperature tolerance need to be investigated. Even though there are several studies related to high temperature on crops that reveal various factors that are up-regulated and down-regulated, there are abundant wild types that are not explored. Touching the untouched wild types can reveal amazing outcomes on how high temperatures can be managed with novel tolerant genes. Advancements in molecular techniques provide rapid detection of traits in wild types. It is also important to implement or follow advanced agricultural practices across the globe towards raising the global temperature, by implementing climate-smart practices, high-throughput phenotyping methods, and revealing the traits that are tolerant can protect, safeguard and defend tomorrow.

Author details

Theivasigamani Parthasarathi[1], Saiyyeda Firdous[1], Einstein Mariya David[1,2], Kuppan Lesharadevi[1,2] and Maduraimuthu Djanaguiraman[3]*

1 VIT School of Agricultural Innovations and Advanced Learning (VAIAL), Vellore Institute of Technology, Vellore, Tamil Nadu, India

2 School of Bio Sciences and Technology, Vellore Institute of Technology, Vellore, Tamil Nadu, India

3 Department of Crop Physiology, Tamil Nadu Agricultural University, Coimbatore, Tamil Nadu, India

*Address all correspondence to: jani@tnau.ac.in

References

[1] Zhao C, Liu B, Piao S, Wang X, Lobell DB, Huang Y, et al. Temperature increase reduces global yields of major crops in four independent estimates. Proceedings of the National Academy of Sciences of the United States of America. 2017;**114**(35):9326-9331. DOI: 10.1073/PNAS.1701762114/SUPPL_FILE/PNAS.1701762114.SAPP.PDF

[2] Hatfield JL, Prueger JH. Temperature extremes: Effect on plant growth and development. Weather and Climate Extremes. 2015;**10**:4-10. DOI: 10.1016/J.WACE.2015.08.001

[3] Schauberger B, Archontoulis S, Arneth A, Balkovic J, Ciais P, Deryng D, et al. Consistent negative response of US crops to high temperatures in observations and crop models. Nature Communications. 2017;**8**(1):1-9. DOI: 10.1038/ncomms13931

[4] Bita CE, Gerats T. Plant tolerance to high temperature in a changing environment: Scientific fundamentals and production of heat stress-tolerant crops. Frontiers in Plant Science. 2013;**4**:1-18. DOI: 10.3389/FPLS.2013.00273

[5] Satake T, Yoshida S. High temperature-induced sterility in indica rices at flowering. Japanese Journal of Crop Science. 1978;**47**(1):6-17. DOI: 10.1626/JCS.47.6

[6] Liu W, Yin T, Zhao Y, Wang X, Wang K, Shen Y, et al. Effects of high temperature on rice grain development and quality formation based on proteomics comparative analysis under field warming. Frontiers in Plant Science. 2021;**12**:2320. DOI: 10.3389/FPLS.2021.746180/BIBTEX

[7] Agnolucci P, Rapti C, Alexander P, de Lipsis V, Holland RA, Eigenbrod F, et al. Impacts of rising temperatures and farm management practices on global yields of 18 crops. Nature Food. 2020;**1**(9):562-571. DOI: 10.1038/s43016-020-00148-x

[8] Shah F, Huang J, Cui K, Nie L, Shah T, Chen C, et al. Impact of high-temperature stress on rice plant and its traits related to tolerance. Journal of Agricultural Science. 2011;**149**(5):545-556. DOI: 10.1017/S0021859611000360

[9] Jagadish SVK. Heat stress during flowering in cereals – Effects and adaptation strategies. New Phytologist. 2020;**226**(6):1567-1572. DOI: 10.1111/NPH.16429

[10] Janni M, Gullì M, Maestri E, Marmiroli M, Valliyodan B, Nguyen HT, et al. Molecular and genetic bases of heat stress responses in crop plants and breeding for increased resilience and productivity. Journal of Experimental Botany. 2020;**71**(13):3780-3802. DOI: 10.1093/JXB/ERAA034

[11] Xu Y, Chu C, Yao S. The impact of high-temperature stress on rice: Challenges and solutions. The Crop Journal. 2021;**9**(5):963-976. DOI: 10.1016/J.CJ.2021.02.011

[12] Iman G. These countries will have the largest populations by 2100 | World Economic Forum. 2020. Available from: https://www.weforum.org/agenda/2020/09/the-world-population-in-2100-by-country/ [Accessed: July 12, 2022]

[13] Fatima Z, Ahmed M, Hussain M, Abbas G, Ul-Allah S, Ahmad S, et al. The fingerprints of climate warming on cereal crops phenology and adaptation options.

Scientific Reports. 2020;**10**:18013. DOI: 10.1038/s41598-020-74740-3

[14] Khan A, Ahmad M, Ahmed M, Iftikhar Hussain M. Rising atmospheric temperature impact on wheat and thermotolerance strategies. Plants. 2020, 2021;**10**(1):43. DOI: 10.3390/PLANTS10010043

[15] van der Merwe R, Labuschagne MT, Herselman L, Hugo A. Effect of heat stress on seed yield components and oil composition in high- and mid-oleic sunflower hybrids. South African Journal of Plant and Soil. 2015;**32**(3):121-128. DOI: 10.1080/02571862.2015.1018354

[16] Canvin DT. The effect of temperature on the oil content and fatty acid composition of the oils from several oil seed crops. Canadian Journal of Botany. 1965;**43**(1):63-69. DOI: 10.1139/B65-008

[17] Yaniv Z, Schafferman D, Zur M. The effect of temperature on oil quality and yield parameters of high- and low-erucic acid Cruciferae seeds (rape and mustard). Industrial Crops and Products. 1995;**3**(4):247-251. DOI: 10.1016/0926-6690(94)00041-V

[18] Hein NT, Ciampitti IA, Jagadish SVK. Bottlenecks and opportunities in field-based high-throughput phenotyping for heat and drought stress. Journal of Experimental Botany. 2021;**72**(14):5102-5116. DOI: 10.1093/JXB/ERAB021

[19] Han S, Jiang S, Xiong R, Shafique K, Zahid KR, Wang Y. Response and tolerance mechanism of food crops under high temperature stress: A review. Brazilian Journal of Biology = Revista Brasleira de Biologia. 2022;**82**:e253898. DOI: 10.1590/1519-6984.253898

[20] Masoomi-Aladizgeh F, Najeeb U, Hamzelou S, Pascovici D, Amirkhani A, Tan DKY, et al. Pollen development in cotton (Gossypium hirsutum) is highly sensitive to heat exposure during the tetrad stage. Plant, Cell & Environment. 2021;**44**(7):2150-2166. DOI: 10.1111/PCE.13908

[21] Urban J, Ingwers MW, McGuire MA, Teskey RO. Increase in leaf temperature opens stomata and decouples net photosynthesis from stomatal conductance in Pinus taeda and Populus deltoides x nigra. Journal of Experimental Botany. 2017;**68**(7):1757-1767. DOI: 10.1093/JXB/ERX052

[22] Wahid A, Gelani S, Ashraf M, Foolad M. Heat tolerance in plants: An overview. Environmental and Experimental Botany. 2007;**61**(3):199-223. DOI: 10.1016/j.envexpbot.2007.05.011

[23] Kumar S, Kaur R, Kaur N, Bhandhari K, Kaushal N, Gupta K, et al. Heat-stress induced inhibition in growth and chlorosis in mungbean (Phaseolus aureus Roxb.) is partly mitigated by ascorbic acid application and is related to reduction in oxidative stress. Acta Physiologiae Plantarum. 2011;**33**(6):2091-2101. DOI: 10.1007/s11738-011-0748-2

[24] Mitra R, Bhatia CR. Bioenergetic cost of heat tolerance in wheat crop. Current Science. 2008;**94**(8):1049-1053

[25] Akter N, Rafiqul Islam M. Heat stress effects and management in wheat. A review. In: Agronomy for Sustainable Development. Vol. 375. France: Springer-Verlag; 2017. DOI: 10.1007/s13593-017-0443-9

[26] Maestri E, Klueva N, Perrotta C, Gulli M, Nguyen HT, Marmiroli N. Molecular genetics of heat tolerance and heat shock proteins in cereals. Plant Molecular Biology. 2002;**48**(5-6):667-681. DOI: 10.1023/A:1014826730024

[27] Shaheen MR, Ayyub CM, Amjad M, Waraich EA. Morpho-physiological evaluation of tomato genotypes under high temperature stress conditions. Journal of the Science of Food and Agriculture. 2016;**96**(8):2698-2704. DOI: 10.1002/jsfa.7388

[28] Omae H, Kumar A, Shono M. Adaptation to high temperature and water deficit in the common bean (Phaseolus vulgaris L.) during the reproductive period. Journal of Botany. 2012;**2012**:1-6. DOI: 10.1155/2012/803413

[29] Hasanuzzaman M, Nahar K, Alam M, Roychowdhury R, Fujita M. Physiological, biochemical, and molecular mechanisms of heat stress tolerance in plants. International Journal of Molecular Sciences. 2013;**14**(5):9643-9684. DOI: 10.3390/ijms14059643

[30] Fahad S, Bajwa AA, Nazir U, Anjum SA, Farooq A, Zohaib A, et al. Crop production under drought and heat stress: Plant responses and management options. Frontiers in Plant Science. 2017;**8**(June):1-16. DOI: 10.3389/fpls.2017.01147

[31] Koevoets IT, Venema JH, Elzenga JTM, Testerink C. Roots withstanding their environment: Exploiting root system architecture responses to abiotic stress to improve crop tolerance. Frontiers in Plant Science. 2016;7(AUG2016):1335. DOI: 10.3389/FPLS.2016.01335/BIBTEX

[32] Fonseca F, de Lima C, Kleine-Vehn J, De Smet I, Feraru E, Quint M. Getting to the root of belowground high temperature responses in plants. Journal of Experimental Botany. 2021;**72**(21):7404-7413. DOI: 10.1093/JXB/ERAB202

[33] Nagel KA, Kastenholz B, Jahnke S, Van Dusschoten D, Aach T, Mühlich M, et al. Temperature responses of roots: Impact on growth, root system architecture and implications for phenotyping. Functional Plant Biology: FPB. 2009;**36**(11):947-959. DOI: 10.1071/FP09184

[34] Bardgett RD, Mommer L, De Vries FT. Going underground: Root traits as drivers of ecosystem processes. In: Trends in Ecology and Evolution. Vol. 29, No. 12. Elsevier; 2014. pp. 692-699. DOI: 10.1016/j.tree.2014.10.006

[35] Luo H, Xu H, Chu C, He F, Fang S. High temperature can change root system architecture and intensify root interactions of plant seedlings. Frontiers in Plant Science. 2020;**11**(February):1-13. DOI: 10.3389/fpls.2020.00160

[36] Sattelmacher B, Marschner H, Kühne R. Effects of the temperature of the rooting zone on the growth and development of roots of potato (Solanum tuberosum). Annals of Botany. 1990;**65**(1):27-36. DOI: 10.1093/oxfordjournals.aob.a087903

[37] Joshi M, Fogelman E, Belausov E, Ginzberg I. Potato root system development and factors that determine its architecture. Journal of Plant Physiology. 2016;**205**:113-123. DOI: 10.1016/j.jplph.2016.08.014

[38] Calleja-Cabrera J, Boter M, Oñate-Sánchez L, Pernas M. Root growth adaptation to climate change in crops. Frontiers in Plant Science. 2020;**11**:544. DOI: 10.3389/fpls.2020.00544

[39] Gavelienė V, Jurkonienė S, Jankovska-Bortkevič E, Švegždienė D. Effects of elevated temperature on root system development of two lupine species. Plants. 2022;**11**(2):192. DOI: 10.3390/plants11020192

[40] Potters G, Pasternak TP, Guisez Y, Palme KJ, Jansen MAK. Stress-induced

morphogenic responses: Growing out of trouble? Trends in Plant Science. 2007;**12**(3):98-105. DOI: 10.1016/J.TPLANTS.2007.01.004

[41] Djanaguiraman M, Perumal R, Ciampitti IA, Gupta SK, Prasad PVV. Quantifying pearl millet response to high temperature stress: Thresholds, sensitive stages, genetic variability and relative sensitivity of pollen and pistil. Plant, Cell & Environment. 2018b;**41**(5):993-1007. DOI: 10.1111/PCE.12931

[42] Narayanan S, Tamura PJ, Roth MR, Prasad PVV, Welti R. Wheat leaf lipids during heat stress: I. High day and night temperatures result in major lipid alterations. Plant, Cell & Environment. 2016;**39**(4):787-803. DOI: 10.1111/PCE.12649

[43] Asthir B. Protective mechanisms of heat tolerance in crop plants. Journal of Plant Interactions. 2015;**10**(1):202-210. DOI: 10.1080/17429145.2015.1067726

[44] Wahid A, Shabbir A. Induction of heat stress tolerance in barley seedlings by pre-sowing seed treatment with glycinebetaine. Plant Growth Regulation. 2005;**46**(2):133-141. DOI: 10.1007/s10725-005-8379-5

[45] Yoo Y-H, Hong W-J, Jung K-H. A systematic view exploring the role of chloroplasts in plant abiotic stress responses. BioMed Research International. 2019;**2019**:1-14. DOI: 10.1155/2019/6534745

[46] Wang QL, Chen JH, He NY, Guo FQ. Metabolic reprogramming in chloroplasts under heat stress in plants. International Journal of Molecular Sciences. 2018;**19**(3):849. DOI: 10.3390/IJMS19030849

[47] Kumar RR, Goswami S, Singh K, Dubey K, Singh S, Sharma R, et al. Identification of putative RuBisCo activase (TaRca1)—The catalytic chaperone regulating carbon assimilatory pathway in wheat (Triticum aestivum) under the heat stress. Frontiers in Plant Science. 2016;**7**(JULY2016):986. DOI: 10.3389/fpls.2016.00986

[48] Wang G, Kong F, Zhang S, Meng X, Wang Y, Meng Q. A tomato chloroplast-targeted DnaJ protein protects Rubisco activity under heat stress. Journal of Experimental Botany. 2015;**66**(11):3027-3040. DOI: 10.1093/JXB/ERV102

[49] Yang X, Han Y, Hao J, Qin X, Liu C, Fan S. Exogenous spermidine enhances the photosynthesis and ultrastructure of lettuce seedlings under high-temperature stress. Scientia Horticulturae. 2022;**291**:110570. DOI: 10.1016/j.scienta.2021.110570

[50] Lohani N, Singh MB, Bhalla PL. High temperature susceptibility of sexual reproduction in crop plants. Journal of Experimental Botany. 2020a;**71**(2):555-568. DOI: 10.1093/jxb/erz426

[51] Djanaguiraman M, Perumal R, Jagadish SVK, Ciampitti IA, Welti R, Prasad PVV. Sensitivity of sorghum pollen and pistil to high-temperature stress. Plant, Cell & Environment. 2018c;**41**(5):1065-1082. DOI: 10.1111/PCE.13089

[52] Liu Y, Li J, Zhu Y, Jones A, Rose RJ, Song Y. Heat Stress in legume seed setting: Effects, causes, and future prospects. Frontiers in Plant Science. 2019;**10**:938. DOI: 10.3389/FPLS.2019.00938/BIBTEX

[53] Hurkman WJ, Vensel WH, Tanaka CK, Whitehand L, Altenbach SB. Effect of high temperature on albumin and globulin accumulation in the endosperm proteome of the developing wheat grain. Journal of Cereal Science.

2009;**49**(1):12-23. DOI: 10.1016/j.jcs.2008.06.014

[54] Wilson ZA, Song J, Taylor B, Yang C. The final split: The regulation of anther dehiscence. Journal of Experimental Botany. 2011;**62**(5):1633-1649. DOI: 10.1093/JXB/ERR014

[55] Jagadish SVK, Craufurd PQ, Wheeler TR. High temperature stress and spikelet fertility in rice (Oryza sativa L.). Journal of Experimental Botany. 2007;**58**(7):1627-1635. DOI: 10.1093/JXB/ERM003

[56] Porch TG, Jahn M. Effects of high-temperature stress on microsporogenesis in heat-sensitive and heat-tolerant genotypes of Phaseolus vulgaris. Plant, Cell & Environment. 2001;**24**(7):723-731. DOI: 10.1046/J.1365-3040.2001.00716.X

[57] Hu Q, Wang W, Lu Q, Huang J, Peng S, Cui K. Abnormal anther development leads to lower spikelet fertility in rice (Oryza sativa L.) under high temperature during the panicle initiation stage. BMC Plant Biology. 2021;**21**(1):428. DOI: 10.1186/s12870-021-03209-w

[58] Sage TL, Bagha S, Lundsgaard-Nielsen V, Branch HA, Sultmanis S, Sage RF. The effect of high temperature stress on male and female reproduction in plants. Field Crops Research. 2015;**182**:30-42. DOI: 10.1016/J.FCR.2015.06.011

[59] Chaudhary S, Devi P, Bhardwaj A, Jha UC, Sharma KD, Prasad PVV, et al. Identification and characterization of contrasting genotypes/cultivars for developing heat tolerance in agricultural crops: Current status and prospects. In: Frontiers in Plant Science. Vol. 11. S.A: Frontiers Media; 2020. p. 1505. DOI: 10.3389/fpls.2020.587264

[60] Tubiello FN, Soussana J-F, Howden SM. Crop and pasture response to climate change. Proceedings of the National Academy of Sciences. 2007;**104**(50):19686-19690. DOI: 10.1073/pnas.0701728104

[61] Djanaguiraman M, Vara Prasad PV, Murugan M, Perumal R, Reddy UK. Physiological differences among sorghum (Sorghum bicolor L. Moench) genotypes under high temperature stress. Environmental and Experimental Botany. 2014;**100**:43-54. DOI: 10.1016/j.envexpbot.2013.11.013

[62] Zhang X, Cai J, Wollenweber B, Liu F, Dai T, Cao W, et al. Multiple heat and drought events affect grain yield and accumulations of high molecular weight glutenin subunits and glutenin macropolymers in wheat. Journal of Cereal Science. 2013;**57**(1):134-140. DOI: 10.1016/j.jcs.2012.10.010

[63] Vasseur F, Pantin F, Vile D. Changes in light intensity reveal a major role for carbon balance in Arabidopsis responses to high temperature. Plant, Cell & Environment. 2011;**34**(9):1563-1576. DOI: 10.1111/J.1365-3040.2011.02353.X

[64] Sethi VP, Dubey RK, Dhath AS. Design and evaluation of modified screen net house for off-season vegetable raising in composite climate. Energy Conversion and Management. 2009;**50**(12):3112-3128. DOI: 10.1016/j.enconman.2009.08.001

[65] Nijabat A, Bolton A, Mahmood-ur-Rehman M, Shah AI, Hussain R, Naveed NH, et al. Cell membrane stability and relative cell injury in response to heat stress during early and late seedling stages of diverse carrot (Daucus carota L.) Germplasm. HortScience. 2020;**55**(9):1446-1452. DOI: 10.21273/HORTSCI15058-20

[66] Sairam RK, Deshmukh PS, Shukla DS. Tolerance to drought and temperature stress in relation to increased antioxidant enzyme activity in wheat. Journal of Agrononmy and Crop Science. 1997;**178**:171-178

[67] Greer DH, Weedon MM. Modelling photosynthetic responses to temperature of grapevine (Vitis vinifera cv. Semillon) leaves on vines grown in a hot climate. Plant, Cell & Environment. 2012;**35**(6):1050-1064. DOI: 10.1111/J.1365-3040.2011.02471.X

[68] Yang Z, Sinclair TR, Zhu M, Messina CD, Cooper M, Hammer GL. Temperature effect on transpiration response of maize plants to vapour pressure deficit. Environmental and Experimental Botany. 2012;**78**:157-162. DOI: 10.1016/J.ENVEXPBOT.2011.12.034

[69] Narayanan S. Effects of high temperature stress and traits associated with tolerance in wheat. Open Access Journal of Science. 2018;**2**(3):177-186. DOI: 10.15406/oajs.2018.02.00067

[70] Ashraf M, Harris PJC. Photosynthesis under stressful environments: An overview. Photosynthetica. 2013;**51**: 163-190. DOI: 10.1007/S11099-013-0021-6

[71] Huang W, Ma HY, Huang Y, Li Y, Wang GL, Jiang Q, et al. Comparative proteomic analysis provides novel insights into chlorophyll biosynthesis in celery under temperature stress. Physiologia Plantarum. 2017;**161**: 468-485. DOI: 10.1111/PPL.12609

[72] Fritschi FB, Ray JD. Soybean leaf nitrogen, chlorophyll content, and chlorophyll a/b ratio. Photosynthetica. 2007;**45**:92-98. DOI: 10.1007/ S11099-007-0014-4

[73] Yuan Z, Cao Q, Zhang K, Ata-Ul-Karim ST, Tan Y, Zhu Y, et al. Optimal leaf positions for SPAD meter measurement in rice. Frontiers in Plant Science. 2016;**7**:719. DOI: 10.3389/ FPLS.2016.00719/BIBT94EX

[74] Li Y, Sun Y, Jiang J, Liu J. Spectroscopic determination of leaf chlorophyll content and color for genetic selection on Sassafras tzumu. Plant Methods. 2019;**15**:1-11. DOI: 10.1186/ s13007-019-0458-0

[75] Barua D, Downs CA, Heckathorn SA. Variation in chloroplast small heat-shock protein function is a major determinant of variation in thermotolerance of photosynthetic electron transport among ecotypes of Chenopodium album. Functional Plant Biology. 2003;**30**: 1071-1079. DOI: 10.1071/FP03106

[76] Mathur S, Agrawal D, Jajoo A. Photosynthesis: Response to high temperature stress. Journal of Photochemistry and Photobiology B: Biology. 2014;**137**:116-126. DOI: 10.1016/J.JPHOTOBIOL.2014.01.010

[77] Fu J, Ristic Z. Analysis of transgenic wheat (*Triticum aestivum* L.) harboring a maize (*Zea mays* L.) gene for plastid EF-Tu: Segregation pattern, expression and effects of the transgene. Plant Molecular Biology. 2010;**73**:339-347. DOI: 10.1007/S11103-010-9622-7

[78] Heckathorn SA, Ryan SL, Baylis JA, Wang D, Hamilton EW, Cundiff L, et al. In vivo evidence from an Agrostis stolonifera selection genotype that chloroplast small heat-shock proteins can protect photosystem II during heat stress. Functional Plant Biology. 2002;**29**: 933-944. DOI: 10.1071/PP01191

[79] Barnabás B, Jäger K, Fehér A. The effect of drought and heat stress on reproductive processes in cereals. Plant, Cell & Environment.

2007;**31**(1):071030190532001.
DOI: 10.1111/j.1365-3040.2007.01727.x

[80] Ceusters N, Valcke R, Frans M, Claes JE, Van den Ende W, Ceusters J. Performance index and PSII connectivity under drought and contrasting light regimes in the CAM orchid phalaenopsis. Frontiers in Plant Science. 2019;**10**:1012. DOI: 10.3389/fpls.2019.01012

[81] Padhi B, Chauhan G, Kandoi D, Stirbet A, Tripathy BC, Govindjee G. A comparison of chlorophyll fluorescence transient measurements, using handy pea and fluorpen fluorometers. Photosynthetica. 2021;**59**:399-408. DOI: 10.32615/ps.2021.026

Chapter 3

Drought Stress in Millets and Its Response Mechanism

Anjali Tiwari, Kapil Kesarwani, Arushi Sharma, Tapan Ghosh, Nisha Bisht and Shailja Punetha

Abstract

Drought is a major abiotic stress that diminishes crop yield and weakens global food security, especially in the current emerging situation of climate change as well as increases in the prevalence and severity of stress elements. Millets are nutrient-dense and capable to resist variety of harsh environmental conditions, including lack of moisture. Millet's crop has evolved dynamically in terms of morphology, physiology, and biochemically that allow them to flee and/or adapt to adverse environmental situations. Drought stress has a significant impact on the vegetative and reproductive phases of plants. Millets generate a very low yield compared to main cereals like wheat and rice, despite their agronomic, nutritional, and health-related benefits. It is necessary to understand how these complex features are regulated and ameliorated the impact of droughts on millet productivity. Keeping this in view, the present work aims to understand the processes used for reducing the negative impacts of droughts in the production of millets varieties using advanced agronomic management strategies (use of information technology) and the biotechnology (improvements in crop genetics).

Keywords: millets, drought, abiotic stress, morphology, physiology

1. Introduction

Millets are the major cereal crop in the developing world, particularly in Africa and Asia's arid and semi-arid tropical regions, where they are utilised in food both for humans and cattle [1]. Millets also play a key role in the subsistence of people living in the mountains. In addition to addressing the issue of the food security at a global level, efforts need to be directed towards such indigenous crops that could flourish amidst the scenario of water scarcity. It is believed that the changing climate will have significant effects on the types of crops cultivated in the next century. Millets are known for their climate-resilient characteristics, such as their ability to adapt to a wide range of ecological conditions, reduced irrigation requirements, improved growth and productivity under low nutrient input conditions, reduced reliance on synthetic fertilisers, and low vulnerability to environmental stresses. Millets might provide alternative climate-smart crops as their adaptation to challenging environments is better than the current major crops of the world [2].

In line with Asia and Africa's food security, millets contribute to over half of the Africa's entire cereal production, together with sorghum. Millets are thus regarded as a poor man's crop because of their major contributions to the diets of resource-constrained producers and consumers. Millets are the primary source of nutrition for small farmer communities in India, Africa, China, and parts of Central America, and they help to ensure food security in Asia and Africa's low-income countries. Millets are ancient food crops which are highly nutritious and can be grown under marginal environmental conditions. Millets are the major energy source and highly nutritious staple foods for sustenance. Millets, often known as 'miscellaneous or coarse cereals,' are a group of small edible grasses in the Poaceae family. In total, millets are distributed around 10 genera and 20 species [3]. Approximately 31,019,370 tonnes of Millets were produced globally in 2018 [4]. Among the summer crops millet's water requirement is much less than those of the two main crops, viz., wheat and rice. Millets are also known as famine crops because these are the only crops assuring yields in famine conditions. About 80% of the millets are used for food and the rest are stocked as feed. Millets have good grain quality that's why it is used in processing industries. Millets have many nutritional, pharmaceutical and nutraceutical properties. They are especially rich in fibre content and starch. They are used in reducing the risk of diabetes, help to lower the cholesterol and are rich in antioxidant activity [5].

Millets are considered to be the essential crops in world's agricultural system due to their resistance to pests and diseases, short growing season and capable to grow in extreme environmental conditions, and because of moderate productivity under enduring water stress conditions when major cereals cannot be relied upon to provide sustainable yields [6]. Millets will stand up to all unwell effects of temperature change apart from doable higher ozone concentrations around urban zones. However, millets would be least struck by increasing greenhouse gas levels within the native micro-climate. Similar to maize and sorghum, millets possess a C4 photosynthesis system, hence they avoid photorespiration. As a result, they efficiently utilise the insufficient moisture present in the semi-arid regions. Since C4 plants are able to close their stomata for long periods, they can considerably reduce moisture loss through the leaves.

2. An overview: small millets

Millets are categorised by a collective term accustomed to see a various group of small-seeded annual C4 Panicoid grasses such as barnyard millet (*Echinochloa frumentacea*), finger millet (*Eleusine coracana*), foxtail (*Setaria italica*), pearl millet (*Pennisetum glaucum* (L.) R. Br.), kodo millet (*Paspalum scrobiculatum* L.), little millet (*Panicum sumatrense* Roth ex Roem. & Schult.) and proso millet (*Panicum miliaceum*) (**Figure 1**) [6]. These are extensively cultivated as food and fodder crops in temperate, sub-tropical and tropical regions across the globe [7, 8] and have outstanding biological process properties. Salient features of the millet's crops are described in **Table 1**.

2.1 Importance of millets crop

Millets can withstand extreme temperatures, droughts, and floods. Millets grow well in arid zones/rain-fed locations with marginal soil fertility and moisture. Because of its effective root system, millets require far less water than other cereal crops for their production. Some crops and their water availability are shown in **Table 2**. Millets are grown with organic inputs and do not require synthetic fertilisers or pesticides.

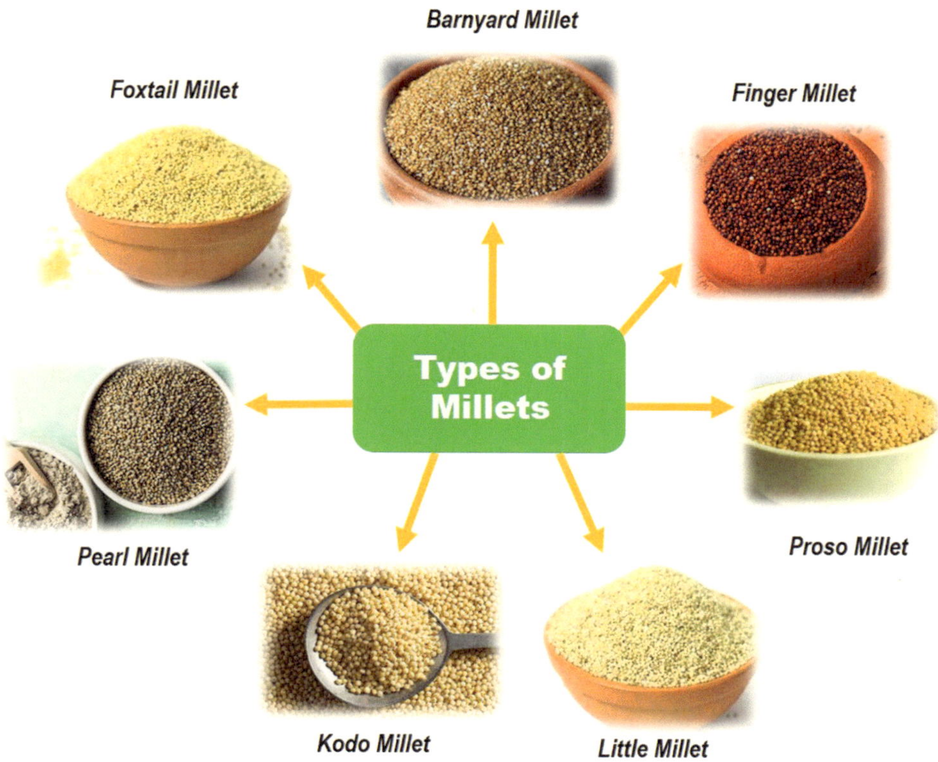

Figure 1.
Representation of different types of small millets grown in the semi-arid regions.

Pesticides are avoided since millets are less susceptible to illnesses and pests. Millets contribute to climate change mitigation by lowering CO_2 levels in the atmosphere [9]. They have a high capacity for carbon sequestration, which aids climate adaptation, particularly in light of global projections of rising methane emissions from rice fields. Millet diets are beneficial for persons with celiac disease and diabetes since they contain no gluten and have a low glycaemic index [1]. Millets crop like pearl millet is most recently used as a low-cost substitute for maize in poultry and animal feed. Millet crops are also used in fermented food and drinks products.

One of the major concerns of the global world is the abrupt changes in the earth's environment, which have had a devastating effect on the earth's ecology [10]. Currently, we are in the age of an agrarian crisis, which has necessitated for crop improvement to tackle the negative effects of climate change. Intensive agriculture of a few crops for food needs has resulted in poor nutrition and genetic degradation, as well as the negligence of locally nutritious crops. Agriculture is the largest consumer of water in India and in the world. In the Central Himalayan part of India, only about 10% of the cultivated land is under irrigation system. It is largely because the millet crops, unlike rice and wheat, have high water use efficiency and the major chunk of the cultivated land in the summer season is devoted to the millet crops. In contemporary times when water is increasingly becoming scarce, we should conserve and make economical use of water resources at each micro- and meso-scale (farmers' field to watershed scale). The millet crops hold great promise for the strategies of conserving water resources and fight against drought

Millet Type	Characteristics							
	Vernacular Names	**Botanical Name**	**Family**	**Ploidy level**	**Chromosome no.**	**Use**	**Nutritional values**	**Agronomic benefits**
Barnyard millet	Barnyard millet (English), Madira, Sawa, Kudraivali, a million-dollar grass and oodalu	*Echinochloa frumentacea* L.	Gramineae	Hexaploid	2n = 6× = 36	Food and Feed	Anti-diabetic and high protein content	Early maturity, anti-fungal
Finger millet	Ragi, koda, mandua, nachani or nagli, African birds' foot, rapoko, koracan, kurakkan.	*Eleusine coracana* (L.) Gaertner	Gramineae	Hexaploid	2n = 6× = 36	Food and Feed	Rich in calcium, methionine and tryptophan	Drought and salt tolerance
Pearl millet	Bajra, gero, hegni, sanyo, dukhon and mahangu	*Pennisetum glaucum*	Gramineae	Diploid	2n = 2× = 14	Food, feed	High protein, starch & minerals	Drought & heat tolerance
Foxtail millet	Kangni (Hindi), Navane (Kannada), Thinai (Tamil), Kang (Gujarati) and Rala (Marathi)	*Setaria italica*	Gramineae	Diploid	2n = 2× = 18	Food, biofuel	—	Drought tolerance
Kodo millet	Kodo millet (English), Varagu, Haraka and Arikalu	*Paspalum scrobiculatum*	Gramineae	Tetraploid	2n = 4× = 40	Food and Feed	High-quality protein	Drought tolerance
Little millet	Little millet (English), Sawai, Samalu	*Panicum sumatrense* L.	Gramineae	Tetraploid	2n = 4× = 36	Food	High in phytochemicals, fibre	Abiotic stress tolerance
Proso millet	Proso, Cheena, Panivaragu, Variga and Baragu	*Panicum miliaceum*	Gramineae	Tetraploid	2n = 4× = 36	Food and feed	Rich in amino acids	Drought tolerance, early maturity

Table 1.
Salient features of millets crop [6].

S. No.	Crop Type	Rainfall Requirement (mm y^{-1})
1.	Sugarcane	2000–2200
2.	Banana	2000–2200
3.	Rice	1200–1300
4.	Sorghum	400–500
5.	Bajra	350–400
6.	Ragi	350–400

Table 2.
Rainfall requirement for various crops [7].

conditions. The millet crops can prove to be helpful in making strategies for conserving water resources and handling drought conditions. Millets are tolerant to the harsh climatic and soil conditions seen in Asia and Africa's semi-arid regions. Millets can be easily grown in harsh environments, particularly in those areas that have insufficient precipitation, low soil fertility, and are unsuitable for the world's primary crops.

Water deficit is one of the most environmental stresses affecting agricultural production and productivity around the world and may result in considerable yield reduction [11]. Hundreds of genes and their products respond to Water stress at transcriptional and translational level [12]. It can be asserted in the context of climate change, millets could be the answer to fighting climate change, poverty and malnutrition. Drought-proofing crops by producing heat-resistant cultivars is one aspect of this adaptation approach for millet crops [13, 14]. In comparison to most of other major crops such as rice and wheat, millets are highly flexible and stress resistant. Water stress affects these millet crops frequently and the problem appears to be becoming worse as weather and climate change is becoming more evident. Millets are agronomically advantageous because they are drought, heat, salt, and biotic stress tolerant, and they may live in marginal lands under rainfed circumstances. Drought-resistance mechanisms in many cereals have been studied by various researchers, but millets drought tolerance has been limited due to a number of factors. The response mechanism of plants in response to drought stress must be assessed. The goal of the present study is to evaluate changes in the millets crop's morphological, physiological, biochemical and molecular features as a result of drought stress, as well as plant tolerance mechanisms to the stress. Elements of drought stress in important millets crops such as finger millet, pearl millet, barnyard millet, foxtail millet, and proso millet were examined in this study. The current methods for discovering drought tolerance genes and metabolic pathways are also discussed. Recent advances in elucidating essential drought stress responses, phenotyping and QTL mapping for drought tolerance, genetic engineering of drought-tolerant crops, and crop management have also been reviewed.

2.2 Drought stress

Drought is outlined as "a temporary reduction in wetness accessibility, in which the amount of available water is significantly below normal for a specified period" and is one of the most common environmental stresses [10]. This condition negatively affects growth and productivity of crops. Water stress is a natural phenomenon in rain-fed (unirrigated) cultivated areas. In general, water stress often causes a series of morphological, physiological, biochemical and molecular changes that unfavourably affect plant growth, development and productivity. Plants are subjected to biotic and abiotic stressors, both of which have a significant impact on their survival [10]. Abiotic stressors are important environmental factors that limit crop productivity and affect the quality and amount of crop yield. Particularly water stress directly affects the physiology of plants, especially photosynthesis. In mountains, summer (*kharif*) crops often encounter water stress. In the face of a global scarcity of water resources, drought has already become a primary factor in limiting crop production worldwide. Water-limited crop production depends on the intensity and the pattern of drought which varies from year to year [15]. The severity of water shortage imposed on field crops also depends on the susceptibility of crops during different stages of their development. The general effects of drought on plant growth are well known. When soil moisture is deficient, crop establishment may be reduced, limited growth of plants, normal development patterns disrupted and eventually, final yield is lowered [16, 17].

2.3 Effects of water stress in millets

Plant cellular activities, growth, development, and economic yield are all affected by water stress. It affects the structure of membranes and organelles at the cellular level, as well as the hydration and structure of proteins and nucleic acids, and the pressure differential across the membrane cell wall complex. Drought causes stomatal closure, which leads to an excessive accumulation of reactive oxygen species (ROS) and oxidative stress. Lipid peroxidation and damage to other biomolecules occur because of this stress [18]. Phytohormones including Abscisic Acid (ABA) and Ethylene (ET) are frequently engaged in drought stress signalling and tolerance. Plant tolerance to drought, salt, and heat stress is improved by salicylic acid (SA) and jasmonic acid (JA) [19]. Plant height, length, biomass, weight, and grain number have all decreased as a result of drought stress in finger millet [20]. Water stress result a decrease in chlorophyll, photosynthesis, and RWC, as well as an increase in proline concentration, in both barnyard and finger millets. Finger millet demonstrated higher levels of tolerance than barnyard millet [21]. Drought-induced oxidative stress caused droopy shoots, curling leaves, increased proline, catalase and malondialdehyde (MDA) content, electrolyte leakage, impaired membrane integrity, and a considerable rise in H_2O_2 in finger and barnyard millets [21]. In finger millet, see [18] found an increased activities of antioxidant enzymes such as glutathione reductase (GR), superoxide dismutase (SOD), ascorbate peroxidase (APX), glutathione peroxidase (GPX) and catalase (CAT) during drought stress. A low photosynthetic and stomatal conductance rates, reduced root respiration, accumulation of protective metabolites (serine, threonine, valine, fructose, glucose, maltose, isomaltose, malate, itaconate) in roots, and better utilisation of carbon and nitrogen were found in a high temperature tolerant variety of foxtail millet [22]. The effect of water stress on different millets crop is given in **Table 3**.

To cope with various environmental challenges, plants have evolved various morphological, biochemical, physiological, and molecular systems. Plant cells detect stress events through a variety of sensors, which activate a variety of signalling pathways. Plant hormones, secondary messengers, transcription regulators, and signal transducers are all involved. Drought stress has a variety of effects on plants vegetative growth, reproductive development and molecular level, and are all affected by these changes (**Figure 2**).

Crop	Drought stress	Effect on plant	References
Finger millet	Drought stress was applied in two regimes (fully irrigated and after drought)	Significant reduction in leaf area, dry matter accumulation, seed weight, radiation use efficiency and yield	[21]
Finger millet	Drought stress was applied in two regimes (fully irrigated and after drought)	Results showed that reduction in plants growth, chlorophyll content due to drought stress	[22]
Pearl millet	After 3 weeks of germination drought stress was induced by ceasing water for 4 weeks	They observed a significant reduction in plant height, biomass, panicle, stalk length, no. of leaves, total grain number and weight	[23]
Finger millet and barnyard millet	Water stress was given (control, mild, medium, and severe condition	A significant reduction in chlorophyll content while MDA, proline and CAT activity increased during stress	[24]
Finger millet and barnyard millet	Water stress was given (control, mild, medium, and severe condition	Significant reduction in protein, carbohydrate, amylase and relative water content	[25]
Finger millet	Stress was given to plant by holding irrigation to 45-day old plants	Their results showed droopy, curling leaves, increased amount of proline, MDA, electrolyte leakage, Hydrogen peroxide and antioxidant activities.	[16]
Proso millet	Well- watered, drought stress at vegetative stage, ear emergence stage, seed filling stage and vegetative and seed filling stages	Drought stress results reduction in grain yield, Water use efficiency and harvest index	[26]
Pearl millet and sorghum	Controlled and stressed condition	Results showed that under water stress condition leaf water potential, rates of stomatal conductance, photosynthesis and transpiration decreased more in sorghum than in pearl millet	[27]

Table 3.
Effect of water stress on millets crop.

2.4 Morphological and physiological adaptations to stress in millets

Drought causes distinct morphological and physiological changes in millets crops, which can be seen at various phases of plant growth. Millets have a short life cycle and plant heights, as well as small leaf regions, thickened cell walls, and dense root systems, all of which aid in stress resistance. The morphological response of millets can often be divided into two categories: shoot and root. Changes in leaf form, leaf expansion, leaf area, leaf size, leaf senescence, leaf pubescence, leaf waxiness, cuticle tolerance, and shoot length are all components of the shoot. Changes in root dry weight, root density, and root length are included in the lower root section. Several studies have found that the relationship between morphological and physiological features including grain yield per plant, grain spike per plant, spike fertility, plant height, root length, shoot length, harvest index, chlorophyll content and relative

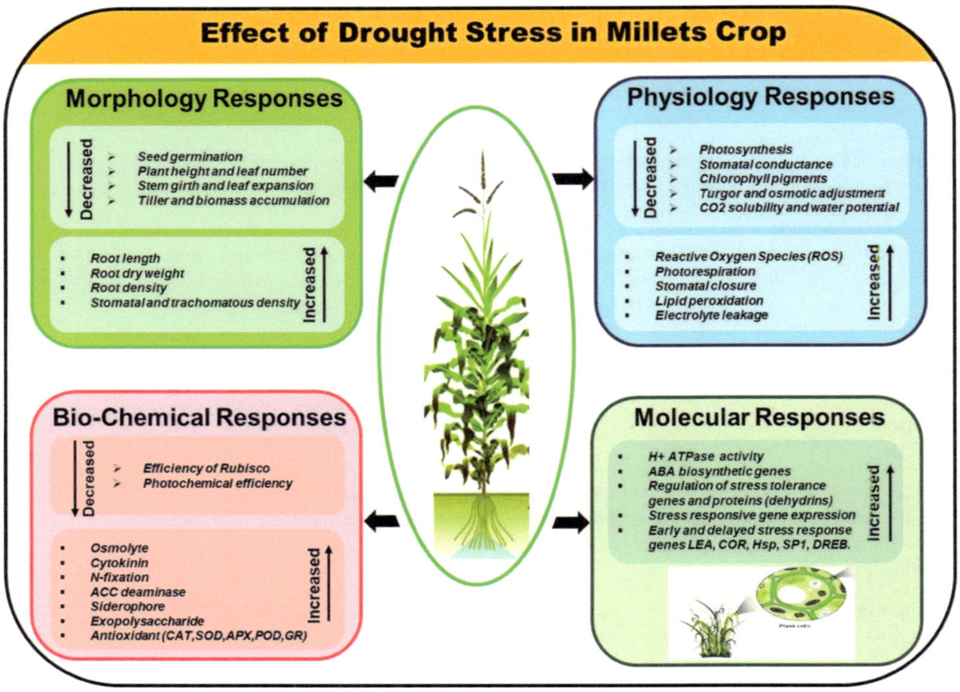

Figure 2.
Effect of drought stress on millets crop and its possible responses.

water content can be used to screen drought-tolerant crops. The various morphological and physiological adaptations of millets to water stresses are summarised in (**Table 4** and **Figure 3**).

2.5 Biochemical adaptations to stress in millets

Drought means water loss and dehydration at normal or even temperatures. Shrinking of cell leads to loss of turgor, osmotic stress and change of membrane potentials, upon severe water loss from the cells, membrane disintegration and abolition of metabolic processes occur [32]. In finger and barnyard millets, the biochemical adaptation response to water stress consists of increased proline content, reduced relative water content, and chlorophyll content [2, 6]. In barnyard and finger millets, some antioxidant enzymes represent adaptive mechanism against water stress. This consists of CAT, phenol and flavonoid content [2]. According to see [33], suggested that during drought stress condition in pearl millet, a higher expression of secondary metabolite genes associated with alkaloid, terpenoid, flavanols, lignin, wax, mevalonic acid (MVA), and Shikimic acid (SA) metabolic pathways were observed in flowering stage than vegetative stage. Following are some stress response mechanisms which takes place in millets during water stress condition (**Table 5** and **Figure 3**).

2.6 Molecular response against water stress in millets

The genome and transcriptome sequences of plants give crucial information for identifying the types of genes involved in the control of drought tolerance, especially in plants that are more resistant to water scarcity. In a study on transcriptome

Crop	Morpho-physiological parameters	Effect on plants	References
Little millet	Root length and shoot length	Root length increased gradually in all drought condition while shoot length decreased	[28]
Foxtail	Root hair density and length	Results showed that under water stressed condition root hair density and length increased.	[29]
Foxtail	Root length and shoot length	A significant reduction in shoot length while root length increased	[30]
Finger millet	Plant height, leaf area total dry matter (TDM), root/shoot ratio by weight and length, grain yield and harvest index	Plant height, leaf area, total dry matter ,root/shoot ratio by weigh, grain yield and harvest index decreased while root/shoot ratio by length increased	[18]
Finger millet and barnyard millet	Relative water content and chlorophyll content	Moderate rate of decline of RWC and chlorophyll	[24, 25]
Proso millet, Foxtail millet and Pearl millet	Seed number per ear and ear number per plant number of tiller, harvest index ear, peduncle and ear length and plant height	Water stress caused reduction in the number of tiller and ear, peduncle and ear length, seed no., harvest index and plant height. Their results showed that foxtail millet showed the greatest yield in both stress and non-stress conditions among other crops.	[31]
Finger millet	Leaf area, dry matter, seed weight and yield	Significant reduction in leaf area, dry matter accumulation, seed weight, radiation use efficiency and yield	[21]
Pearl millet	Plant height, biomass, panicle, stalk length, no. of leaves, total grain no. and weight	They observed a significant redcution in plant height, biomass, panicle, stalk length, no. of leaves, total grain number and weight	[23]

Table 4.
Morphological and physiological adaptations to stress in millets.

sequencing in finger millet provides information of 2824 genes under water stress condition [39]. According to see [40], SiLEA14 gene from foxtail millet, increased the tolerance in transgenic Arabidopsis plants to salt and osmotic stress. The induction of AKR1 gene (Aldo Keto Reductases) in roots and leaves of finger millets is studied with increasing water stress and salt stress. The up-regulated AKR1 gene shows physical defence against oxidative stress (**Table 6** and **Figure 3**).

A single gene known as -carbonic anhydrase (PgCA) was continuously up-regulated in pearl millet exposed to several abiotic stimuli such as drought, salinity, and heat [57]. There are other genes which are known to be involved in drought response or tolerance in millets were EcDehydrin 7 [49], Ec-apx1 [50], EcbHLH57 [51], EcbZIP60 [52], EcGBF3 [53], EcbZIP17 [54], mt1D [55], Metallothionein, Farnesylated protein ATFP6, Farnesyl pyrophosphate synthase and Protein phosphatase 2A, RISBZ4 [56] from finger millet, SiARDP [41], SiCDPK24 [42], SiLTP [43],

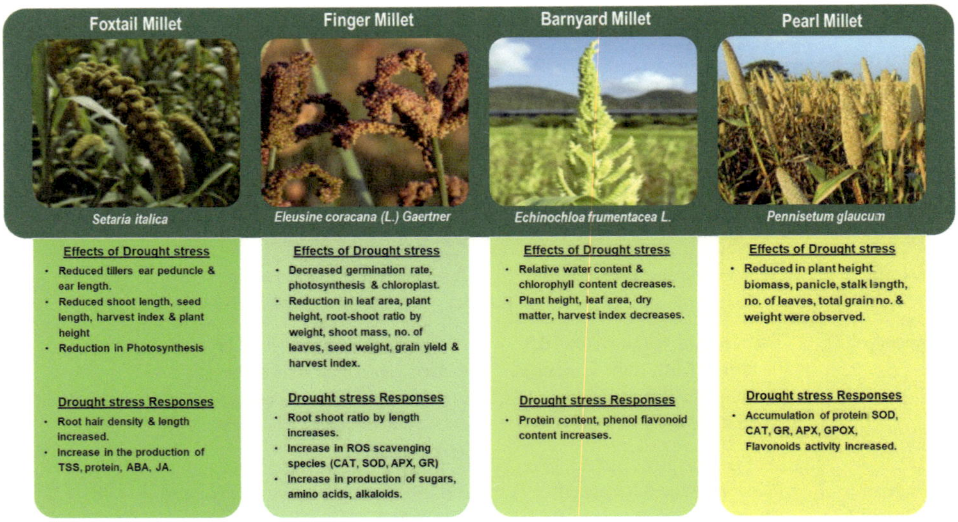

Figure 3.
Effect and response mechanism of drought tolerant millets.

SiATG8a [44], SiNF-YA1 and SiNFYB8 [45], SiASR1 [46], SiASR4 [47] and SiMYB56 [48] wild foxtail millet and PgGPx [58], PgRab7 [59] and PgeIF4A [60] from wild foxtail millet.

3. Management techniques for water stress in millets

In order to manage water stress, genetic advancements must be combined with appropriate cultural behaviours. To cope water stress, several cultural practices should be done. In recent year, conventional and molecular breeding techniques have been evolved to improve stress tolerance in plants.

3.1 Improvement of water use efficiency in millets

To improve water, use efficiency of millets, multiple factors such as physiological characteristics, time of planting, soil characteristics, meteorological conditions frequency of tillage and application of herbicide, should be measured properly [20, 6].

3.2 Conventional breeding

In conventional breeding, old plant breeding techniques such as introduction, selection and hybridization were used for the identification of stress-tolerant genetic traits in crops [24]. The major phases in conventional breeding are the artificial introduction of water stress and the selection of stress tolerant genotypes. Through hybridization, stress tolerant genes can be transferred into commercially growing types for improved performance under water stress conditions. Drought resistance selection can be done in the field or in a greenhouse. The field environment is particularly ideal for selection work. When it comes to determining which features to utilise to improve selection efficiency, those that contribute to productivity will be more valuable than

Crop	Biochemical parameters	Plants adaptation mechanism	References
Finger millet and barnyard millet	MDA, proline, CAT, phenol and flavonoids	A significant increase in MDA, proline, CAT, phenol and flavonoids activity during stress condition in both the millets crop	[24]
Finger millet and barnyard millet	Protein, carbohydrates and amylase	Significant reduction in protein, carbohydrate and amylase content	[25]
Finger millet	Proline, MDA, electrolyte leakage, Hydrogen peroxide and antioxidant activities.	Significant accumulation of proline, MDA, electrolyte leakage, Hydrogen peroxide and increased antioxidant activities.	[16]
Finger millet	Proline, glycine betaine and TSS and antioxidant enzymes (SOD, CAT, APX, GPX)	Accumulation of Proline, glycine betaine and TSS and antioxidant enzymes (SOD, CAT, APX, GPX) increased under stress.	[34]
Foxtail millet	TSS, proline, ABA and JA phytohormones	Significant increase in TSS, TSS, proline, ABA and JA phytohormones	[35]
Pearl Millet	Proline, superoxide dismutase (SOD), catalase (CAT), glutathione reductase (GR), ascorbate peroxidase (APX) and guaiacol peroxidase (GPOX) activities	Accumulation of Proline, superoxide dismutase (SOD), catalase (CAT), glutathione reductase (GR), ascorbate peroxidase (APX) and guaiacol peroxidase (GPOX) activities increased	[36]
Pearl Millet	ABA and water potential	Accumulated, higher ABA content while water potential decreased in stress condition	[37]
Pearl millet	Flavonoids, lignin, terpenoids	A higher accumulation of flavonoids, lignin and terpenoids under water stress condition	[38]

Table 5.
Biochemical adaptations to stress in millets.

those that contribute to survival. Using managed field-based stress around the blooming period, when crop is particularly susceptible to water stress has been the key to enhanced pace of advancement in breeding and selection.

3.3 Molecular breeding

Plant genome improvement is insufficient for the development of novel plant varieties using traditional breeding approaches. Since the 1990s, molecular markers have been utilised to identify superior hybrid lines to overcome this barrier in plant breeding procedures. Molecular breeding is evolving a new breeding technology, which has the potential to improve crops dramatically. The majority of attributes including stress tolerance, are quantitative and influenced heavily by the environment. With the help of molecular breeding techniques, many biofortified crops and plant types with high yielding variety with new characteristics like pest and disease resistant were developed. Marker-assisted breeding (known as gene stacking) is a more efficient and cost-effective method than traditional breeding, which makes it impossible to transfer many resistance genes into a single agricultural plant at

Genes	Plants response	References
Foxtail millet		
SiLEA 14	It enhances tolerance in Arabidopsis plants to drought stress	[40]
SiARDP	Increased drought and salt tolerance in transgenic Arabidopsis	[41]
SiCDPK24	Improved drought resistance in transgenic Arabidopsis	[51]
SiLTP	Enhanced drought and salt tolerance in transgenic tobacco	[52]
SiATG8a	It enhances tolerance in Arabidopsis plants to drought stress and nitrogen starvation	[53]
SiNF-YA1, SiNFYB8	Increased drought, salinity and osmotic stress tolerance in tobacco	[54]
SiASR1	Improved tolerance to drought and oxidative stress tolerance in transgenic tobacco	[55]
SiASR4	Increased drought and salt stress tolerance in transgenic foxtail millet and Arabidopsis	[56]
SiMYB56	It enhances drought tolerance in transgenic rice plants	[57]
Finger millet		
EcDehydrin7	Overexpression of EcDehydrin7 increased drought tolerance in transgenic tobacco	[42]
Ec-apx1	Enhanced drought tolerance in plant	[43]
EcbHLH57	Conferred drought, salt and oxidative tolerance in transgenic tobacco plants	[44]
EcbZIP60	Displayed tolerance to drought stress with enhanced photosynthesis in tobacco	[45]
EcGBF3	Enhanced drought, Osmotic and salinity tolerance in Arabidopsis	[46]
EcbZIP17	Improved tolerance to various environmental stresses via ER signalling pathways	[47]
Mt1D	Overexpression of mt1D in finger millet showed better osmotic adjustment and chlorophyll retention under drought	[48]
Metallothionein	Increased drought tolerance	[49]
Farnesylated protein ATFP6	Improved tolerance to drought stress	[49]
Farnesyl pyrophosphate synthase	Enhanced tolerance to drought stress	[49]
Protein phosphatase 2A	Improved tolerance to drought stress	[49]
RISBZ4	Improved tolerance to drought stress	[49]
Pearl millet		
PgCA	Upregulate PgCA when it exposed to drought condition	[50]
PgGPx	Improved drought and salt in transgenic rice	[58]
PgRab7	Transgenic rice plants displayed to drought and salinity stress	[59]
PgeIF4A	Enhanced tolerance to drought, salinity and oxidative stress in groundnut	[60]

Table 6.
Drought-tolerance genes and their key characteristics in millets.

the same time [61]. Every generation in conventional breeding necessitates time-consuming and costly progeny testing [62]. Resistance genes were introduced into the recurrent parent from the donor parent using marker-assisted breeding. This method can also be used to find the recurrent parent genotype utilising markers that are dispersed over the genome and aren't intimately connected to the target trait [63]. With the use of molecular markers, genes driving the majority of variation in stressed situations like heat and drought have been identified, and field testing has been augmented with marker-assisted selection. The genomic regions that control stress tolerance in plants is known as quantitative trait loci (QTL), and their identification and mapping aids in the early selection or screening of genotypes with stress resistance.

Markers can be used for QTL identification, mapping, and screening as well. Different approaches in molecular breeding for producing water stress-tolerant genotypes include Marker Assisted Selection (MAS), Marker Assisted Backcrossing, Marker Assisted Stacking, and so on (**Figure 4**).

3.4 Transgenic approach

Transgenic techniques require modification of both qualitative and quantitative characters by transferring desired genes [42]. The identification of a specific gene of interest that contributes to water stress tolerance, as well as the transfer of that gene through transgenic breeding, expands the possibilities for crop improvement in drought conditions. Water stress tolerance is induced when a transgene is overexpressed. The engineering of genes that encode growth regulators, suitable solutes, and antioxidants involved in stress tolerance have received the most attention.

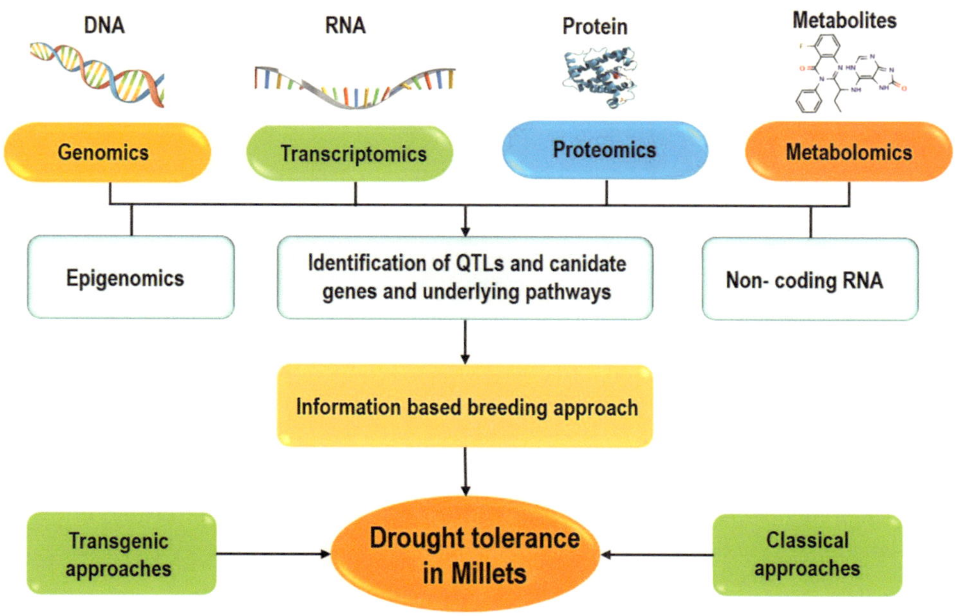

Figure 4.
Schematic representation of the molecular breeding approach for drought tolerance in millets crop.

3.5 Microbe-plant interactions

Microorganisms live in a wide range of environmental conditions, from sub-zero temperatures to desert extremes [64]. Plant-supported phylogenetically varied microbial communities to withstand drought by altering phytohormone levels in the rhizosphere and creating water-sequestering biofilms [65]. Rhizospheric bacteria that lives near plants roots produce a variety of biocontrol chemicals as well as plant growth promoters [66]. These microorganisms are also modifying soil structure, fertility, pH, and oxygen availability [67]. Some other examples of microbes assisted drought tolerance in plants have been listed in **Table 7**.

3.6 Drought resistance in millets: Potential characteristics

Water scarcity and recurrent drought spells in agricultural ecosystems have resulted in considerable yield losses for numerous crops around the world.

Crop	Microbes	Plant part used for isolation	Benefits	References
Foxtail	Actinobacter	Roots	The findings showed that host plants in the rhizoplane enrich specific bacteria and functions.	[68]
Foxtail	*Acinetobacter calcoaceticus* EU- LRNA-72 and *Penicillium sp*	Roots	Increasing glycine betaine, proline, and sugar buildup while lowering lipid peroxidation.	[69]
Foxtail	*Pseudomonas fluorescens*	Roots	Increased soil moisture and improved the root adhering soil/root tissue ratio by efficiently colonising the root adhering soil.	[70]

Table 7.
Drought tolerance in plants were improved by microbes.

Millets Crop	Trait	References
Foxtail	Root structure, grain weight, Dry root weight and root length, root thickness, shoot biomass, root depth, leaf area index, TSS, proline, ABA and JA phytohormones	[29, 30, 31, 35]
Pearl millet	Plant height, biomass, panicle, stalk length, no. of leaves, total grain no. and weight, leaf water potential, rates of stomatal conductance, photosynthesis and transpiration Seed number per ear and ear number per plant number of tiller, harvest index ear, peduncle and ear length and plant height	[23, 27, 31]
Finger millet	Plant height, leaf area total dry matter (TDM), root/shoot ratio by weight and length, grain yield and harvest index	[18, 21, 22, 23]
Barnyard millet	Plant height, leaf area total dry matter (TDM), root/shoot ratio by weight and length, grain yield and harvest index	[24, 25]
Proso millet	Seed number per ear and ear number per plant number of tiller, harvest index ear, peduncle and ear length and plant height	[26, 31]

Table 8.
Potential traits/characters for screening millets crop for drought resistance.

Innovative research results and the rapid development of several unique tools and methodologies in drought-resistance breeding have resulted in significant improvement. However, our understanding of drought resistance in millets crop is still limited, and we know very little about the complex genetic architecture of drought tolerance. We need to uncover the genetic bases of any trait associated with drought resistance in crops that can be used in crop breeding. There are various traits have been used to screen for drought tolerance, including smaller leaf area, leaf area maintenance, water use efficiency, root and shoot biomass, osmotic adjustment, pod number per plant, grain weight, biochemical parameters in millets by different researchers **Table 8** [25, 33, 35, 37, 38, 71].

4. Conclusions

Abiotic stresses are a significant barrier to crop productivity around the world. Plants respond to drought in a variety of ways, the most common of which are changes in plant development and morphology. Crop yields in the past were drastically reduced, according to historical evidence. Drought stress slows crop growth and development, resulting in changes in the crop's morphological, physiological, and biochemical characteristics [10].

Millets play an important role in the livelihood of the developing world's people; on the other hand, have a wide range of coping mechanisms to deal with these challenges. Mostly foxtail millet, pearl millet, proso millet and finger millets perform better tolerance against drought stress [2, 6, 18, 20, 23, 24, 26, 27, 29, 37, 42]. So far, we have studied stress tolerance mechanisms, adaptations, genetic modification, targeted expression of enzymes and transporters, and the role of proline, among other things, in millets [3, 5, 7, 24]. Both traditional and new methods of improvement have yet to be fully adopted. Climate change is expected to have a substantial impact on the types of crops farmed in the coming century.

In future, the key to successful crop improvement will be the ability to identify and access genetic diversity including new or improved variability for target traits by selecting parental germplasm proven to be resilient under likely climate change, including extreme events such as high temperatures [72]. Understanding how millets crops respond to drought stress is therefore essential for drought tolerance breeding. The traditional breeding strategy has demonstrated its ability to sustain productivity growth in numerous crops over the previous century. Meanwhile, current technical breakthroughs have hastened the production of novel cultivars and their impact.

An understanding of the genetic basis of drought tolerance in millets is prerequisite for plant breeders to evolve superior genotypes by adopting biotechnological approaches. There is an urgent need to improve the efficiency of molecular breeding and transgenic approach to develop new and proficient varieties with boosted natural osmolytes and raised tolerance for crops. Additional molecular studies are required to enhance knowledge on proteomic and metabolic activities on millets crop in response to drought stress. The current study's goal was to combine multiple drought tolerance mechanisms and improve these processes in millets crop. Drought stress causes plants to go through morphological, physiological, biochemical, and molecular changes. Changes in leaf structure, root growth, and stomata regulation are all morpho-physiological processes. Changes in phytohormonal levels, such as ABA, JA, Auxins, Ethylene, Gibberellins, Cytokinin, and Brassinosteroids, are biochemical processes. Plant phytohormone levels rise in response to drought, resulting in the activation of morphophysiological and other biochemical processes [73]. Furthermore, emerging 'omics' sciences such as genomics, transcriptomics, proteomics, and metabolomics could greatly improve our current understanding of the underlying drought-tolerant candidate genes as well as deciphering the complex gene networks and signalling pathways involved in drought tolerance in millets crop. Importantly, novel strategies like as GE tools and 'speed breeding' will aid in a better understanding and effectively speed up the creation of DS-resistant millets crop to reduce the risk of global food instability.

Acknowledgements

The authors are thankful to the Director, GBP-NIHE (Almora, Uttarakhand) for providing the necessary support to conduct this study. The work was supported by project no. GBPI/In-House Project-03/2020-2021/9094 funded by Ministry of Environment, Forest & Climate Change, Govt. of India (New Delhi).

Author details

Anjali Tiwari, Kapil Kesarwani*, Arushi Sharma, Tapan Ghosh, Nisha Bisht and Shailja Punetha
G.B. Pant National Institute of Himalayan Environment, Almora, Uttarakhand, India

*Address all correspondence to: kapilgbpi@gmail.com

References

[1] Tadele Z. Drought adaptation in millets. In: Shanker AK, Shanker C, editors. Abiotic and Biotic Stress in Plants—Recent Advances and Future Perspectives [Internet]. London: IntechOpen; 2016

[2] Tiwari A, Rastogi A, Singh V, Arunachalam A. Water stress effects on nutritional values and relative water content of barnyard and finger millet crops. International Journal on Agricultural Sciences. 2019;**10**(1):23-28

[3] Bennetzen JL, Dida MM, Manyera NWM, Devos KM: Characterization of Genetic Diversity in Finger Millet (Eleusine coracana). 2003. Available from: http://www.cerealsgenomics.org/

[4] FAO, FAOSTAT database Rome: Food and Agriculture Organization [Internet]. 2018. Available from: https://www.fao.org/faostat/en/#data/QC. [Accessed: 2020-07-04]

[5] Chandrasekara A, Shahidi F. Effect of food preparation methods on the antioxidant properties of finger millet grain phenolics. Journal of Agricultural and Food Chemistry. 2010;**58**:6706-6714

[6] Tiwari A, Rastogi A, Singh V, Arunachalam A. Effect of water stress on oxidative damage and antioxidant enzyme activity in finger millet and barnyard millet. Indian Journal of Hill Farming. 2020;**33**(1):36-45

[7] Dwivedi S, Upadhyaya H, Senthilvel S, Hash C, Fukunaga K, Diao X, et al. Millets: Genetic and genomic resources. In: Janick J, editor. Plant Breeding Reviews. Hoboken, NJ: John Wiley & Sons; 2012;**35**:247-375. DOI: 10.1002/9781118100509.ch5

[8] Lata C, Gupta S, Prasad M. Foxtail millet: A model crop for genetic and genomic. Critical Reviews in Biotechnology. 2013;**33**(3):328-343

[9] FAO, The World Sorghum and Millet Economies: Facts, Trends and Outlook Part II Millets Introduction [Internet]. 1996. Available from: http://www.fao.org/docrep/W1808E/w1808e0c.htm. [Accessed: 2021-09-18]

[10] Tiwari A, Punetha S, Kesarvani K. Drought stress and its impact on plant mechanism. International Journal of Plant Sciences. 2021;**16**(AAEBSSD):95-112

[11] Alexieva V, Sergiev I, Mapelli S, Karanov E. The effect of drought and ultraviolet radiation on growth and stress markers in pea and wheat. Plant, Cell & Environment. 2001;**24**:1337-1334

[12] Chinnusamy V, Schumaker K, Zhu J. Molecular genetic perspectives on cross- talk and specificity in abiotic stress signalling in plants. Journal of Experimental Botany. 2004;**55**:225-236

[13] Kurukulasuriya P, Benhin JKA, Deressa T. Will African agriculture survive climate change? World Bank Economic Review. 2006;**20**(3):367-388

[14] Benhin JKA. South African crop farming and climate change: An economic assessment of impacts. Global Environmental Change. 2008;**18**:666-678

[15] Ashraf M, Foolad MR. Roles of glycine betaine and proline in improving plant abiotic stress resistance. Environmental and Experimental Botany. 2006;**59**:206-216

[16] Manivannan P, Jaleel CA, Kishorekumar A, Sankar B,

Somasundaram R, Sridharan R, et al. Changes in antioxidant metabolism of Vigna unguiculata (L.) Walp. By propiconazole under water deficit stress. Colloids and Surfaces. B, Biointerfaces. 2007;**57**(1):69-74

[17] Sankar B, Jaleel CA, Manivannan P, Kishorekumar A, Somasundram R, Panneerselvam R. Drought-induced biochemical modifications and proline metabolism in Abelmoschus esculentus (L.) Moench. Acta Botanica Croatica. 2007;**66**:43-56

[18] Bhatt D, Negi M, Sharma P, Saxena SC, Dobriyal AK. Responses to drought induced oxidative stress in five finger millet varieties differing in their geographical distribution. Physiology and Molecular Biology of Plants. 2011;**17**(4):347-353

[19] Tiwari S, Lata C, Singh Chauhan P, Prasad V, Prasad M. A functional genomic perspective on drought signalling and its crosstalk with phytohormone-mediated signalling pathways in plants. Current Genomics. 2017;**18**(6):469-482

[20] Khatoon H, Singh V. Impact of water stress on physiological and biochemical parameters of finger millet (Eleusine coracana L.). Research in Environment and Life Science. 2016;**9**(12):1474-1477

[21] Tiwari A, Kesarwani K, Ghosh T, Rai S. Declining nutrients from our plates. Research Biotica. 2020;**2**(4):135-140

[22] Aidoo MK, Bdolach E, Fait A, Lazarovitch N, Rachmilevitch S. Tolerance to high soil temperature in foxtail millet (Setaria italica L.) is related to shoot and root growth and metabolism. Plant Physiology and Biochemistry. 2016;**106**:73-81. DOI: 10.1016/j.plaphy.2016.04.038

[23] Maqsood M, Azam Ali SN. Effects of environmental stress on growth, radiation use efficiency and yield of finger millet. Pakistan Journal of Botany. 2006;**39**(2):463-474

[24] Assefa A, Fetene M. Agro-morphological, physiological and yield related performances of finger millet [Eleusine coracana (L.) Gaertn.] accessions evaluated for drought resistance under field condition. Asian Journal of Agriculture and Rural Development. 2013;**3**:709-720

[25] Debieu M, Sine B, Passot S, Grondin A, Akata E, Gangashetty P, et al. Response to early drought stress and identification of QTLs controlling biomass production under drought in pearl millet. PLoS One. 2018;**13**(10):e0201635. DOI: 10.1371/journal.pone.0201635

[26] Seghatoleslami MJ, Kafi M, Majidi E. Effect of drought stress at different growth stages on yield and water use efficiency of five proso millet (Panicum miliaceum L.) genotypes. Pakistan Journal of Botany. 2008;**40**:1427-1432

[27] Subramanian VB, Maheshwari M. Comparison of physiological responses of pearl millet and sorghum to water stress. Proceedings of Indian Academy of Sciences (Plant Sciences). 1989;**99**(6):517-522

[28] Ajithkumar IP, Panneerselvam R. Osmolyte accumulation, photosynthetic pigment and growth of Setaria italica (L.) P. Beauv. Under drought stress. Asian Pacific Journal of Reproduction. 2013;**2**(3):220-224

[29] Liu TY, Ye N, Song T, Cao Y, Gao B, Zhang D, et al. Rhizosheath formation and involvement in foxtail millet (Setaria italica) root growth under drought stress.

Journal of Integrative Plant Biology. 2019;**61**(4):449-462

[30] Ajithkumar IP, Panneerselvam R. ROS scavenging system, osmotic maintenance, pigment and growth status of Panicum sumatrense Roth. Under drought stress. Cell Biochemistry and Biophysics. 2014;**68**(3):587-595

[31] Seghatoleslami MJ, Kafi M, Majidi E. Effect of drought stress at different growth stages on yield and water use efficiency of five proso millet (Panicum miliaceum L.) genotypes. Pakistan Journal of Botany. 2008;**40**:1427-1432

[32] Mahajan S, Tuteja N. Cold, salinity and drought stresses. An overview. Archives of Biochemistry and Biophysics. 2006;**444**:139-158

[33] Shivhare R, Asif MH, Lata C. Comparative transcriptome analysis reveals the genes and pathways involved in terminal drought tolerance in pearl millet. Plant Molecular Biology. 2020;**103**(6):639-652. DOI: 10.1007/s11103-020-01015-w

[34] Mundada PS, Nikam TD, Kumar SA, Umdale SD, Ahire ML. Morpho-physiological and biochemical responses of finger millet (Eleusine coracana (L.) Gaertn.) genotypes to PEG-induced osmotic stress. Biocatalysis and Agricultural Biotechnology. 2020;**23**:101488

[35] Xu BQ, Gao XL, Gao JF, Li J, Yang P, Feng BL. Transcriptome profiling using RNA-seq to provide insights into foxtail millet seedling tolerance to short-term water deficit stress induced by PEG-6000. Journal of Integrative Agriculture. 2019;**18**(11):2457-2471

[36] Vijayalakshmi T, Varalaxmi Y, Jainender S, Yadav SK, Vanaja M. Physiological and biochemical basis of water-deficit stress tolerance in pearl millet hybrid and parents. American Journal of Plant Science. 2012;**3**:1730-1740

[37] Henson IE, Mahalakshmi V, Alagarswamy G, Bidinger FR. Leal abscisic acid content and recovery from water stress in pearl millet (Pennisetum americanum (L.) Leeke). Journal of Experimental Botany. 1984;**35**:99-109

[38] Shivhare R, Lata C. Assessment of pearl millet genotypes for drought stress tolerance at early and late seedling stages. Acta Physiologiae Plantarum. 2019;**41**(3):39

[39] Qi X, Xie S, Liu Y, Yi F, Yu J. Genome-wide annotation of genes and noncoding RNAs of foxtail millet in response to simulated drought stress by deep sequencing. Plant Molecular Biology. 2013;**83**:459-473

[40] Wang M, Li P, Li C, Pan Y, Jiang X, Zhu D, et al. SiLEA14, a novel atypical LEA protein, confers abiotic stress resistance in foxtail millet. BMC Plant Biology. 2014;**14**(1):290

[41] Li C, Yue J, Wu X, Xu C, Yu J. An ABA-responsive DRE-binding protein gene from Setaria italica, SiARDP, the target gene of SiAREB, plays a critical role under drought stress. Journal of Experimental Botany. 2014;**65**(18):5415-5427

[42] Yu TF, Zhao WY, Fu JD, Liu YW, Chen M, Zhou YB, et al. Genome-wide analysis of CDPK family in foxtail millet and determination of SiCDPK24 functions in drought stress. Frontiers in Plant Science. 2018;**9**:541

[43] Pan Y, Li J, Jiao L, Li C, Zhu D, Yu J, et al. A non-specific Setaria italica L. lipid transfer protein gene plays a critical

role under abiotic stress. Frontiers. Plant Science. 2016;7:1752

[44] Li WW, Chen M, Zhong L, Liu JM, Xu ZS, Li LC, et al. Overexpression of the autophagy-related gene SiATG8a from foxtail millet (Setaria italica L.) confers tolerance to both nitrogen starvation and drought stress in Arabidopsis. Biochemical and Biophysical Research Communications. 2015;**468**(4):800-806

[45] Feng ZJ, He GH, Zheng WJ, Lu PP, Chen M, Gong YM, et al. Foxtail millet NF-Y families: Genome-wide survey and evolution analyses identified two functional genes important in abiotic stresses. Frontiers in Plant Science. 2015;**6**:1142

[46] Feng ZJ, Xu ZS, Sun J, Li LC, Ming C, Yang GX, et al. Investigation of the ASR family in foxtail millet and the role of ASR1 in drought/oxidative stress tolerance. Plant Cell Reports. 2016;**35**:115-128

[47] Li J, Dong Y, Li C, Pan Y, Yu J. SiASR4, the target gene of SiARDP from Setaria italica, improves abiotic stress adaption in plants. Frontiers in Plant Science. 2017;7:2053

[48] Xu W, Tang W, Wang C, Ge L, Sun J, Qi X, et al. SiMYB56 confers drought stress tolerance in transgenic rice by regulating lignin biosynthesis and ABA signaling pathway. Frontiers in Plant Science. 2020;**11**:785

[49] Singh RK, Singh VK, Raghavendrarao S, Phanindra MLV, Raman KV, Solanke AU, et al. Expression of finger millet EcDehydrin7 in transgenic tobacco confers tolerance to drought stress. Applied Biochemistry and Biotechnology. 2015;**177**(1):207-216

[50] Bhatt D, Saxena SC, Jain S. Cloning, expression and functional validation of drought inducible ascorbate peroxidase (Ec-apx1) from Eleusine coracana. Molecular Biology Reports. 2013;**40**(2):1155-1165

[51] Babitha KC, Vemanna RS, Nataraja KN, Udayakumar M. Overexpression of EcbHLH57 transcription factor from Eleusine coracana L. in tobacco confers tolerance to salt, oxidative and drought stress. PLoS One. 2015;**10**:e0137098

[52] Babitha KC, Ramu SV, Nataraja KN, Sheshshayee MS, Udayakumar M. EcbZIP60, a basic leucine zipper transcription factor from Eleusine coracana L. improves abiotic stress tolerance in tobacco by activating unfolded protein response pathway. Molecular Breeding. 2015;**35**:181

[53] Ramegowda V, Gill US, Sivalingam PN, Gupta A, Gupta C, Govind G, et al. GBF3 transcription factor imparts drought tolerance in Arabidopsis thaliana. Scientific Reports. 2017;7:9148

[54] Chopperla R, Singh S, Sangala R, Padaria JC, Mohanty S, Sharma TR, et al. The membrane tethered transcription factor EcbZIP17 from finger millet promotes plant growth and enhances tolerance to abiotic stresses. Scientific Reports. 2018;**8**:2148

[55] Hema R, Vemanna RS, Sreeramulu S, Reddy CP, Senthil KM, Udayakumar M. Stable expression of mtlD gene imparts multiple stress tolerance in finger millet. PLoS One;**9**:e99110

[56] Sreekumar PM, Nataraja KN, Yashoda BK, Ramegowda HV, Mamrutha HM, Rama N. Expression analysis of stress responsive pathway genes linked to drought hardiness in an adapted crop, finger millet (Eleusine coracana). Journal of Plant Biochemistry and Biotechnology. 2013;**22**(2):193-201

[57] Kaul T, Reddy PS, Mahanty S, Thirulogachandar V, Reddy RA, Kumar B, et al. Biochemical and molecular characterization of stress-induced beta-carbonic anhydrase from a C (4) plant, Pennisetum glaucum. Journal of Plant Physiology. 2011;**168**(6):601-610

[58] Islam T, Manna M, Reddy MK. Glutathione peroxidase of Pennisetum glaucum (PgGPx) is a functional Cd2+ dependent peroxiredoxin that enhances tolerance against salinity and drought stress. PLoS One. 2015;**10**:e0143344

[59] Tripathy MK, Tiwari BS, Reddy MK, Deswal R, Sopory SK. Ectopic expression of PgRab7 in rice plants (Oryza sativa L.) results in differential tolerance at the vegetative and seed setting stage during salinity and drought stress. Protoplasma. 2017;**254**:109-124

[60] Rao TSRB, Naresh JV, Reddy S, Reddy MK, Mallikarjuna G. Expression of Pennisetum glaucum eukaryotic translational initiation factor 4A (PgeIF4A) confers improved drought, salinity, and oxidative stress tolerance in groundnut. Frontiers in Plant Science. 2017;**8**:453

[61] Varshney RK, Singh VK, Kumar A, Powell W, Sorrells ME. Can genomics deliver climate-change ready crops. Current Opinion in Plant Biology. 2018;**45**:1-7

[62] Lema M. Marker assisted selection in comparison to conventional plant breeding: Review article. Agricultural Research and Technology. 2018;**14**(2):1-10

[63] Collard BCY, Jahufer MZZ, Brouwer JB, Pang ECK. An introduction to markers, quantitative trait loci (QTL) mapping and marker-assisted selection for crop improvement: The basic concepts. Euphytica. 2005;**142**(2):169-196

[64] Xue J, Yu Y, Bai Y, Wang L, Wu Y. Marine oil-degrading microorganisms and biodegradation process of petroleum hydrocarbon in marine environments: A review. Current Microbiology. 2015;**71**:220-228

[65] Bhattacharyya A, Pablo CHD, Mavrodi OV, Weller DM, Thomashow LS, Mavrodi DV. Rhizosphere plant-microbe interactions under water stress. Advances in Applied Microbiology. 2021;**115**:65-113. DOI: 10.1016/bs.aambs.2021.03.001

[66] Ullah A, Manghwar H, Shaban M, Khan AH, Akbar A, Ali U, et al. Phytohormones enhanced drought tolerance in plants: A coping strategy. Environmental Science and Pollution Research. 2018;**25**:33103-33118

[67] Finkel OM, Castrillo G, Paredes SH, González IS, Dangl JL. Understanding and exploiting plant beneficial microbes. Current Opinion in Plant Biology. 2017;**38**:155-163

[68] Jin T, Wang Y, Huang Y, Xu J, Zhang P, Wang N. Taxonomic structure and functional association of foxtail millet root microbiome. Gigascience. 2017;**6**:1-12. DOI: 10.1093/gigascience/gix089

[69] Kour D, Rana KL, Yadav AN, Sheikh I, Kumar V, Dhaliwal HS, et al. Amelioration of drought stress in foxtail millet (Setaria italica L.) by P-solubilizing drought-tolerant microbes with multifarious plant growth promoting attributes. Environmental Sustainability. 2020;**3**:23-34

[70] Niu X, Song L, Xiao Y, Ge W. Drought-tolerant plant growth promoting rhizobacteria associated with foxtail millet in a semiarid agroecosystem

and their potential in alleviating drought stress. Frontiers in Microbiology. 2018;**8**:2580

[71] Yu A, Zhao J, Wang Z, Tian G, Liu X, Guo E, et al. Transcriptome and metabolite analysis reveal the drought tolerance of foxtail millet significantly correlated with phenylpropanoids-related pathways during germination process under PEG stress. BMC Plant Biology. 2021;**20**:274. DOI: 10.1186/s12870-020-02483-4

[72] Newton AC, Johnson SN, Gregory PJ. Implications of climate change for diseases, crop yields and food security. Euphytica. 2011;**179**:3-18

[73] Ilyas M, Nisar M, Khan N, Hazrat A, Khan AM, Hayat K, et al. Drought tolerance strategies in plants: A mechanistic approach. Journal of Plant Growth Regulation. 2021;**40**(3):926-944

Chapter 4

Abiotic Stresses and Their Management in Vegetable Crop Production

Khursheed Hussain, Sameena Lone, Faheema Mushtaq, Ajaz Malik, Sumati Narayan, Majid Rashid and Gazala Nazir

Abstract

The stress concept, first proposed by Hans Selye in 1936, has also been applied to plants to describe adverse and environmental restrictions. The notion of plant stress, differs significantly from that of animals and humans. Due to ever fluctuating climatic circumstances and variables, the crop-environment interaction in horticultural crops leading to losses in yields and quality of produce occurs and thus climate change with respect to horticulture industry is attracting more attention. Abiotic stress is the leading cause of crop yield loss globally, lowering average yields by more than half for most main crop plants. Abiotic stressors are highly correlated and connected, causing morphological, biochemical, physiological and molecular changes in vegetable crops, leading in a significant profit drop. Water stress is the most common abiotic stress that causes significant losses in vegetable production, especially because it is often coupled by additional stresses like as salt, high temperatures, and nutritional deficits. Increased CO_2 and temperature in the atmosphere, variation in amounts of precipitation causing more frequent droughts and floods, widespread runoffresulting in soil nutrient leaching and a loss in fresh-water availability are all contributing factors. Efforts to mitigate various pressures should be focused both throughout the growing season and after harvest. Stress-tolerant cultivars are being developed using a variety of methods, including traditional breeding and transgenic technology. Instead of genetic engineering, using vegetable breeding procedures or directed breeding is one the best options to improve stress tolerance in vegetables. Besides, post-harvest treatments, application of growth regulators, antioxidants, germplasm and *in vitro* selection, and modified environment packaging with different plastics may all help to improve tolerance and hence increase the shelf and nutritive life of vegetables.

Keywords: abiotic stress, vegetable crops, yield, growth regulators, tolerance

1. Introduction

Crops are increasingly commonly subjected to abiotic stressors in today's climate change scenarios. Abiotic stress, such as drought, salt, and severe temperatures, which typically cause main crop losses around the world, are predicted to produce a yield loss of more than 50% in agricultural crop plants [1]. We should also focus on increasing the food production and supplies by two-fold till 2050 so as to fulfill the requirement and demand of human population. This can be achieved by a basic comprehension of mechanisms underlying abiotic stresses. It is for this reason that development of stress-tolerant plants has received gotten a lot of attention in recent years for these reasons. During production, processing, storage, and distribution, harvested vegetables might be subjected to a variety of abiotic stressors. When there is a moderate or severe abiotic stress, quality losses nearly always occur at market [2, 3]. Moreover, there are abiotic stressorswhich ultimately decreases the defense mechanisms of plants and increases their susceptibility to infection by pathogens. Understanding the nature andorigins of abiotic stressors that impact vegetables is critical. In addition, increased understanding opens up possibilities for better control or resistance [3, 4]. As a result, as postharvest problems limit the storage and shelf-life potential of vegetables, understanding the effects of field abiotic stresses on postharvest stress susceptibility will become increasingly important [4]. Why there are studies in relation to advances in physiology, molecular biology, and genetics is because of the fact that our comprehension of plants' responses to various stresses, as well as the basis for varietal tolerance variances is a big solution to the very big problem of abiotic stresses.

As the best way to solve these issues is to focus on both pre-harvest and post-harvest abiotic stress reduction, it's crucial to understand the relationship between pre-harvest and post-harvest abiotic stresses that occur during vegetable crop production and handling, storage, and distribution, respectively.

2. Pre-harvest stresses during vegetable production

2.1 Temperature extremes

Plants are vulnerable to low temperature stress if the temperature dips below 15°C, and high temperature stress if the temperature rises over 45°C. Plants are affected by high temperature stress in a variety of ways, including physiology, biochemistry, and gene regulation mechanisms. High temperatures during the reproductive period of plants can increase senescence, diminish fruit set and lower yield. Furthermore, temperature stress makes the plant vulnerable to pests and other environmental issues besides limitingor preventing seed germination, depending on the species and stress level. Exposure to high temperatures throughout the growing season might also affect antioxidants in vegetable crops. Susceptibility to postharvest chilling injury can be exacerbatedif the preharvest temperature causes chilling induced harm in the field. As a result, the magnitude of the preharvest temperature extreme will determine whether the exposure has a favorable or negative impact on postharvest stress sensitivity.

2.2 Drought

We know that a third of the world's population resides in areas that are having water-stress condition which may become more severe due to increasing carbon-dioxide concentrations in the atmosphere. The climatic changes are therefore more expected with severe droughts in furture. Water scarcity is expected to remain a major abiotic issue influencing worldwide crop output. Reduced canopy absorption of photosynthetically active sunlight, decreased radiation-use efficiency, and reduced harvest indexare all effects of soil moisture deficit on crop output. Drought circumstances during the development of vegetable crops are becoming increasingly common as a result of climate change patterns [5]. In root crops, field water deficiency (stress) has been found to have both positive and negative effects. Water stress prior to harvest (Irrigating to 25–75 percent of soil water field capacity) may weaken the cells, resulting in increased membrane leakage (cell damage) and, as a result, more weight loss in storage for root crops like carrots.In response to mechanical stresssuch as bruisingpotato cells undergo decompartmentationresulting in black spot conditions [6]. Water stress, especially during the tuber-forming stage, can make potatoes more susceptible to the black spot condition after harvest [6].

2.3 Light

Tomatoes grow smaller when cultivated in low light environments, such as early spring in northern latitudes [7], and because theratio of surface area to volume is higher in smaller fruits, vulnerability to postharvest desiccation stress increases [8]. It has been reported that when lettuce is cultivated under less intense lights, due to less number of photons, photosynthetic efficiency as well as quality traits like vitamin C is suboptimal and thus decreasing shelf life after harvest.

2.4 Salinity

Excessive quantities of soluble salts in the soil water (soil solution) are known as saline soil, and they can significantly affect plant growth, resulting in lower crop yields and even plant mortality in extreme cases. Salts are substances that dissolve into ions, such as NaCl, MgSO4, KNO3, and sodium bicarbonate. Electrical conductivity (ECe), exchangeable sodium percentage (ESP) or sodium adsorption ratio (SAR), and pH of soil paste (saturated) extractare used to calculate it. As a result, saline soils have saturated soil paste extracts with an ECe of more than 4 dSm^{-1}, an ESPof less than 15%, and a pH of less than 8.5 [9]. Tomato crop grown inhigh salinity generate smaller fruits with a greater soluble solids content. It is a matter of understanding that fruits having smaller size have more surface area than their volumes (Known as Surface-area to Volume ratios), making them more vulnerable to postharvest water loss (desiccation stress) [8].

2.5 Flooding stress

Crops are also subjected to severe physiological stress as a result of sudden inundation following heavy rainfall events. Plants must adapt to a distinct, but equally challenging, flooding environmentthat occurs in a more regular cycle of seasonal fluctuations in river levels and concomitant slow flooding of crop lands. Waterlogging is the term for soil condition when there is flooding that creates hypoxia which also affects stems causing wilting with other physiological conditions.

2.6 Plant nutrition

When plants are unable to complete the reproductive stage of their life cycle due to a shortage of mineral components, they are considered essential. In several crops, calcium supplementation during production has been associated to postharvest issues [10]. Calcium has been proposed as a possible signaling molecule involved in the development of abiotic stress cross tolerance [11]. As a result, the effect of preharvest calcium nutrition on postharvest stress resistance is likely to be multifaceted, and it will depend on whether the vegetable is also exposed to abiotic environmental difficulties. Preharvest nitrogen levels are frequently linked to poor postharvest vegetable quality. Excessive nitrogen fertilization causes large zinc and aluminum accumulations in cabbage, as well as nitrate-induced manganese deficiency [10]. Nitrogen fertilizer affects black spot susceptibility in potatoes. Nitrogen deficit or lower-than-recommended nitrogen treatment rates, on the other hand, will almost always result in higher vitamin C concentration in plants. Vitamin C concentration has been connected to storage life potential [12], which is likely due to the antioxidant nutrient's usefulness in preventing oxidative damage, which leads to quality losses in storage. Potassium deficit in carrots is linked to increased weight loss during storage. At potassium levels below 1 mM in the soil media, weight loss was directly linked to increased membrane leakage (i.e., damaged cells) in carrot tissues.

3. Post-harvest stresses during handling, storage and distribution of vegetable crops

3.1 Temperature extremes

The relevance of temperature in determining a harvest index cannot be overstated. Many products, particularly those delivered by air or ocean container, face persistent postharvest temperature abuse during distribution [13]. Heat treatments limit respiration and ethylene generation, diminish protein synthesis, and accelerate protein degradation in the short term. Fruit vegetables, root and tuber crops are all susceptible to chilling [14]. The ability to produce flavor has been shown to be a sensitive early indicator of cold stress effects, andChilling injury is related with visible (surface pitting, interior browning) and texture (accelerated softening and development of mealiness) alterations.

3.2 Oxygen and carbon-dioxide level

The difficulty is worsened when processing products in modified atmosphere (MA) containers rather than controlled atmosphere (CA) systems, because temperature is generally not as easily managed in MA packages as produce goes through a distribution chain [4]. High CO_2 stress can cause a diverse array of physiological problems, such as black heart in potatoes and brown stain in lettuce [15]. Chilling stress, ethylene-induced diseases, and vulnerability to pathogenic attack can all be influenced by high CO_2 levels [15].

Low oxygen levels are known to cause stress-induced changes in metabolism and metabolite accumulations [16], but acute low oxygen injury does not show up until the tissue is re-aerated and an uncontrolled oxygen burst (consisting of hydrogen peroxide and other radicals) occurs, causing lipid peroxidation, protein denaturation, and membrane injury [17]. Varied vegetables have different low O_2 stress thresholds depending on architecture, temperature, physiological age, the presence of supplementary gases (e.g., CO_2, CO, SO_2), and the duration of exposure.

3.3 Mechanical injury

Injury from impact is linked to product loading for transportation, incidents during transportation, unloading, and throughout the packaging and processing lines. Cuts can produce brief increases in respiration, ethylene synthesis, phenolics generation, and cell degradation near the injury site [18]. The severity of the reaction to cutting is highly dependent on tissue properties, the maturity of the vegetable of interest, the coarseness or sharpness of the cutting object employed, and the cutting temperature.

Fresh-cut vegetable products have the highest rate of cut injuries. Many vegetables suffer from cut injuries during the harvesting process, which are more severe in machine-harvested produce than in hand-harvested produce. Cut-edge browning or blackening is the most typical symptom of cutting-related diseases, however yellowing in green tissues and whitening on carrots can also occur [18]. The degree and amount of bruising received is influenced by maturity, tissue or cellular orientation at the region of the injury, water potential, form of the object imparting the bruising force, energy and angle of impact, and product temperature. Internal black patches appear in potato tubers as a result of impact trauma.

3.4 Desiccation

Water loss in vegetable tissues causes degradation, which is a significant problem inpostharvest processing and distribution [19]. Water stress can cause rapid senescence, which manifests as tissue weakening, membrane degradation, and yellowing in addition to wilting [19]. The vapor pressure deficit, which is the connection that explains the difference in water activity of the vegetable and the water activity of the atmosphere surrounding it [19], is thedriving factor for water loss. The more of a vapor pressure deficit there is, the more water is lost. For limiting water loss in any vegetable, there are a few postharvest handling principles to follow:

a. Delays in chilling will result in extended exposure to higher vapor pressure deficit circumstances, therefore cooling after harvest should be done as soon as possible.

b. When a warm product is placed in a cool area, it loses water faster than a cool product, hence rapid precooling before storage is crucial; and

c. Water loss can be minimized by storing products at the coolest storage temperature and with the highest relative humidity achievable.

4. Abiotic stress response Mechanismsat the biochemical and molecular levels

Acute and sub-acute reactions to abiotic stresses exist; acute responses reflect circumstances where cell death is a direct resultof the stress, whereassub-acute responses represent cases where the stress induces adaptive changes in biochemistry and gene expression [6]. Many reactive oxygen species (ROS), especially hydrogen peroxide, operate as signaling molecules that cause biochemical changes in gene expression. Abiotic stresses disrupt vegetable cellular homeostasis, resulting in increased production of reactive oxygen species (ROS) in the apoplast, mitochondria, peroxisomes, cytoplasm, chloroplasts, and endoplasmic reticulum [20]. The cell's ability to cope at first will be largely determined by its endogenous free radical scavenging capacity [20]. When free radical production surpasses endogenous scavenging capability, ROS interact with sensors whose full natureis unknown, triggeringmitogen activated protein kinase (MAPK) cascade events and up-regulating transcription factors and calcium/calmodulin kinases directly. The MAPK cascade reactionactivates a number oftranscription factors that enable de novo ROS formation, ROS scavenging systems, heat shock protein accumulation, and NADPH supply modulation in the cell [20]. Some MAPK cascade pathways have also been connected to ethylene production particularly, which is likely why ethylene production appears to be intrinsic to most stress reactions. Heat shock proteins (HSPs) accumulation, whichismediated by transcription factor activation downstream of the MAPK cascade [20], has been shown to improve long-term stress resistance in afflicted tissues [21]. Under normal, stress-free conditions, HSPs are thought to be an essential element inprotein folding, assembly, translocation, and degradation [22]. HSPs have also been linked to protein stabilization, membrane stability, and protein refolding under stress. As a result, HSPs are assumed to play akey role in protecting plant tissues from stress by ensuring cellular homeostasis. Metabolic Responses to Abiotic Stress Signaling, physiological regulation, and defense responses are all heavily influenced by metabolism when the environment is hostile and plant development is harmed. Abiotic stressors alter the production, concentration, transport, and storage of primary and secondary metabolites in a feedback loop.

5. Drought stress and plant metabolomics

Water deficiency stress causes a variety of physiological and biochemical changes in plants, including cell development and photosynthesisarrest, as well as increased respiration. Thegenome's expression is modified extensively, activating and suppressing a wide range of genes with various roles. Abscisic acid (ABA) accumulates inhydric-stressed plant tissues and increases stomatal closure, which reduces transpiration. Plants use this technique to reduce water loss and reduce stress injury.

6. Temperature extremes and plant metabolomics

In sensitive vegetables, chilling stress causes the formation of lipid peroxidation products, superoxide anions, and hydrogen peroxide, as well as a reduction in flavor volatile synthesis. It interferes with ethylene metabolism [23], causing the softening process to speed up. Cell wall metabolism is also altered by chilling stress, with up-regulation of cell wall breakdown enzymes as pectin methyl esterase and endopoly-galacturonase. Heat stress causes metabolic changesthat lead to the buildup of heat shock proteins, which are known to give long-term stress resistance in heat-exposed animals [22]. It can alsostop lycopene from being produced and accumulated.

7. Wounding stress and plant metabolomics

The upregulation of phenylalanine ammonia lyase (PAL) by wounding stress causes phenolic buildup [24]. The upregulation of PAL was linked to the synthesis of ethylene by wounds. Theinitial reaction to wounding stress is characterized by a progressive accumulation of ACC synthase, ACC, and ethylene synthesis in tomatoes, which can last up to 2 hours, butethylene production reduces if the sliced tomatoes are kept for longer periods of time. ACC synthase and ACC, on the other hand, continue to accumulate, showing that there is a capability to generate ethylene, but ACC to ethylene transition is prevented or inhibited. Other metabolites, such asisocoumarin in carrots [25], anthocyanins in red-pigmented lettuce midribs [25], methanethiol, allylisothiocyanates, and dimethyl disulfide in cabbage [26], and six-carbon aldehydes and alcohols in cut peppers [27], rise in response to wounding stress.

8. Oxidative stress and plant metabolomics

When exposed to anaerobic or anoxic environments, the body goes through two separate phases. The first phase is characterized by a metabolic shift caused by a limitation of the principal electron acceptor, molecular O2, in the mitochondrial electron transport chain. ATP levels, pyruvate dehydroxylase activity, and cytoplasmic pH have all decreased as a result of this impairment [15]. The activity ofpyruvate decarboxylase, alcohol dehydrogenase, and lactate dehydrogenase also rise under these conditions [15]. Anaerobic respiration is induced, and acetaldehyde, ethanol, ethyl acetate, and/or lactateaccumulate. Damage to the mitochondrial electron transport chain also causes electron leakage in the cells, resulting in the production of superoxide anions and hydrogen peroxide, which are destroyed by existing cellular antioxidant systems. During an anoxic or hypoxic incident, ascorbate and glutathione levels can also rise. Any or all of these alterations in vegetable tissues are signs of oxygen stress and have an impact on qualitative qualities. However, it isnot until the second phase, when the vegetable is returned to greater O_2 atmospheres, that true tissue harm occurs. When cells are exposedto aerobic environments, rates of oxygen radical production in the impaired electron transport chainincrease, resulting in largeaccumulations of superoxide anion, hydrogen peroxide, and hydroxyl radical that cannot be fully decomposed by existing antioxidant protection systems, resulting in membrane damage, enzymatic browning, and cell death [17].

9. Salinity stress and plant metabolomics

Salt stress causes the production of abscisic acid, which is transferred to guard cells and seals stomata, resulting in impaired photosynthesis, photoinhibition, and oxidative damage. This results in an instantaneous halt tocell expansion, which manifests as slowed plant growth, rapid development, and senescence. Plants use strategies including reduced photosynthesis, stomatal conductance, and transpiration rates to cope with salt stress. Because sodium ion has the same chemical structure as potassium ion, it competes with potassium uptake and suppresses it. Potassium deficiency inhibits growth since it is involved in the capacitance of a wide range of enzyme activities, as well as controlling membrane potential and cell turgor.

10. Desiccation stress and plant metabolomics

Under extreme handling conditions, desiccation stress increased the osmotic potential of carrots, which is a function of free sugars in the roots. (i.e., at 13°C). The elevation in osmotic potential in response to water loss was most likely explained by increased polysaccharide hydrolyzing enzyme activity in response to stress. As a result, enzymes such polygalacturonase and pectin esterase may become more active, resulting in a loss of cell wall integrity and a rise in soluble sugars. This could account for at least some of the loss of stiffness observed in carrots as they lose water.

11. Management of Abiotic stresses

11.1 Abiotic stress tolerance and the role of plant growth regulators

Applications of growth regulators may also improve stress resistance, particularly invegetables that are prone to rapid senescence in response to stress. As a result, anti-ethylene products such amino vinyl glycine (AVG) and 1-methylcyclopropene (1-MCP) may help to extend storage or shelf life if ethylene synthesis in reaction to stress is a major problem. Bell peppers and zucchini squash can benefit fromother growth hormones, such as methyl jasmonate (which increases leaf senescence). In some crops, abscisic acid has been shown to decrease chilling-induced damage. Other growth regulators (e.g. 2, 4-D) have been proposed for use in avoiding senescence in leafy vegetables, but their practical utility is limited.

11.2 Postharvest treatments to enhance stress resistance

A variety of postharvest treatments have been tested to improve vegetable abiotic stress tolerance [3]. Temperature manipulation (including intermittent warming), extreme atmospheres (high O_2, CO_2, and low O_2), growth regulators, anti-transpirants, antioxidant dips, growth regulators, nitric oxide, and ethanol haveall been put to the test [3]. In fresh-cut items, hot or warm water treatments have been demonstrated to reduce cutting-induced damage. Treatments like these can also be utilized to prevent chilling harm by inducing heat shock proteins [21]. In cut and packed lettuce, a warm water treatment has also been demonstrated to lower irradiation susceptibility [28]. Gradual cooling (2°C per day) has been reported to lower the tomato chilling injury susceptibility, most likely by enabling the intrinsic stress resistance systems to mature before true chilling conditions [29]. Atmospheric treatments, such as modified or controlled atmospheres, have been demonstrated to aid in the reduction ofchilling injury in a variety of vegetable crops.

11.3 Use of molecular probes for marker-assisted breeding

As there are many genes and proteins linked to stress tolerance in plants, applying the stress of interest and doing quantitative trait loci (QTL) analysis is the best way to find stress tolerant lines. This method canbe employed with intact plants and/or harvested plant parts, with the plant component of interest in the breeding improvement strategybeing used in most cases [30]. The method necessitates the examination of adaptive changes in QTL expression rather than constitutive expression. To distinguish between resistant and susceptible lines, a stress protocol must be created to which the target vegetable will be exposed. However, because the stress response is complicated, successful use of QTLs will necessitate an interdisciplinary effort that integratesbiochemistry, gene mapping, and phenotyping activities to allow for reliable interpretation and successful application of adaptive QTLs for stress resistance selection.

11.4 Molecular engineering

Because of two major factors, molecular engineering for stress resistance in vegetables is limited:

1. Due to the general complexity of the stress response network, single gene insertions are unlikely to modify stress resistance, and.

2. Many key vegetable crops have yet to be successfully transformed, and ways to do so have yet to be devised. Insertion of anti-freeze genes to defend against low temperatureharm is one area where progress has been made. Future developments, on the other hand, will demand a deeper molecular knowledge of the stress response network and regulatory points.

11.5 Germplasm selection

Germplasm selection and cultivarsdeveloped through breeding programmes will be more resistant to postharvest stress and so have improved storage capacity. Hodges *et al.* [12] were able to establish that variations in the balance of antioxidant systems in the tissues caused larger accumulations of ROS, notably hydrogen peroxide, in a cultivar that was more prone to yellowing. They theorized that greater ROS levels were directly responsible forthe yellowing of the chlorophyll in spinach leaves. *In vitro* selection isa technique in which plant cells from a target vegetable are tissue-cultured and subjected to a stressor, with the surviving cells used to regenerate new plants with higher stress resistance [31]. It is a far less expensive technique than molecular engineering, and laboratories may be built up practically everywhere in the globe with basic utilities and utilizing low-cost technology. This method has proven to be particularly effective in regenerating germplasm from a variety of crop plants that can be regenerated using tissue culture techniques.

11.6 Postharvest handling

Simple changes to postharvest handling methods can occasionally result in a significant reduction in stress exposure, allowing for longer storage and/or shelf life. Because most produce is refrigerated as a required step to minimize rotting and protect food safety, avoiding low temperature stress is typically impossible. While quick cooling is normally suggested to maintain quality, delaying or gradually chilling sensitive crops to allow them to acclimate to storage and handling conditions may be beneficial. Slow cooling of tomatoes from 12–4°C at a rate of 2°C per day has recently been proven to decrease chilling harm when stored at the lower temperature [19]. In many circumstances, modified environment packaging is thought tohelp minimize moisture loss in fresh-cut and whole vegetables by controlling humidity surrounding the product. Many vegetables have improved their shelf life and quality by using plastic film packaging or wraps to prevent desiccation [19]. Anti-transpiration coatings have also been demonstrated to be useful in preserving quality by reducingwater loss. A single stress resistance-enhancing treatment may notprovide enough resistance to all postharvest stresses [3]. To obtain optimal levels of resistance, it may be beneficial to explore using a combination of two or more stress tolerance increasing treatments.

12. Conclusion

Drought, excessive watering, severe temperatures, salt, and mineral toxicityall have a negative impact on thegrowth, development, yield, and quality of vegetables on and off the farm until they reach the customer. Furthermore, climate change has introduced new environmental variablesthat may influence the vulnerability of vegetables to postharvest stress. Crop management can have a substantial impact on stress susceptibility. Adapting horticulture crops to changing surroundings could be the single most essential action we can take to prevent climate change's negative consequences. The management steps must be followed from the field circumstances to the point where the product reaches the consumer. While many crops are being bred for stress resistance to helpthem adapt to climate change, it is unclear if in the field breeding for stress resistance will also transmit stress resistance qualities to the harvestedcomponent. To properly evaluate the benefits that abiotic stress during production may offer for postharvest abioticchallenges, it's critical tounderstand the basis of molecular and biochemical response networks to diverse stresses faced in the field and throughout the postharvest continuum. Theuse directed plant breeding to improve the toleranceto stress in vegetablesshould probably be the focus of attention. Temperature modulation, usage of growth regulators, anti-transpirants, antioxidants, and other sorts of postharvest management can improve tolerance and hence extend the keeping quality of vegetables. The use of various plastic sheets to generate tailored environment packaging is one of the best promising technologies. As a result, new vegetable types that are resistant to abiotic stressors are urgently needed to assure food security and safety for many years to come.

Author details

Khursheed Hussain*, Sameena Lone, Faheema Mushtaq, Ajaz Malik, Sumati Narayan, Majid Rashid and Gazala Nazir
Division of Vegetable Science, SKUAST-Kashmir, Shalimar (J&K), India

*Address all correspondence to: khussainskuast@gmail.com

References

[1] Bray EA, Bailey-Serres J, Weretilnyk E. Responses to abiotic stresses. In: Gruissem W, Buchannan B, Jones R, editors. Biochemistry and Molecular Biology of Plants. Rockville, MD: American Society of Plant Physiologists; 2000. pp. 1158-1249

[2] Toivonen PMA. Effects of storage conditions and postharvest procedures on oxidative stress in fruits and vegetables. In: Hodges DM, editor. Postharvest Oxidative Stress in Horticultural Crops. New York: Food Products Press; 2003a. pp. 69-90

[3] Toivonen PMA. Postharvest treatments to control oxidative stress in fruits and vegetables. In: Hodges DM, editor. Postharvest Oxidative Stress in Horticultural Crops. New York: Food Products Press; 2003b. pp. 225-246

[4] Toivonen PMA. Postharvest storage procedures and oxidative stress. HortScience. 2005;**39**(5):938-942

[5] Morris LL. Field and transit chilling of fall-grown tomatoes. In: Proceedings of the Conference on Transportation of Perishables. Davis, CA, USA: University of California; 1954. pp. 101-105

[6] Shibairo SI, Upadhyaya MK, Toivonen PMA. Potassium nutrition and postharvest moisture loss in carrots (*Daucus carota* L.). Journal of Horticultural Science and Biotechnology. 1998;**73**(6):862-866

[7] Stevens LH, Davelaar E. Biochemical potential of potato tubers to synthesize blackspot pigments in relation to their actual blackspot susceptibility. Journal of Agricultural and Food Chemistry. 1997;**145**(11):4221-4226

[8] Hamouz K, Becka D, Morava J. Effect of environmental conditions on the susceptibility to mechanical damage in potatoes. In: Agricultura-Scientia-Prosperitas: Prosperity of Agriculture is Science, Prague (Czech Republic), 11-12 Dec 2001. Ceska Zemedelska Univerzita; 2001

[9] Shibairo SI, Upadhyaya MK, Toivonen PMA. Postharvest moisture loss characteristics of carrot (Daucus carota L.) cultivars during short-term storage. Scientia Horticulturae. 1997;**71**(1):1-12

[10] Witkowska I, Woltering EJ. Pre-harvest light intensity affects shelf-life of fresh-cut lettuce. Acta Horticulturae. 2010;**877**:223-227

[11] Abrol IP. Salt-affected soils: An overview. In: Chopra VL, Paroda SL, editors. Approches for Incorporating Drought and Salinity Resistance in Crop Plants. New Delhi, India: Oxford and IBH Publishing Company; 1986. pp. 1-23

[12] Sams CE, Conway WS. Preharvest nutritional factors affecting postharvest physiology. In: Bartz JA, Brecht JK, editors. Postharvest Physiology and Pathology of Vegetables. 2nd ed. New York: Marcel Dekker, Inc; 2003. pp. 161-176

[13] Bowler C, Fluhr R. The role of calcium and activated oxygens as signals for controlling cross-tolerance. Trends in Plant Science. 2000;**5**(6):241-246

[14] Hodges DM, Wismer WV, Forney CF. Antioxidant responses in leaves of two cultivars of spinach (Spinacia oleracea L.) differing in their senescence rates. Journal of the American Society for Horticultural Science. 2001;**126**(5):611-617

[15] Toivonen PMA. Postharvest physiology of vegetables. In: Hui YH, Sinha N, Ahmed J, Evranuz EÖ, Siddiq M,

editors. Handbook of Vegetables and Vegetable Processing. Ames, IA: Wiley-Blackwell Publishing; 2010. pp. 199-220

[16] Saltveit ME. Temperature Extremes. In: Bartz JA, Brecht JK, editors. Postharvest Physiology and Pathology of Vegetables. 2nd ed. New York: Marcel Dekker, Inc; 2003. pp. 457-483

[17] Kanellis AK, Tonutti P, Perata P. Biochemical and molecular aspects of modified and controlled atmospheres. In: Yahia E, editor. Modified and Controlled Atmospheres for Storage, Transportation and Packaging of Horticultural Commodities. Boca Raton, FL: CRC Press; 2009. pp. 553-567

[18] Blokhina OB, Chirkova TV, Fagerstedt KV. Anoxic stress leads to hydrogen peroxide formation in plant cells. Journal of Experimental Botany. 2001;**52**(359):1179-1190

[19] Blokhina O, Virolainen E, Fagerstedt KV. Antioxidants, oxidative injury and oxygen deprivation stress: A review. Annals of Botany. 2003;**91**(2):179-194

[20] Miller AR. Harvest and handling injury: Physiology, biochemistry, and detection. In: Bartz JA, Brecht JK, editors. Postharvest Physiology and Pathology of Vegetables. 2nd ed. New York: Marcel Dekker, Inc; 2003. pp. 177-208

[21] Toivonen PMA, DeEll JR. Physiology of fresh-cut fruits and vegetables. In: Lamikanra O, editor. Fresh-Cut Fruits and Vegetables: Science, Technology and Market. Boca Raton, FL: CRC Press; 2002. pp. 91-123

[22] Ben-Yehoshua S, Rodov V. Transpiration and water stress. In: Bartz JA, Brecht JK, editors. Postharvest Physiology and Pathology of Vegetables. 2nd ed. New York: Marcel Dekker, Inc; 2003. pp. 111-159

[23] Jaspers P, Kangasjärvi J. Reactive oxygen species in abiotic stress Signalling. Physiologia Plantarum. 2010;**138**(4):405-413

[24] Panja P, Kuchi VS, Adhikary S. Abiotic stresses and its management for quality production of vegetables. Research Trends in Horticulture Sciences. 2018:37-56. DOI: 10.22271/ed.book07a03

[25] Sabehat A, Weiss D, Lurie S. The correlation between heat-shock protein accumulation and persistence and chilling tolerance in tomato fruit. Plant Physiology. 1996;**110**(2):531-537

[26] Wang W, Vinocur B, Shoseyov O, Altman A. Role of plant heat-shock proteins and molecular chaperones in the abiotic stress response. Trends in Plant Science. 2004;**9**(5):244-252

[27] Obenland DM, Vensel WH, Hurkman WJ. Alterations in protein expression associated with the development of Mealiness in peaches. Journal of Horticultural Science & Biotechnology. 2008;**83**(1):85-93

[28] Wu CM, Liou SE. Effect of tissue disruption on volatile constituents of bell peppers. Journal of Agricultural and Food Chemistry. 1986;**34**(4):770-772

[29] Shibairo SI, Upadhyaya MK, Toivonen PMA. Changes in water potential, osmotic potential, and tissue electrolyte leakage during mass loss in carrots stored under different conditions. Scientia Horticulturae. 2002;**95**:13-21

[30] Inari T, Yamauhi R, Kato K, Takeuchi T. Changes in Pectic polysaccharides during the ripening of cherry tomato fruits. Food Science and Technology Research. 2002;**8**(1):55-58

[31] Wang CY. Reduction of chilling by methyl Jasmonate. Acta Horticulturae. 1994;**368**:901-907

Chapter 5

Copper Toxicity in Plants: Nutritional, Physiological, and Biochemical Aspects

Flávio José Rodrigues Cruz, Raphael Leone da Cruz Ferreira, Susana Silva Conceição, Edson Ugulino Lima, Cândido Ferreira de Oliveira Neto, Jessivaldo Rodrigues Galvão, Sebastião da Cunha Lopes and Ismael de Jesus Matos Viegas

Abstract

Copper (Cu) is an essential micronutrient for plants because it participates in several redox reactions and the structural constitution of the Fe–Cu cluster. Although it is required in small concentrations at toxic levels, Cu triggers physiological and biochemical disorders that reduce plant growth. In higher plants, the normal range of Cu concentration is in the range of 2–20 mg Cu kg^{-1} DW. Above the upper limit of this range, Cu toxicity may occur if the plants are not tolerant to the stress caused by toxic levels of Cu. In view of the growing agricultural and industrial activity that are the main sources of Cu addition in nature, understanding the physiological and biochemical mechanisms of Cu toxicity in plants constitutes an important tool for the selection of more tolerant genotypes based on biochemical and physiological indicators to heavy metal stresses. In this chapter, we propose a systematic review of plants grown under toxic levels of Cu, based on the responses of physiological, biochemical, and nutritional variables. Understanding these responses will contribute to improving the understanding of the basic mechanisms of stress tolerance by toxic levels of Cu in higher plants, providing valuable information for the improvement of genotypes resistant to toxic levels of Cu in the plant culture medium.

Keywords: nutritional disorder, gas exchange, micronutrient, plant growth

1. Introduction

The micronutrient copper (Cu) is a transition metal with atomic number 29, an atomic mass of 63.5 g mol^{-1}, and a density of 8.96 g cm^{-3}. It is the 25th most abundant chemical component in the Earth's crust and the third most used worldwide [1].

Cu occurs naturally in soils with contents ranging from 60 to 125 mg kg^{-1} [2]. It is an essential micronutrient for plant development and, under physiological conditions, it exists in the form Cu^{+} and Cu^{2+}. Cu acts as a structural element in regulatory proteins and participates in the electron transport chain of photosynthesis and respiration, oxidative metabolism, cell wall metabolism, and hormonal signaling [3, 4]. Among the proteins, ascorbate oxidase, Zn/Cu superoxide dismutase, and Cu amino oxidase are those that have more than one Cu atom in their structures (8, 2, and 2 Cu atoms, respectively) [5].

Cu is absorbed in the form of Cu^{2+} or Cu chelate and, despite being poorly mobile in plants, it can be translocated from old leaves to new leaves. Its concentration in the dry mass of plants is small and generally ranges from 2 to 20 mg kg^{-1}. However, concentrations in the dry mass of plants varying between 20 and 100 mg kg^{-1} are toxic to most plants [6].

Soil contamination by Cu is mainly caused by human action represented by industrial, mining, and agricultural activities. Intensive use of Cu-containing agrochemicals or swine manure is the main source of Cu entry into the agricultural soils [7, 8]. This scenario is worrying because the world population is expected to reach nine billion inhabitants in 2050 [9], which suggests an increase in the area of agricultural crops to meet the world demand for plant and animal foods with the consequent increase in the consumption of Cu-containing agrochemicals to phytosanitary purposes.

At toxic levels in the soil, Cu reduces the absorption of water and mineral nutrients [10], promotes oxidative stress [11] and affects photosynthesis [12], causing reduced growth [13] and plant production [14].

This chapter aims to address the impact of copper toxicity on plant growth, emphasizing key physiological, biochemical, and nutritional variables in studies of heavy metal toxicity in higher plants.

2. Copper absorption mechanism

Mineral nutrients are absorbed from the soil matrix by plants through the cells of the root epidermis and then transferred to its center through the parenchyma, endoderm, and xylem. This unidirectional pathway of transition-metal absorption is supported by different metal transporters that act in coordination with other metal transport molecules that sequester/chelate so that adequate absorption and transport of ions occurs in all plant tissues throughout the metagenesis of plants [15].

Cu absorption occurs through three types of transporters present in the plasma membrane of root cells—P-type ATPase copper transports, COPT copper transports, ZIP family transports, and NRAMP family transports. P-type metal transporters (P-type ATPase copper transports) are responsible for the transmembrane transport of toxic metals, such as divalent cations (Cu^{2+}, Zn^{2+}, Cd^{2+}, and Pb^{2+}). These transporters use ATP to pump charged molecules across cell membranes [15, 16]. The family of proteins responsible for the transport of Cu in the reduced form is the COPT transport protein (copper transporter protein). These proteins are formed by five members, that is, COPT1, COPT2, COPT3, COPT4, and COPT5 [17]. The zinc–iron-regulated transporter-like protein (ZIP) are involved in the absorption of Cu ions. Depending on the concentration of Cu in the plant-growth medium, ZIP2 or ZIP4 can act as a Cu carrier [18]. The natural resistance-associated macrophage protein (NRAMP) is responsible for the reallocation of ions, such as Fe, Ni, Mn, Zn, and Cu, from the root and shoot against cellular and vacuolar membranes [19].

After being absorbed by the roots, Cu can be transported in the xylem to the shoot in the form of Cu^{+} and Cu^{2+}. But usually, the transport of Cu from the root to the shoot is done in the form of Cu^{-} complex. In plants, the xylem is the main source of

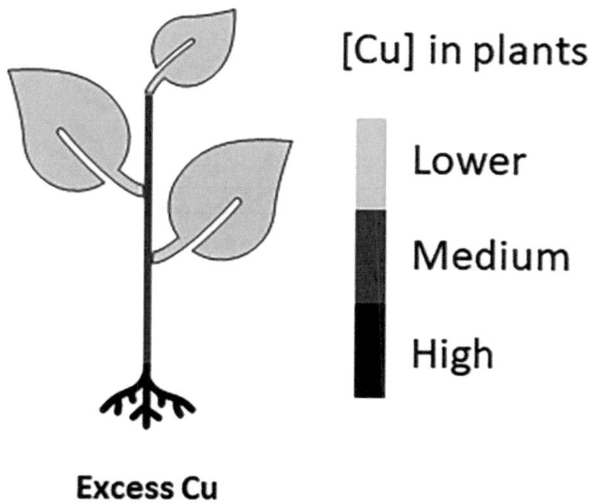

Figure 1.
Variation in Cu content in different plant structures. Because it is not very mobile in plants, Cu accumulates to a greater extent in the root, stem, and leaves.

Cu for the shoot [20, 21]. However, Cu is poorly mobile in plants, accumulating to a greater extent in the root system compared to the shoots of plants [3, 22] (**Figure 1**).

3. Effect of copper toxicity on antioxidant metabolism

Cu is an essential micronutrient for plants because it is a component of several enzymes that act in electron transport and catalysis of redox reactions in mitochondria and chloroplasts [3]. However, at toxic levels, Cu generates oxidative stress that damages cellular structures and molecules, such as DNA, proteins, and lipids. Cu has specific chemical characteristics that generate oxidative stress through the catalysis of oxidation–reduction reactions that form reactive oxygen species, such as singlet oxygen (O_2^-), hydrogen peroxide (H_2O_2), and hydroxyl radical (OH^-). These reactive oxygen species promote lipid peroxidation [23].

The biochemical process of free radical formation triggered by Cu involves three mechanisms—participation in Fenton-type reactions (1), reduced glutathione depletion (2), and substitution of Fe in the Fe–S cluster (3) [24].

$$Cu^{+} + H_2O_2 \rightarrow Cu^{2+} + OH^{-} + OH \quad (1)$$

$$2Cu^{2+} + 2GSH \rightarrow Cu^{+} + GSSG + 2H \quad (2)$$

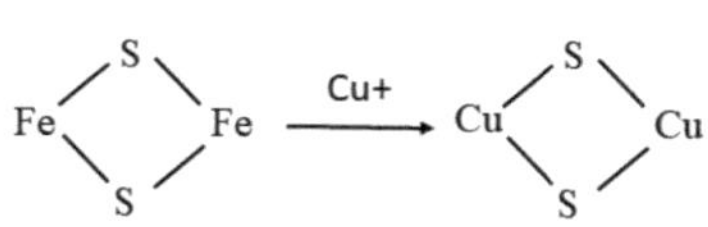

(3)

However, the production of free radicals can be potentiated by the Haber–Weiss reactions (4) and (5) in which superoxide and hydroxyl radicals are formed [25].

$$H_2O_2 + OH \rightarrow H_2O + O_2^{-} + H^{+} \quad (4)$$

$$H_2O_2 + O_2^{-} \rightarrow O_2 + OH^{-} + OH \quad (5)$$

Plants under heavy metal stress conditions have enzymatic (superoxide dismutase, catalase, ascorbate peroxidase, and glutathione reductase) and nonenzymatic (reduced glutathione and ascorbate) mechanisms that minimize the negative impact of oxidative stress triggered by free radicals.

Cu toxicity (100–500 μM Cu) increases the activity of important antioxidant enzymes, such as superoxide dismutase, ascorbate peroxidase, catalase, and glutathione reductase [26, 27], because, in excess, Cu increases the production of O_2^{-} and H_2O_2 by modulating the Haber–Weiss reaction [25]. In addition, nonenzymatic mechanisms are involved in the attenuation of oxidative stress triggered by Cu toxicity, such as reduced glutathione (GSH) and ascorbate. GSH is a tripeptide widely distributed in plant tissues. It reacts with harmful oxidants to protect thiol groups of proteins [28, 29]. The concentration of GSH can be reduced in plants treated with toxic levels of Cu because it is used to neutralize free radicals directly or indirectly [28, 30].

Iron–sulfur (Fe–S) groups are versatile cofactors formed by inorganic iron and sulfide atoms because various metabolic pathways and proteins require Fe–S groups for proper functioning [31–33]. The main function of Fe–S-containing proteins is the transfer of electrons to produce redox potential in chloroplasts and mitochondria [34]. Under Cu toxicity conditions, iron (Fe) can be replaced by Cu in Fe–S groups, impairing the functioning of the electron transport chain in mitochondria and chloroplasts and, consequently, reducing plant growth [5].

4. Effect of copper toxicity on gas exchange

Cu is an essential micronutrient element in the transport of electrons between photosystems II and I because it enters the structural constitution of plastocyanin, an important component of the electron transport chain along photosystems [3–5]. However, under toxic levels of Cu, photosystem II is the most sensitive site of Cu action [5, 35]. The most apparent toxic effect of Cu is the inhibition of the oxygen evolution complex and the quenching of photochemical variables [36–38]. Cu^{2+} ions inhibit both the acceptor and the donor site of photosystem II. However, the oxidizing site of photosystem II is more sensitive to toxic Cu [39, 40].

Cu at toxic levels affects photosynthesis because it decreases Rubisco enzymatic activity and stomatal conductance, in addition to increasing the intercellular concentration of CO_2 in plants [12, 41, 42]. Therefore, Cu can affect photosynthesis due to stomatal (reduction in stomatal conductance) and non-stomatal factors (damage to Rubisco and the electron transport chain). For example, toxic levels of Cu (100–1000 mg L^{-1}) impose a stomatal limitation on photosynthesis because the decrease in photosynthetic activity and intercellular concentration of CO_2 occurs in response to the decrease in stomatal conductance caused by Cu [43, 44]. On the other hand, Cu toxicity can promote non-stomatal limitation of photosynthesis due to a decrease in photosynthesis and stomatal conductance in parallel with an increase in the intercellular concentration of CO_2 [12]. This suggests biochemical damage to enzymatic components of photosynthesis, such as Rubisco. Toxic Cu (700 mg kg^{-1}) negatively affects photosynthesis, transpiration, stomatal conductance, and internal cell concentration in plants [45]. Toxic levels of Cu (800 mg kg-1) cause a considerable reduction in photosynthesis and stomatal conductance, with deleterious effects on plant height and stem diameter [13].

5. Effect of copper toxicity on nutritional status

Depending on the concentration and plant species/genotype, heavy metals can induce toxicity that manifests itself through a decrease in chlorophyll concentration, reduced nitrate reductase activity, nutritional disorder, and, consequently, reduced plant growth [11, 46]. Cu at toxic levels promotes changes in root cell membrane permeability, expression of phosphorus membrane transporters (P), volume, and root area, which result in lower P absorption [47].

The negative impact of toxic levels of Cu on the mineral metabolism of plants may originate from morphological changes in the root system that decrease the surface for nutrient uptake. Thus, there is a close relationship between nutritional disorders and plant root growth. Toxic levels of Cu (50 μM) reduce the diameter, length, area, and root biomass, coinciding with lower levels of P, Ca, Mg, Mn, and S in plants [47].

Toxic levels of Cu negatively affect the nitrogen metabolism of plants by decreasing the reduction of nitrate and its assimilation into organic compounds. The toxicity of 10.3 μM reduces the activity of nitrate reductase in roots and leaves, contributing to a decrease in root nitrate content [48]. High concentrations of Cu (5–20 μM) inhibit the activity of nitrate reductase and genes encoding the synthesis of low-affinity nitrate transporters (NRT.1), resulting in lower nitrogen uptake and accumulation in plants [42]. Cu (20–100 μmol) considerably reduces the activity of nitrate reductase and the accumulation of nitrate in leaves and roots of seedlings, with negative repercussions on plant growth [49].

Regarding sulfur metabolism in plants, Cu toxicity (5 –10 μM) induced an increase in total sulfur and glucosinolate levels as a defense against Cu-induced stress (antioxidant role and/or Cu chelating agent) [50]. The concentration of thiols and the activity of the enzyme O-acetylserine (thiol)lyase are increased in the aerial part of plants in the presence of 20 μM of Cu [51]. These changes suggest that Cu at toxic levels modulates changes in the concentration and activity of components of sulfur metabolism in plants to minimize the deleterious effects of Cu.

Despite the toxic effect of Cu, root architecture can be remodeled by plants. For example, the total density of lateral roots, the density of lateral roots less than 0.3 cm, and the density of lateral roots greater than 0.3 cm show an increase in their values when subjected to toxic levels of Cu (10–75 μM). Despite this, root remodeling lacks molecular studies to highlight the mechanisms involved in this process. Lateral root elongation and primary root mitotic activity are inhibited in the toxicity range of 50–75 μM Cu. These changes in the root system occur together with a reduction in the concentration of essential nutrients, such as P, K, Ca, Mn, and Fe, in the aerial part of plants [52].

Plant roots show a reduction in the concentration of essential nutrients (Ca, K, and F) and a decrease in root length and biomass in the presence of toxic levels of 4–80 μM of Cu [53], suggesting that morphological changes in roots and shoots induced by Cu are linked to nutritional disorders in higher plants exposed in the medium and long term to toxic levels of Cu.

Specie	Root Cu concentration	Shoot Cu concentration	References
Oryza sativa	91–3380 μg g^{-1} DM	17.5–508 μg g^{-1} DW	[46]
Solanum lycopersicum	0.6–3.5 μg g^{-1} DM	0.18–0.4 μg g^{-1} DM	[11]
Cucumis sativus	89.91–5575.10 μg g^{-1} DM	20.10–81.65 μg g^{-1} DM	[47]
Brassica pekinensis	50–500 mg kg^{-1} DM	20–40 mg g^{-1} DM	[48]
Swingle citrumelo	20–270 mg kg^{-1} DM	4–7 mg kg^{-1} DM	[43]
Zea mays	—	0.15–0.25 μg g^{-1} DM	[42]
Amaranthus tricolor	—	8.3–43.5 mg kg^{-1} DM	[28]
Cannabis sativa	10.0–470 μg g^{-1} DM	5.0–20 μg g^{-1} DM	[51]
Arabidopsis thaliana	14.7–205.3 mg kg^{-1} DM	7.0–10.6 mg kg^{-1} DM	[52]
Spinacea oleracea	33–4727 μg g^{-1} DM	25–729 g^{-1} DM	[53]
Citrus grandis	8–31 μg g^{-1} DM	17–690 μg g^{-1} DM	[41]
Grapevine ("14/ rute/1")	0.012–1372 mg g^{-1} DM	0.016–0.101 mg g^{-1} DM	[12]
Spinacea oleracea	50–1700 mg g^{-1} DM	nd*–40 mg g^{-1} DM	[13]

**Not detectable.*

Table 1.
Ranges of Cu concentration in roots and shoots of plants grown under copper toxicity.

The Cu is poorly mobile in plants, with higher levels in the root, stem, and leaves, respectively. This pattern of Cu accumulation is evident in studies of Cu toxicity in plants, in which the highest ranges of Cu contents are observed in the root system followed by the shoot (**Table 1**).

6. Effect of copper toxicity on growth

Cu is an essential oligonutrient for plant growth, but it is lethal when it exceeds the permissible limit, leading to poor plant-growth performance with loss of production. Thus, the reduction in plant growth induced by heavy metals is a final consequence of changes that initially occur at the biochemical, physiological, and mineral levels of plants.

The physiological functions of essential plant nutrients are disturbed when the concentration of these nutrients is below an adequate limit due to the presence of toxic levels of Cu in the culture medium. Thus, the reduction in the concentration of Fe, Zn (in leaves), and Mg (roots) modulated by toxic levels of Cu coincides with the decrease in leaf area, root length, and SPAD index in plants. This suggests that Cu at toxic levels affects the physiological functions of these nutrients, consequently decreasing plant growth [54]. Natural populations of plants not tolerant to the toxicity of 100 μM of Cu (*Rumex japonicus*) have a reduction in shoot and root dry mass together with a decrease in P, Mg, and Fe contents in the root and P and Fe in the shoot [55]. Thus, disorders in the mineral metabolism of plants are accompanied by a reduction in plant growth and/or production when the plant culture medium has toxic levels of Cu.

In the context of net carbon assimilation, photosynthesis is a vital process, because it allows the carboxylation of CO_2 and the synthesis of phosphate trioses that will constitute the different structural and nonstructural components of plants. However, this process is affected under toxic concentrations of heavy metals due to damage to the photochemical and/or biochemical apparatus, culminating in reduced plant growth. Cu toxicity (700 mg kg^{-1}) has a negative impact on gas exchange (photosynthesis, transpiration, stomatal conductance, and internal cell concentration), chloroplast pigments (chlorophylls a and b, and carotenoids), and photochemical parameters (Fv/Fm, qP, and ETR). These changes promoted by Cu contribute decisively to the reduction of vegetative growth [45]. Furthermore, Cu has a specific action in inhibiting the oxygen evolution complex in photosystem II, which is associated with the oxidation of cytochrome b559 [5, 38]. In soils naturally rich in Cu (3050 g g^{-1}), height and seed production are strongly reduced under wheat-growing conditions. These results were accompanied by lower photosynthetic activity and lower concentration of chlorophylls [14]. All these disorders in the photochemical apparatus of photosynthesis have the ultimate effect of reducing shoot growth, roots, and plant production.

7. Morphological symptoms of copper toxicity in plants

Toxicity due to toxic levels of Cu manifests itself in the root system, which tends to lose its vigor with longer exposure to Cu, acquiring a dark color and thickening, culminating in reduced growth. In the aerial part of the plants, the morphological symptoms of Cu toxicity are evidenced by the chlorosis of the leaves and a marked reduction in growth (leaf area, height, and stem diameter). In advanced stages, leaf edges may become necrotic [10, 56]. **Figure 2** shows the general aspects of Cu toxicity

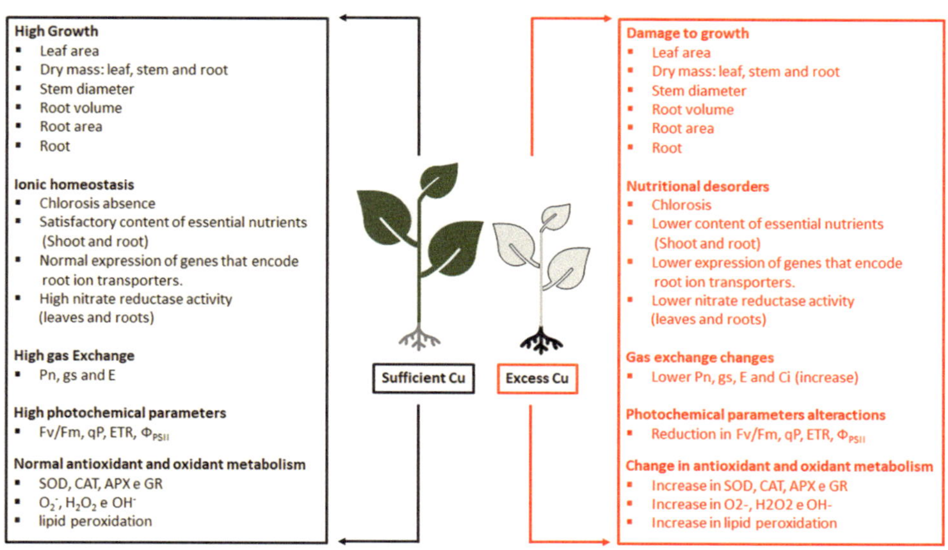

Figure 2.
Overview of copper toxicity in plants at the level of gas exchange, oxidant-to-antioxidant metabolism, mineral metabolism, and growth.

in plants, which affects photosynthetic, antioxidant, and mineral metabolism, culminating in reduced plant growth.

8. Conclusion

Contamination of soil and plants with trace elements is one of the most severe ecological problems in many industrialized countries due to industrial, mining, and agricultural activities. However, important progress has been made in understanding the biochemical, physiological, nutritional, and morphological mechanisms of Cu toxicity associated with higher plants. Cu toxicity reduces plant growth because Cu at toxic levels acts as a prooxidant, increasing the production of free radicals that cause damage to cellular and subcellular structures, causing protein oxidation and lipid peroxidation. In another toxic mechanism, excess Cu can replace S in F-S groups, forming Fe–Cu, which affects electron transport in chloroplasts. In addition, Cu toxicity affects the oxygen evolution complex and cytochrome b559. These changes together imply a reduction in the synthesis of phosphate triose and, consequently, in the production of dry matter in plants. In the context of mineral metabolism, Cu has a strong impact on the reduction of P, Ca, and Fe nutrient concentrations. Despite the existence of numerous studies involving the toxicity of Cu, little is reported in the literature about the prooxidant role of Cu at the physiological and molecular levels. For example, plants accumulate most of Cu in the root, but there is a need for further understanding of whether enzymatic and nonenzymatic antioxidant mechanisms in the root contribute to the tolerance of accumulator plants to toxic levels of Cu. Furthermore, understanding how toxicity modulates sulfur metabolism is crucial because sulfur is a key element in the antioxidant activity of glutathione in plants. This suggests the need for further studies to demonstrate the toxic role of Cu and its relationship with the production of oxidative stress in plants because antioxidant metabolism is one of the key mechanisms in inducing tolerance to trace element toxicity.

Author details

Flávio José Rodrigues Cruz[1*], Raphael Leone da Cruz Ferreira[1], Susana Silva Conceição[2], Edson Ugulino Lima[3], Cândido Ferreira de Oliveira Neto[2], Jessivaldo Rodrigues Galvão[2], Sebastião da Cunha Lopes[4] and Ismael de Jesus Matos Viegas[2]

1 Federal Institute of Education, Science and Technology of Amapá, Laranjal do Jari, AP, Brazil

2 Federal Rural University of Amazonia, Belém, PA, Brazil

3 State University of Pará, Belém, PA, Brazil

4 Federal University of Pará, Belém, PA, Brazil

*Address all correspondence to: fjrc@bol.com

References

[1] Karlin KD, Tyeklár Z. Bioinorganic Chemistry of Copper. Springer Science & Business Media; 2012. DOI: 10.1007/978-94-011-6875-5

[2] Kabata-Pendias A. Trace Elements in Soils and Plants. 4th ed. Boca Raton, FL: CRC Press; 2010. p. 548

[3] Marschner H. Mineral Nutrition of Higher Plants. San Diego: Academic Press; 1995. p. 889

[4] Yruela I. Copper in plants: Acquisition, transport and interactions. Functional Plant Biology. 2009;**36**:409-430

[5] Maksymiec W. Effect of copper on cellular processes in higher plants. Photosynthetica. 1997;**34**:321-342

[6] Kabata-Pendias A, Pendias H. Trace Elements in Soils and Plants. 3rd ed. Boca Raton: CRC Press; 2001. p. 413

[7] Van-Zwieten L, Merrington G, Van-Zwieten M. Review of impacts on soil biota caused by copper residues from fungicide application. Environmental Consultants. 2004;**1**:5-9

[8] Gräber I, Hansen JF, Olesen SE, Petersen J, Østergaard HS, Krogh L. Copper and zinc in danish agricultural soils in intensive pig production areas. Danish Journal of Geography. 2005;**105**:15-22

[9] FAO. United Nations Food and Agriculture Organization - How to Feed the World in 2050. pp. 1-35. Available from: http://www.fao.org/fileadmin/templates/wsfs/docs/expert_paper/How_to_Feed_the_World_in_2050.pdf [Accessed: November 28, 2021]

[10] Shabbir Z, Sardar A, Shabbir A, Abbas G, Shamshad S, Khalid S, et al. Copper uptake, essentiality, toxicity, detoxification and risk assessment in soil-plant Environment. Chemosphere. 2020;**259**:127436

[11] Nazir F, Hussain A, Fariduddin Q. Hydrogen peroxide modulate photosynthesis and antioxidant systems in tomato (*Solanum lycopersicum* L.) plants under copper stress. Chemosphere. 2019;**230**:544-558

[12] Cambrollé J, García JL, Figueroa ME, Cantos M. Evaluating wild grapevine tolerance to copper toxicity. Chemosphere. 2015;**120**:171-178

[13] Marques DM, Júnior VV, da Silva AB, Antovani JR, Magalhães PC, de Souza TC. Copper toxicity on photosynthetic responses and root morphology of Hymenaea courbaril L. (Caesalpinioideae). Water Air & Soil Pollution. 2018;**229**:138

[14] Moustakas M, Ouzounidou G, Symeonidis L, Karataglis S. Field study of the effects of excess copper on wheat photosynthesis and productivity. Soil Science and Plant Nutrition. 1997;**43**:531-539

[15] Kumar V, Pandita S, Sidhu GPS, Sharma A, Khanna K, Kaur P, et al. Copper bioavailability, uptake, toxicity and tolerance in plants: A comprehensive review. Chemosphere. 2021;**262**:127810

[16] Williams LE, Pittman JK, Hall JL. Emerging mechanisms for heavy metal transport in plants. Biochimica et Biophysica Acta Biomembrane. 2000;**1465**:104-126

[17] Puig S. Function and regulation of the plant COPT family of high-affinity copper transport proteins. Advances in Botany. 2014;**2014**:1-9

[18] Kavitha PG, Kuruvilla S, Mathew MK. Functional characterization of a transition metal ion transporter, OsZIP6 from rice (*Oryza sativa* L.). Plant Physiology and Biochemistry. 2015;**97**:165-174

[19] Krämer U, Talke IN, Hanikenne M. Transition metal transport. FEBS Letters. 2007;**581**:2263-2272

[20] Deng F, Yamaji N, Xia J, Ma JF. A member of the heavy metal P-type ATPase OsHMA5 is involved in xylem loading of copper in rice. Plant Physiology. 2013;**163**:1353-1362

[21] Lange B, van der Ent A, Baker AJM, Echevarria G, Mahy G, Malaisse F, et al. Copper and cobalt accumulation in plants: A critical assessment of the current state of knowledge. New Phytologist. 2017;**213**:537-551

[22] Siedlecka A. Some aspect of interaction between heavy metals and minerals nutrients. Acta Societats Botanicorum Poloniae. 1997;**64**: 365-272

[23] Aust SD, Morehouse LA, Thomas CE. Role of metals in oxygen radical reactions. Free Radical Biology and Medicine. 1985;**1**:3-25

[24] Solioz M. Copper and bacteria. In: Solioz M, editor. Copper and Bacteria Evolution, Homeostasis and Toxicity. Springer; 2018. pp. 11-19

[25] Apel K, Hirt H. Reactive oxygen species: Metabolism, oxidative stress, and signal transduction. Annual Review of Plant Biology. 2004;**55**:373-379

[26] Lombardi L, Sebastiani L. Copper toxicity in Prunus cerasifera: Growth and antioxidante enzymes responses of in vitro grown plants. Plant Science. 2005;**168**:797-802

[27] Peng HY, Yang XE, Yang MJ, Tian SK. Responses of antioxidant enzyme system to copper toxicity and copper detoxification in the leaves of *Elsholtzia splendens*. Journal of Plant Nutrition. 2006;**29**:1619-1635

[28] Shi-shen K. Effects of copper on the photosynthesis and oxidative metabolism of *Amaranthus tricolor* Seedlings. Agricultural Sciences in China. 2007;**6**:1182-1192

[29] Hasanuzzaman M, Nahar K, Anee TI, Fujita M. Glutathione in plants: Biosynthesis and physiological role in environmental stress tolerance. Physiology and Molecular Biology of Plants. 2017;**23**:249-268

[30] Foyer CH, Noctor G. Ascorbate and glutathione: The heart of the redox hub. Plant Physiology. 2011;**155**:2-18

[31] Johnson DC, Dean DR, Smith AD, Johnson MK. Structure, function, and formation of biological iron-sulfur clusters. Annual Review in Biochemistry. 2005;**74**:247-281

[32] Balk J, Schaedler TA. Iron cofactor assembly in plants. Annual Review of Plant Biology. 2014;**65**:125-153

[33] Fonseca JP, Lee HK, Boschiero C, Griffiths M, Lee S, Zhao P, et al. Iron–sulfur cluster protein nitrogen fixations-like1 and its interactor frataxin function in plant immunity. Plant Physiology. 2020;**184**:1532-1548

[34] Balk J, Lobréaux S. Biogenesis of iron–sulfur proteins in plants. Trends in Plant Science. 2005;**10**:324-331

[35] Droppa M, Horváth G. The role of copper in photosynthesis. Plant Science. 1990;**9**:111-123

[36] Hsu BD, Lee JY. Toxic effects of copper on Photosystem II of spinach

chloroplasts. Plant Physiology. 1988, 1998;**87**:116-119

[37] Arellano JB, Lazaro JJ, Lopez-Gorge J, Baron M. The donor side of Photosystem II as the copper-inhibitory binding site. Photosynthesis Research. 1995;**45**:127-134

[38] Burda K, Kruk J, Schmid G, Strzalka K. Inhibition of oxygen evolution in Photosystem II by Cu(II) ions is associated with oxidation of cytochrome b559. Biochemical Journal. 2003;**371**:597-601

[39] Haberman HM. Reversal of copper inhibition in chloroplasts reactions by manganese. Plant Physiology. 1969;**44**:331-336

[40] Vierke G, Stuckmeier P. Binding of copper(II) to proteins of the photosynthetic membrane and its correlation with inhibition of electron transport in class II chloroplasts of spinach. Zeitschrift für Naturforschung. 1977;**32**:605-610

[41] Li XY, Lin ML, Hu PP, Lai NW, Huang ZR, Chen LS. Copper toxicity differentially regulates the seedling growth, copper distribution, and photosynthetic performance of *Citrus sinensis* and *Citrus grandis*. Journal of Plant Growth Regulation. 2021;**40**. DOI: 10.1007/s00344-021-10516-x

[42] Yusuf M, Almehrzi ASS, Alnajjar AJN, Alam P, Elsayed N, Khalil R, et al. Glucose modulates copper induced changes in photosynthesis, ion uptake, antioxidants and proline in *Cucumis sativus* plants. Carbohydrate Research. 2021;**501**:108271

[43] Hippler FWR, Boaretto RM, Dovis VL, Quaggio JA, Azevedo RA, Mattos-Jr D. Oxidative stress induced by Cu nutritional disorders in Citrus depends on nitrogen and calcium. Scientific Reports. 2018;**8**:1641

[44] Gong Q, Wanga L, Daia T, Zhoua J, Kanga Q, Chena H, et al. Effects of copper on the growth, antioxidant enzymes and photosynthesis of spinach seedlings. Ecotoxicology and Environmental Safety. 2019;**171**:771-780

[45] Gong Q, Li ZH, Wang L, Zhou JY, Kang Q, Niu DD. Gibberellic acid application on biomass, oxidative stress response, and photosynthesis in spinach (*Spinacia oleracea* L.) seedlings under copper stress. Environmental Science and Pollution Research. 2021;**28**:53594-53604

[46] Lindon FC, Henriques FS. Effects of copper toxicity on growth and the uptake and translocation of metals in rice plants. Journal of Plant Nutrition. 1993;**16**:1449-1464

[47] Feil SB, Pii Y, Valentinuzzi F, Tiziani R, Mimmo T, Cesco S. Copper toxicity affects phosphorus uptake mechanisms at molecular and physiological levels in *Cucumis sativus* plant. Plant Physiology and Biochemistry. 2020;**157**:138-147

[48] Xiong ZT, Liu C, Geng B. Phytotoxic effects of copper on nitrogen metabolism and plant growth in Brassica pekinensis Rupr. Ecotoxicology and Environmental Safety. 2006;**64**:273-280

[49] Pourakbar L, Khayami M, Khara J, Farbodnia T. Phisiological effect of copper on some biochemical parameters in Zea mays L. seedlings. Pakistan Journal of Biological Sciences. 2007;**10**:4092-4096

[50] Aghajanzadeh TA, Prajapatib DH, Burow M. Copper toxicity affects indolic glucosinolates and gene expression of key enzymes for their biosynthesis in Chinese

cabbage. Archives of Agronomy and Soil Science. 2020;**66**:1288-1301

[51] Quagliata G, Celletti S, Coppa E, Mimmo T, Cesco S, Astolfi S. Potential Use of Copper-Contaminated Soils for Hemp (Cannabis sativa L.) Cultivation. Environments. 2021;**8**:111

[52] Lequeux H, Hermans C, Lutts S, Verbruggen N. Response to copper excess in Arabidopsis thaliana: Impact on the root system architecture, hormone distribution, lignin accumulation and mineral profile. Plant Physiology and Biochemistry. 2010;**48**:673-682

[53] Ouzounidou G, Ilias I, Tranopoulou H, Karataglis S. Amelioration of copper toxicity by iron on spinach physiology. Journal of Plant Nutrition. 1998;**21**:2089-2101

[54] Martins LL, Mourato MP. Effect of excess copper on tomato plants: Growth parameters, enzyme activities, chlorophyll, and mineral content. Journal of Plant Nutrition. 2006;**29**:2179-2198

[55] Ke W, Xiong ZT, Chen S, Chen J. Effects of copper and mineral nutrition on growth, copper accumulation and mineral element uptake in two *Rumex japonicus* populations from a copper mine and an uncontaminated field sites. Environmental and Experimental Botany. 2007;**59**:59-67

[56] Salmani MS, Khorsandi F, Yasrebi J, Karimian N. Biochar effects on copper availability and uptake by sunflower in a copper contaminated calcareous soil. International Journal of Plant, Animal and Environmental Sciences. 2014;**4**:389-395

Chapter 6

Metal Nanoparticles and Abiotic Stress Tolerance

Maryam Dahajipour Heidarabadi

Abstract

Nanotechnology opens a large scope of novel applications in the fields of biotechnology and agricultural industries. Nanoparticles (NPs) are described as particles with at least one dimension in the 1–100 nm size range. They have unique physicochemical properties, i.e., high surface area, high reactivity, tunable pore size, and particle morphology. Abiotic stresses like drought, salinity, flooding, heat, heavy metals, etc. are major constraints that affect the growth and productivity of plants. To overcome the impact of these abiotic stresses, many strategies could be considered to support plant growth including the use of metal nanoparticles. Several metal nanoparticles (such as Zn, Fe, Ti, Ag, Mn, Cu, and Mo NPs) are being studied to assess their potential in protecting plants from abiotic stresses, improving plants, and modulating various plant processes. The present review has investigated the beneficial role of metal nanoparticles in alleviation of various abiotic stresses in some plants.

Keywords: abiotic stress tolerance, antioxidant activity, metal nanoparticles, secondary metabolites

1. Introduction

Nanotechnology, the fourth wave of the industrial revolution, is one of the new technologies that is developing rapidly [1]. In the last decade, nanotechnology has been considered as an important tool to increase agricultural production. Nanotechnology has the potential to transform the agricultural industry through the development of new formulations for pesticides and fertilizers, the identification and diagnosis of plant diseases, water supply for agriculture, and soil management [2]. The agricultural industry is considered as one of the important economic pillars of developed countries. As the world's population grows, the need for food and agricultural products is increasing. Factors such as climate change, limited soil and water resources, increased environmental pollution, and plant diseases create problems in agriculture and the production of sufficient and healthy food [3]. In general, nanotechnology can make a significant contribution to the growing prosperity of the industry by optimizing the consumption of agricultural inputs such as water, fertilizers, and pesticides and reducing effluents and pollution [4]. Nanotechnology can improve the overall use efficiency of agricultural inputs such as water, light, and chemicals. Health and functions of both soil and plant improve through microbiome enhancement and decreasing losses by managing crop disease better, which leads to less collateral damage to the environment. Therefore, nanotechnology has a promising potential to develop sustainable agriculture [5].

The environmental factors that restrict plant growth, vitality, and fertility are known as abiotic stresses. Plants are naturally exposed to a variety of abiotic stresses such as drought, salinity, heavy metals, chilling, and heat [6]. Plants as sessile organisms have numerous mechanisms to cope with changes in their growth conditions to show the necessary flexibility in responding to environmental stresses, without affecting cellular, physiological and developmental processes [7]. One of the main concerns in sustainable agriculture is increasing tolerance to abiotic stresses. Based on this, researchers have been able to take a big step in global sustainable agriculture by reducing the harmful effects of abiotic stresses [8]. When plants are exposed to abiotic stresses, reactive oxygen species (ROS) accumulate at the toxicity level in the cell. Overproduction of ROS causes the degradation of membrane lipids and proteins, cell toxicity, and reduction of plant growth. The antioxidant defense system scavenges ROS to alleviate oxidative stress [9]. In the last years, the use of nanoparticles in technology has been considered due to their properties such as small size, high surface area, higher solubility, and reactivity compared to bulk materials [10]. Metallic nanoparticles (MNPs) including Zn, Fe, Ti, Ag, Mn, Cu, and Mo NPs have earned significant attention due to their environmentally friendly implementations in the agricultural sector [11]. They have recently been used for seed germination, plant growth, and stress tolerance of a number of plants [12, 13]. The goal of this review was to better understand the stress resistance mechanisms and MNP-mediated plant tolerance increase via antioxidant activity regulation.

2. Types of nanoparticles

Nanoparticles (NPs) have dimensions between 1 and 100 nm. They have unique physical and chemical properties such as high surface vitality, large surface-to-volume ratio, and high reactivity [14]. NPs are generally classified into different groups: metal-based NPs, metalloid NPs, metal magnetic NPs, metal oxide NPs, dendrimers, and carbon-based NPs (**Table 1**). In the last decade, metal and metal oxide-based NPs are comprehensively studied in agriculture fields for the improvement of crop productivity and increasing the plant flexibility and tolerance under abiotic stress

Types of Nanoparticles	Example
Metal-based NPs	Gold, copper, aluminum, iron, silver, platinum, palladium
Metalloid NPs	Selenium, silicon, boron, arsenic, tellurium
Metal magnetic NPs	Cobalt, manganese, nickel, iron
Metal oxide NPs	Titanium dioxide, cerium oxide, iron oxide, aluminum oxide, zinc oxide, copper oxide
Dendrimers	Hybrid, tecto, micellar, chiral, liquid crystalline, triazine
Carbon-based NPs	Carbon nanotubes, carbon nanohorn, nanodiamond, fullerene, graphite, graphene, graphene oxide, carbon dot

Table 1.
Types of nanoparticles [15].

conditions [16]. Metal-based NPs and their oxides including nanomaterials of gold, silver, copper, aluminum, iron, titanium dioxide (TiO_2), cerium oxide (CeO_2), iron oxide (FeO), aluminum oxide (Al_2O_3), and zinc oxide (ZnO) are gaining so much attention of scientists to modulate abiotic stress [12, 13, 17].

3. Effect of metal nanoparticles on antioxidant defense system improving of plants during abiotic stress exposure

Reactive oxygen species (ROS) are generated in various plant cell compartments such as plasma membranes, endoplasmic reticulum, peroxisomes, chloroplasts, mitochondria, and cell wall in natural and stress conditions [18]. ROS such as singlet oxygen ($^1O_2$), superoxide (O^{2-}), hydrogen peroxide (H_2O_2), and hydroxyl radicals ($OH^{\cdot}$) are accumulated in all the abiotic stresses that result in oxidative stress. Increased ROS act as a signal, and ROS scavengers are one of the defense mechanisms in plants [8]. As the level of ROS is elevated in plants in response to abiotic stress exposure, an antioxidant defense system that is capable of scavenging ROS is activated [19]. Antioxidant defense system of plants includes a number of antioxidant enzymes such as catalase (CAT), ascorbate peroxidase (APX), dehydroascorbate reductase (DHAR), guaiacol peroxidase (GPX), glutathione reductase (GR), monodehydroascorbate reductase (MDHAR), and superoxide dismutase (SOD) and non-enzyme antioxidant compounds such as proline, glycine betaein, anthocyanins, flavonoids, carotenoids, glutathione, and ascorbate [20–22]. As shown in **Figure 1**, MNPs enter the plant cell through penetration or transportation via specific channels in the plasma membrane. Then, as stress signaling molecules, by inducing the expression of regulatory factors in the activation of the defense system, they result in stress tolerance.

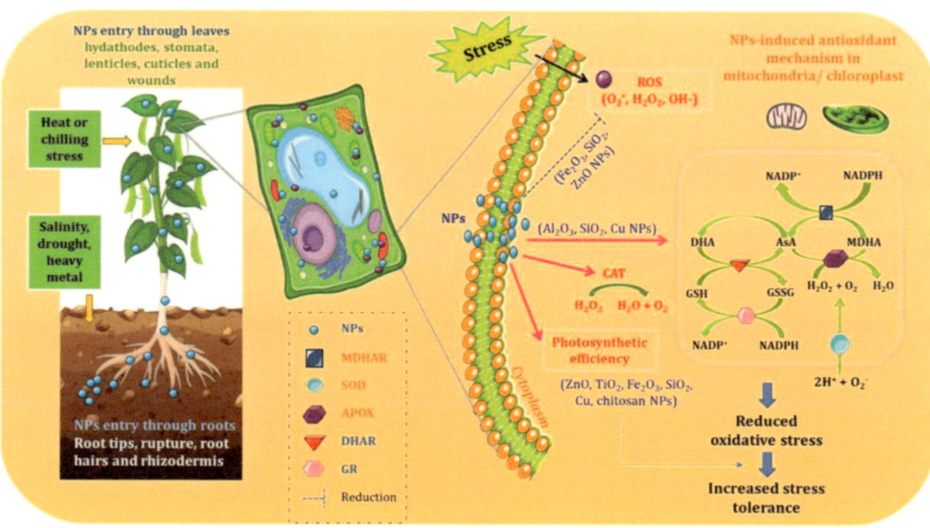

Figure 1.
Antioxidative mechanism of action of MNPs in plants under abiotic stress (NPs: Nanoparticles; MDHAR: Monodehydroascorbate reductase; SOD: Superoxide dismutase; APOX: Ascorbate peroxidase; DHAR: Dehydroascorbate reductase; GR: Glutathione reductase; ROS: Reactive oxygen species) [17].

In addition, MNPs can activate the plant defense system under stress by maintaining ROS levels to induce the ROS signaling network. MNPs treatment alters biological pathways involved in defense mechanisms by upregulation of genes that encode proteins that play a key role in ROS balance such as peroxidases (POX), NADPH oxidase, glutathione S-transferase (GST), and superoxide dismutase (SOD) [23]. MNPs upregulate the genes responsible for the activation of antioxidant enzymes. For instance, Laware and Raskar experiments on onion seedlings showed that TiO2 NPs enhanced SOD enzyme's activity in NP-supplemented plants [24]. Also, results from transcriptomic studies showed that the expression of Cu/Zn SOD, Fe/Mn SOD, catalase, and ascorbate peroxidase in plants that were treated with ZnO NPs under drought was notably enhanced [25]. Concrete evidence provided by Thakur et al. showed that an increase in GPX and SOD activities in wheat plants treated with ZnO NPs under heat stress improved heat tolerance by further reducing H_2O_2 levels and establishing membrane stability [26]. Studies have also reported that MNPs participate in the induction of Ca^{2+}-binding protein expression, resulted in stress tolerance by launching a cascade of intracellular signaling and upregulation of associated genes [27].

4. Secondary metabolites role induced by MNPs in abiotic stress tolerance

Metabolites are essential molecules for growth, adaptation to stress, and defense of a living organism. Metabolic pathways leading to the synthesis of molecules including carbohydrates, proteins, amino acids, fatty acids, and nucleotides are considered as the primary metabolism, and the compounds produced in these pathways, which are necessary for plant survival, are called primary metabolites [28]. Primary metabolites are involved in various life functions in plants, such as cell division, growth and development, photosynthesis, respiration, and reproduction [29]. Plants produce a diverse group of organic compounds called secondary metabolites that do not have a role in processes such as photosynthesis, respiration, metabolism, protein synthesis, and nutrient accumulation directly [30].

MNPs	Species	Concentration	Secondary metabolite	Refs
CuNPs	*Cucumis sativus*	10 and 20 mg L^{-1}	Acetyl glucosamine, Phenyl lactate, 4-aminobutyrate	[36]
AgNPs	*Artemisia annua*	900 mg L^{-1}	Artemisinin	[37]
	Arabidopsis thaliana	0.2–25 μg mL^{-1}	Anthocyanin and Flavonoid	[38]
	Corylus avellana	0–10 mg L^{-1}	Taxol and Baccatin II	[39]
CeNPs	*Solanum lycopersicum*	0–500 mg kg^{-1}	lycopene	[40]
ZnNPs	*Thymus vulgaris*	100 and 150 mg L^{-1}	Thymol and Carvacrol	[41]
TiNPs	*Salvia officinalis*	0–1000 mg L^{-1}	Monoterpenes and Camphene	[42]
FeNPs	*Hypericum perforatum*	0–150 mg L^{-1}	Hypericin and Hyperforin	[43]

Table 2.
MNPs as an elicitor for secondary metabolites.

A significant number of secondary metabolites, such as terpenoids, steroids, phenolics, flavonoids, and alkaloids, by the removal of ROS in cellular stress and defense response, function as an adaptation mechanism to stress conditions [31]. Evidence has shown that secondary metabolites are involved in the non-enzymatic defense of plants against stress [32]. The MNPs are commonly found in agrochemicals such as pesticides, fungicides, herbicides, and fertilizers [33]. A few studies have demonstrated that treatment of plants with MNPs resulted in increased production of secondary metabolites, which might act as antioxidants to scavenge the ROS [34, 35]. Several studies have shown that MNPs have the potential to induce plant secondary metabolites' production (**Table 2**).

5. Alleviation of abiotic stress by metal-based and metal-oxide NPs application

Abiotic stresses including drought, salinity, heat, chilling, heavy metal toxicities, etc. are major obstacles to plant growth and productivity [7]. Plants adapt to and alleviate abiotic stresses by alterations in morphological, physiological, biochemical, and molecular levels. Researchers have shown that MNPs help plants to overcome abiotic stresses by their concentration-dependent impact on plant growth and development [15]. MNPs can be supplied to plants in form of seed coating, soil, or foliar application according to their mode of action. Extensive researches have elucidated the positive effects of some MNPs on some plant species under different abiotic stress conditions (**Table 3**). For instance, MNPs were effective in ameliorating the detrimental effects of abiotic stresses by increasing flavonoid, anthocyanin, phenolic, and photosynthetic pigment contents, upregulating the antioxidant enzymes, reducing the stress markers (MDA and H_2O_2), water balance, ion accumulation, improvement of the nutrient absorption, and the Na^+/K^+ ratio [17, 44–60].

MNPs	Characters and Size (nm)	Species	Abiotic stress	Mode of application	Morphophysiological responses	Refs
Silver NPs	Hexagonal and spherical (17–34)	*Solanum melongena*	Drought	Foliar spray	Upregulating the antioxidant enzymes	[17]
	Spherical (27–33)	Triticum aestivum	Salinity	Foliar spray	Upregulating the antioxidant enzymes, proline metabolism, and ion accumulation	[44]
	Cubic to rectangular (8–28)	*T. aestivum*	Heat	Soil application	Balanced relative water content, and improved chlorophyll content	[45]
	Spherical (~10 nm)	*Phaseolus vulgaris*	Chilling	Seed treatment	Increased seedling height, fresh and dry weight, and net photosynthesis	[46]
	Mono nanopowder (15)	*Glycine max*	Flooding	Silica sand	Improve the growth by accumulation of calnexin/calreticulin and glycoproteins	[47]

Zinc NPs	NR	*Cucumis sativus*	Drought	Foliar application	Reduction in ROS accumulation and lipid peroxidation, improve antioxidant defense system, nutrient absorption, and osmolytes accumulation	[48]
	Spherical and hexagonal (~20 nm)	*Brassica napus*	Salinity	Seed treatment	Seedling development through the biosynthesis of pigments, osmotic protection, reduction of ROS accumulation, adjustment of antioxidant enzymes, and improvement of the nutrient absorption	[49]
	Spherical (80)	wheat cultivars	Heat	Foliar spray	Increasing antioxidant enzymes activities	[50]
	Spherical (30)	*Oryza Sativa*	Chilling	Foliar application	Upregulation of the chilling-induced gene expression of the antioxidant system and chilling response transcription factors	[51]
	Hexagonal, square and spherical (2–64)	*Leucaena leucocephala*	Heavy metals	Hydroponic application	Reduced MDA content and the elevated level of antioxidant enzyme activities	[52]
Copper NPs	Crystalline powder (30–40)	*Zea mays*	Drought	Seed treatment	Water balance, photosynthesis pigment, ROS-scavenging enzyme activities and anthocyanin biosynthesis	[53]
	Spherical (20–50)	*Lycopersicon esculentum*	Salinity	Foliar application	Improving the Na^{+}/K^{+} ratio and stimulates the plant's antioxidant mechanism	[54]
Iron NPs	Spherical (20–40)	*Dracocephalum moldavica*	Salinity	Foliar application	Increasing non-enzymatic system as phenolic compounds and flavonoid content	[55]
	NR	*T. aestivum*	Drought	Seed treatment	increases photosynthetic pigments, proline, reduced lipid peroxidation, electrolyte leakage, and improved antioxidative defense system	[56]
	Spherical (50)	wheat cultivars	Heat	Foliar spray	Increasing antioxiodant enzymes activities, appearance of new bands in some isozymes and decreasing of lipid peroxidation product malondialdehyde	[50]
	Spherical (16)	*L. esculentum*	Heat	Foliar spray	enhanced photosynthesis by regulating energy dissipation, caused cooling of leaves through inducing stomatal opening	[57]

Titanium NPs	Crystalline and nearly spherical (15–25)	*T. aestivum*	Drought	Foliar application	Increased growth, yield, gluten, and starch content	[58]
	Spherical (16)	*L. esculentum*	Heat	Foliar spray	Enhanced photosynthesis by regulating energy dissipation, caused cooling of leaves through inducing stomatal opening	[57]
	Spherical-like shapes (70–90)	*D. moldavica*	Salinity	Hydroponic application	Positive impact on agronomically important attributes by increased antioxidant enzyme activity	[59]
Aluminum NPs	Crystalline powder (30–60)	*G. max*	Flooding		Upregulated the AsA/GSH pathway (ROS scavenger) and increased ribosomal proteins	[60]

NR: not reported.

Table 3.
Application of some metal NPs induce abiotic stress tolerance in different plant species.

6. Conclusion

Plant production globally is subjected to various environmental stress challenges. Today, the application of nanotechnology in various scientific fields is expanding. Recent studies have highlighted the potential applications of nanotechnology in improving plant growth and performance. MNPs due to their small size and having large surface area, as compared to their bulk chemical forms penetrate and absorb in relatively shorter period of time into plant cells. It has been found that MNPs have a multitude of beneficial effects on morphological, physiological, and biochemical characteristics of plants and enhance their tolerance under a variety of abiotic stresses. Accordingly, the application of MNPs in abiotic stress improvement has been noticed by agricultural researchers. MNPs enhance ROS level in plants that is associated with the amplification of a stress signal that can efficiently activate defense systems of them. It may be concluded that MNPs alleviate the abiotic stress-caused damage by activating the defense system in plants. In addition to, MNPs can regulate photosynthetic efficiency, water balance, nutrient absorption, and osmolytes accumulation, thereby enhancing growth and productivity of plants. However, to investigate the exact action of MNPs in improve plant stress, further research is needed at molecular and subcellular levels. Although MNPs have many advantages that deserve to be explored for alleviation of abiotic stress in plants; it must be noticed that its application without care can lead to a series of issues to the plants, animals, and finally to humankind. The toxicity level of MNPs is related to their concentration, size, number, surface activity, modification, and aggregation. Thus, ambiguities about the risk of use and fate of MNPs in plants and soil, as well as their interaction with the environment, should not be overlooked. In addition, green synthesized MNPs in comparison with chemically synthesized MNPs can be efficiently used due to the lack of limitation of use and toxicity in modulating various abiotic stresses in plants. Finally, developing a comprehensive database, an alert system, as well as international cooperation in regulation and legislation is essential for the use of this technology.

Author details

Maryam Dahajipour Heidarabadi
Department of Genetics and Plant Production, Agriculture College, Vali-e-Asr University of Rafsanjan, Rafsanjan, Iran

*Address all correspondence to: m.dahaji@vru.ac.ir

References

[1] Handford CE, Dean M, Henchion M, Spence M, Elliott CT, Campbell K. Implications of nanotechnology for the Agri-food industry: Opportunities, benefits and risks. Trends in Food Science and Technology. 2014;**40**(2):226-241

[2] Acharya A, Pal PK. Agriculture nanotechnology: Translating research outcome to field applications by influencing environmental sustainability. NanoImpact. 2020;**19**:100232

[3] Foley JA, Ramankutty N, Brauman KA, Cassidy ES, Gerber JS, Johnston M, et al. Solutions for a cultivated planet. Nature. 2011;**478**(7369):337-342

[4] Sekhon BS. Nanotechnology in Agri-food production: An overview. Nanotechnology, Science and Applications. 2014;**7**:31-53

[5] Tipu MMH, Baroi A, Rana J, Islam S, Jahan R, Miah MS, et al. Potential applications of nanotechnology in agriculture: A smart tool for sustainable agriculture. In: Agricultural Development in Asia: Potential Use of Nano-Materials and Nano-Technology. IntechOpen; 2021

[6] Kumar AA, Mishra P, Kumari K, Panigrahi K. Environmental stress influencing plant development and flowering. Frontiers in Bioscience. 2012;**4**:1315-1324

[7] Yang L, Wen K-S, Ruan X, Zhao Y-X, Wei F, Wang Q. Response of plant secondary metabolites to environmental factors. Molecules. 2018;**23**(4):762

[8] Khan N, Ali S, Shahid MA, Mustafa A, Sayyed R, Curá JA. Insights into the interactions among roots, rhizosphere, and rhizobacteria for improving plant growth and tolerance to abiotic stresses: A review. Cell. 2021;**10**(6):1551

[9] Das K, Roychoudhury A. Reactive oxygen species (ROS) and response of antioxidants as ROS-scavengers during environmental stress in plants. Frontiers in Environmental Science. 2014;**2**:53

[10] Baig N, Kammakakam I, Falath W. Nanomaterials: A review of synthesis methods, properties, recent progress, and challenges. Materials Advances. 2021;**2**(6):1821-1871

[11] Rai PK, Kumar V, Lee S, Raza N, Kim K-H, Ok YS, et al. Nanoparticle-plant interaction: Implications in energy, environment, and agriculture. Environment International. 2018;**119**:1-19

[12] Abdel Latef AAH, Abu Alhmad MF, Abdelfattah KE. The possible roles of priming with ZnO nanoparticles in mitigation of salinity stress in lupine (*Lupinus termis*) plants. Journal of Plant Growth Regulation. 2017;**36**(1):60-70

[13] Taran N, Storozhenko V, Svietlova N, Batsmanova L, Shvartau V, Kovalenko M. Effect of zinc and copper nanoparticles on drought resistance of wheat seedlings. Nanoscale Research Letters. 2017;**12**(1):1-6

[14] Sukhanova A, Bozrova S, Sokolov P, Berestovoy M, Karaulov A, Nabiev I. Dependence of nanoparticle toxicity on their physical and chemical properties. Nanoscale Research Letters. 2018;**13**(1):1-21

[15] Sarraf M, Vishwakarma K, Kumar V, Arif N, Das S, Johnson R, et al. Metal/

metalloid-based nanomaterials for plant abiotic stress tolerance: An overview of the mechanisms. Plants. 2022;**11**(3):316

[16] Paramo LA, Feregrino-Pérez AA, Guevara R, Mendoza S, Esquivel K. Nanoparticles in agroindustry: Applications, toxicity, challenges, and trends. Nanomaterials. 2020;**10**(9):1654

[17] Alabdallah NM, Hasan MM, Salih AM, Roushdy S, Al-Shammari AS, Alsanie SI, et al. Silver nanoparticles improve growth and protect against oxidative damage in eggplant seedlings under drought stress. Plant, Soil and Environment. 2021;**67**(11):617-624

[18] Mansoor S, Ali Wani O, Lone JK, Manhas S, Kour N, Alam P, et al. Reactive oxygen species in plants: From source to sink. Antioxidants. 2022;**11**(2):225

[19] Bowler C, Mv M, Inze D. Superoxide dismutase and stress tolerance. Annual Review of Plant Biology. 1992;**43**(1):83-116

[20] Liu Y, Xiao Z, Chen F, Yue L, Zou H, Lyu J, et al. Metallic oxide nanomaterials act as antioxidant nanozymes in higher plants: Trends, meta-analysis, and prospect. Science of The Total Environment. 2021;**780**:146578

[21] Rico CM, Peralta-Videa JR, Gardea-Torresdey JL. Chemistry, biochemistry of nanoparticles, and their role in antioxidant defense system in plants. In: Nanotechnology and Plant Sciences. Springer, Champions; 2015

[22] Hasan MM, Gong L, Nie Z-F, Li F-P, Ahammed GJ, Fang X-W. ABA-induced stomatal movements in vascular plants during dehydration and rehydration. Environmental and Experimental Botany. 2021;**186**:104436

[23] Ruotolo R, Maestri E, Pagano L, Marmiroli M, White JC, Marmiroli N. Plant response to metal-containing engineered nanomaterials: An omics-based perspective. Environmental Science & Technology. 2018;**52**(5):2451-2467

[24] Laware S, Raskar S. Effect of titanium dioxide nanoparticles on hydrolytic and antioxidant enzymes during seed germination in onion. International Journal of Current Microbiology and Applied Sciences. 2014;**3**(7):749-760

[25] Sun L, Song F, Guo J, Zhu X, Liu S, Liu F, et al. Nano-ZnO-induced drought tolerance is associated with melatonin synthesis and metabolism in maize. International Journal of Molecular Sciences. 2020;**21**(3):782

[26] Thakur S, Asthir B, Kaur G, Kalia A, Sharma A. Zinc oxide and titanium dioxide nanoparticles influence heat stress tolerance mediated by antioxidant defense system in wheat. Cereal Research Communications. 2021:1-12

[27] Linh TM, Mai NC, Hoe PT, Lien LQ, Ban NK, Hien LTT, et al. Metal-based nanoparticles enhance drought tolerance in soybean. Journal of Nanomaterials. 2020;**2020**:4056563

[28] Piechulla B, Heldt H-W. Plant Biochemistry. Academic Press; 2010

[29] Bourgaud F, Gravot A, Milesi S, Gontier E. Production of plant secondary metabolites: A historical perspective. Plant Science. 2001;**161**(5):839-851

[30] Irchhaiya R, Kumar A, Yadav A, Gupta N, Kumar S, Gupta N, et al. Metabolites in plants and its classification. World Journal of Pharmacy and Pharmaceutical Sciences. 2015;**4**(1):287-305

[31] Isah T. Stress and defense responses in plant secondary metabolites production. Biological Research. 2019;**52**:39

[32] Edreva A, Velikova V, Tsonev T, Dagnon S, Gürel A, Aktaş L, et al. Stress-protective role of secondary metabolites: Diversity of functions and mechanisms. General and Applied Plant Physiology. 2008;**34**(1-2):67-78

[33] DeRosa MC, Monreal C, Schnitzer M, Walsh R, Sultan Y. Nanotechnology in fertilizers. Nature Nanotechnology. 2010;**5**(2):91-91

[34] Dixon RA, Paiva NL. Stress-induced phenylpropanoid metabolism. The Plant Cell. 1995;**7**(7):1085

[35] Franklin G, Conceição LF, Kombrink E, Dias AC. Xanthone biosynthesis in Hypericum perforatum cells provides antioxidant and antimicrobial protection upon biotic stress. Phytochemistry. 2009;**70**(1):60-68

[36] Zhao L, Huang Y, Hu J, Zhou H, Adeleye AS, Keller AA. 1H NMR and GC-MS based metabolomics reveal defense and detoxification mechanism of cucumber plant under nano-Cu stress. Environmental Science and Technology. 2016;**50**(4):2000-2010

[37] Zhang B, Zheng LP, Yi Li W, Wen WJ. Stimulation of artemisinin production in *Artemisia annua* hairy roots by Ag-SiO_2 core-shell nanoparticles. Current Nanoscience. 2013;**9**(3):363-370

[38] García-Sánchez S, Bernales I, Cristobal S. Early response to nanoparticles in the Arabidopsis transcriptome compromises plant defence and root-hair development through salicylic acid signalling. BMC Genomics. 2015;**16**(1):1-17

[39] Jamshidi M, Ghanati F. Taxanes content and cytotoxicity of hazel cells extract after elicitation with silver nanoparticles. Plant Physiology and Biochemistry. 2017;**110**:178-184

[40] Barrios AC, Medina-Velo IA, Zuverza-Mena N, Dominguez OE, Peralta-Videa JR, Gardea-Torresdey JL. Nutritional quality assessment of tomato fruits after exposure to uncoated and citric acid coated cerium oxide nanoparticles, bulk cerium oxide, cerium acetate and citric acid. Plant Physiology and Biochemistry. 2017;**110**:100-107

[41] Mosavat N, Golkar P, Yousefifard M, Javed R. Modulation of callus growth and secondary metabolites in different thymus species and *Zataria multiflora* micropropagated under ZnO nanoparticles stress. Biotechnology and Applied Biochemistry. 2019;**66**(3):316-322

[42] Ghorbanpour M. Major essential oil constituents, total phenolics and flavonoids content and antioxidant activity of *Salvia officinalis* plant in response to nano-titanium dioxide. Indian Journal of Plant Physiology. 2015;**20**(3):249-256

[43] Sharafi E, Fotokian MH, Loo H. Improvement of hypericin and hyperforin production using zinc and iron nano-oxides as elicitors in cell suspension culture of John'swort (*Hypericum perforatum* L.). Journal of Medicinal Plants and By-products. 2013;**2**(2):177-184

[44] Wahid I, Kumari S, Ahmad R, Hussain SJ, Alamri S, Siddiqui MH, et al. Silver nanoparticle regulates salt tolerance in wheat through changes in ABA concentration, ion homeostasis, and defense systems. Biomolecules. 2020;**10**(11):1506

[45] Iqbal M, Raja NI, Wattoo FH, Hussain M, Ejaz M, Saira H. Assessment of AgNPs exposure on physiological and biochemical changes and antioxidative defence system in wheat (*Triticum aestivum* L.) under heat stress. IET Nanobiotechnology. 2019;**13**(2):230-236

[46] Prażak R, Święciło A, Krzepiłko A, Michałek S, Arczewska M. Impact of Ag nanoparticles on seed germination and seedling growth of green beans in normal and chill temperatures. Agriculture. 2020;**10**(8):312

[47] Hashimoto T, Mustafa G, Nishiuchi T, Komatsu S. Comparative analysis of the effect of inorganic and organic chemicals with silver nanoparticles on soybean under flooding stress. International Journal of Molecular Sciences. 2020;**21**(4):1300

[48] Ghani MI, Saleem S, Rather SA, Rehmani MS, Alamri S, Rajput VD, et al. Foliar application of zinc oxide nanoparticles: An effective strategy to mitigate drought stress in cucumber seedling by modulating antioxidant defense system and osmolytes accumulation. Chemosphere. 2022;**289**:133202

[49] El-Badri AM, Batool M, Mohamed IA, Khatab A, Sherif A, Wang Z, et al. Modulation of salinity impact on early seedling stage via nano-priming application of zinc oxide on rapeseed (*Brassica napus* L.). Plant Physiology and Biochemistry. 2021;**166**:376-392

[50] Hassan NS, Salah El Din TA, Hendawey MH, Borai IH, Mahdi AA. Magnetite and zinc oxide nanoparticles alleviated heat stress in wheat plants. Current Nanomaterials. 2018;**3**(1):32-43

[51] Song Y, Jiang M, Zhang H, Li R. Zinc oxide nanoparticles alleviate chilling stress in rice (*Oryza sativa* L.) by regulating antioxidative system and chilling response transcription factors. Molecules. 2021;**26**(8):2196

[52] Venkatachalam P, Jayaraj M, Manikandan R, Geetha N, Rene ER, Sharma N, et al. Zinc oxide nanoparticles (ZnONPs) alleviate heavy metal-induced toxicity in *Leucaena leucocephala* seedlings: A physiochemical analysis. Plant Physiology and Biochemistry. 2017;**110**:59-69

[53] Van Nguyen D, Nguyen HM, Le NT, Nguyen KH, Nguyen HT, Le HM, et al. Copper nanoparticle application enhances plant growth and grain yield in maize under drought stress conditions. Journal of Plant Growth Regulation. 2022;**41**(1):364-375

[54] Pérez-Labrada F, López-Vargas ER, Ortega-Ortiz H, Cadenas-Pliego G, Benavides-Mendoza A, Juárez-Maldonado A. Responses of tomato plants under saline stress to foliar application of copper nanoparticles. Plants. 2019;**8**(6):151

[55] Moradbeygi H, Jamei R, Heidari R, Darvishzadeh R. Investigating the enzymatic and non-enzymatic antioxidant defense by applying iron oxide nanoparticles in *Dracocephalum moldavica* L. plant under salinity stress. Scientia Horticulturae. 2020;**272**:109537

[56] Noor R, Yasmin H, Ilyas N, Nosheen A, Hassan MN, Mumtaz S, et al. Comparative analysis of iron oxide nanoparticles synthesized from ginger (*Zingiber officinale*) and cumin seeds (*Cuminum cyminum*) to induce resistance in wheat against drought stress. Chemosphere. 2022;**292**:133201

[57] Qi M, Liu Y, Li T. Nano-TiO_2 improve the photosynthesis of tomato leaves

under mild heat stress. Biological Trace Element Research. 2013;**156**(1):323-328

[58] Jaberzadeh A, Moaveni P, Moghadam HRT, Zahedi H. Influence of bulk and nanoparticles titanium foliar application on some agronomic traits, seed gluten and starch contents of wheat subjected to water deficit stress. Notulae Botanicae Horti Agrobotanici Cluj-Napoca. 2013;**41**(1):201-207

[59] Gohari G, Mohammadi A, Akbari A, Panahirad S, Dadpour MR, Fotopoulos V, et al. Titanium dioxide nanoparticles (TiO_2 NPs) promote growth and ameliorate salinity stress effects on essential oil profile and biochemical attributes of *Dracocephalum moldavica*. Scientific Reports. 2020;**10**(1):1-14

[60] Mustafa G, Komatsu S. Insights into the response of soybean mitochondrial proteins to various sizes of aluminum oxide nanoparticles under flooding stress. Journal of Proteome Research. 2016;**15**(12):4464-4475

Section 2

Mechanisms of Plant Genomic Responses to Stresses

Chapter 7

Heat Shock Proteins (HSP70) Gene: Plant Transcriptomic Oven in the Hot Desert

Fatima Batool, Batcho Anicet Agossa, Zainab Y. Sandhu, Muhammad Bilal Sarwar, Sameera Hassan and Bushra Rashid

Abstract

Heat stress is considered to induce a wide range of physiological and biochemical changes that cause severe damage to plant cell membrane, disrupt protein synthesis, and affect the efficiency of photosynthetic system by reducing the transpiration due to stomata closure. A brief and mild heat shock is known to induce acquired thermo tolerance in plants that is associated with concomitant production of heat shock proteins' (HSPs) gene family including HSP70. The findings from different studies by use of technologies have thrown light on the importance of HSP70 to heat, other abiotic stresses and environmental challenges in desserts. There is clear evidence that under heat stress, HSP70 gene stabilized the membrane structure, chlorophyll and water breakdown. It was also found that under heat stress, HSP70 decreased the malondialdehyde (MDA) content and increased the production of superoxide dismutase (SOD) and peroxidase (POD) in transgenic plants as compared to non-transgenic plants. Some reactive oxygen species (ROS) such as superoxide, hydrogen peroxide and hydroxyl radical are also synthesized and accumulated when plants are stressed by heat. Hence HSP70 can confidently be used for transforming a number of heat tolerant crop species.

Keywords: heat stress, genomic approaches, heat shock protein gene family, plant transformation, plant physiology

1. Introduction

Heat Shock Proteins (HSPs) are a family of functionally associated proteins which regulates their expression when cells are subjected to high temperatures or other stimuli [1]. As intracellular chaperones, HSP genes are an evolutionarily conserved class of proteins in all living organisms, from bacteria to humans. They are important components that contribute to cellular homeostasis under favorable and harmful growth conditions in prokaryotic and eukaryotic cells [2]. They are responsible for the folding, assembly, translocation and degradation of proteins during normal cell growth and development. Many HSP members perform critically important chaperone roles, such as three-dimensional folding of newly formed proteins and/or proteins weakened by cell stress [3]. For this reason, many chaperones are known to be HSPs due to their existence as aggregates when denatured by heat stress. Under high temperature conditions, their expression levels are increased by transcription of heat shock transcription factors (HSTFs) which are enabled by trimerization of their monomeric forms. This gene regulation system of HSPs is one of the most defined response systems known at molecular level for organisms exposed to extreme temperature conditions [4]. Based on their respective molecular weights, HSPs are referred to small HSP (HSP18, HSP20 and HSP40) and large HSP (HSP60, HSP70, HSP90 and HSP100) [5].

HSP70s are an essential part of the cellular protein folding mechanism and help the cells to defend against stress. Strongly regulated by heat stress, their mechanism interacts with expanded protein peptide segments as well as partly folded proteins to resist aggregation, reshape folding pathways, and regulate activity [6]. When there is no interaction with peptide substrate, HSP70 is normally ATP (Adenosine 5′-triphosphate) bound, distinguished by very poor ATPase activity. When newly synthesized proteins emerge from the ribosomes, the HSP70 substrate binding domain identifies and interacts with hydrophobic amino acid residue sequences [7]. Ubiquitously found in all living species, they inhibit aggregation and assist in the reproduction of nonactive proteins in both normal and stress conditions and provide heat-tolerance in plants that are under heat stress. They also inhibit protein folding in mitochondria/chloroplast during post-translational importation. They are also involved in the import and translocation mechanism of proteins and promote the proteolytic degradation of defective proteins by transferring these proteins to lysosomes or proteasomes [8]. The key role is to manage protein folding and quality control in a crowded cell environment. It also plays a crucial role in signal transduction networks, cell cycle regulation, protein degradation and protein trafficking. In addition, it may also play a role in the morphological evolution and adaptation to stress. They were isolated from the cytosol, the endoplasmic reticulum and the plastid of many plants [9]. Recent studies have shown that HSP70 interacts with the 26S proteasome and plays a crucial function in its assembly and maintenance. As the most abundant proteins in the cells, HSP70 genes have been recorded in many plant species, such as Arabidopsis, soybean, tobacco, rice, maize wheat and Agave [10, 11]. They are also constitutively expressed in plants but their expression is developmentally regulated and caused by various environmental factors such as drought, cold, high temperature and salt. Studies have shown their involvement in cytosols, mitochondria and chloroplasts, play an important role in remodeling machines that contribute to preserving the integrity of the cell proteome by promoting protein remodeling, disaggregation, reactivation and degradation of malformed and inactive proteins [12]. The mechanism for the recovery of proteins from aggregation often requires the assistance of another ATP-dependent chaperone system. The HSP70 family solubilizes the aggregated protein and extracts it in a process that can be repeated with the aid of a specific HSP family of genes [13]. This review focuses on recent discoveries of molecular and cellular mechanisms of HSP70 that govern the tolerance of plants in unfavorable environmental conditions.

2. Biochemical and physiological responses of plants against heat stress

Heat inducible genes can be categorized in two agencies. The primary organization consists of proteins that maximum probably characteristic in abiotic stress tolerance.

These encompass molecules along with antifreeze proteins, chaperones osmotin, late embryogenesis abundant (LEA) proteins, mRNA-binding proteins, key enzymes involved in osmolyte biosynthesis, proteins of water channel, transporters like sugars and proline, detoxification enzymes and numerous proteases. The second institution consists of regulatory proteins i.e. protein factors concerned in addition law of signal transduction and stress-responsive gene expression [14, 15]. These encompass diverse transcription factors, protein kinases, protein phosphatases, enzymes worried in phospholipid metabolism, and different signaling molecules inclusive of calmodulin-binding protein. Many transcription factor genes have been stress inducible, suggesting that numerous transcriptional regulatory mechanisms can also function in regulating heat, drought, cold, or excessive salinity stress signal transduction pathways. Those transcription factors could govern expression of stress-inducible genes both cooperatively and independently [16].

3. Methods to study gene identification

Many techniques have been carried out to evaluate the gene expression for the improvement of plants such as subtractive hybridization of cDNA (Deoxyribonucleic Acid) libraries, homology searching, differential display, genome-wide identification and third generation sequencing [17]. Differential display reverse transcriptase polymerase chain reaction (DDRT-PCR) is a delicate, easy and significant technique to evaluate cDNA [18]. Differential display has benefit as compare to other techniques because big quantity of RNA (Ribonucleic acid) is not needed for analysis. It has been used with great success to identify several differentially expressed genes from plants [19]. The main thing is to do research by using oligonucleotide primers, of which one will be used as anchored primer to the poly-adenylate tail of mRNAs subgroup and the second will be used as arbitrary that will be short in sequence length so that it may combine at different positions as compare to the first primers [20]. The resulted mRNA after using these primer can then be manipulate by using RT-PCR (Reverse transcription polymerase chain reaction) and checked on agarose gel. Multiple primer's pairs can be used to obtain the complementary DNA fragments that depend on strong link with sequence specificity of respective primer. Novel HSP70 genes expressed under stressful conditions have been identified and isolated from tomato, Arabidopsis and wheat by using the DDPCR [21–23]. However, number of genes studied in one attempt of experiment are low as compared to the other advanced techniques like high-throughput expression profiling as qPCR (quantitative polymerase chain reaction) and Northern blotting [24, 25]. Advancements of second-generation sequencing technology offers opportunities for the discovery of millions of novel markers in non-model crop organisms as well as the detection of genes for agronomic traits [26]. Identification of genes within a population gives an understanding that how essential is to control the agronomic traits. The ability to produce sequence data is being supported by increasingly high throughput technologies such as next or second-generation sequencing. It identify the systems that yield vast number (usually millions) of short DNA sequence reads between 25 and 400 bp [27].

The first model plant genome sequenced was *Arabidopsis thaliana* (Arabidopsis Genome, 2000) and first crop genome sequenced was the Rice [28–30]. Current crop genome sequencing programmes are quickly shifting pace with emerging technology that's why second-generation sequencing is adopted to obtain insight into their chosen genome. Although the sequencing and assembly of large and complex crop genomes remains an important task, a considerable amount of information can be obtained from low-coverage shotgun sequencing of these genomes. Short paired reading data generated using second generation technologies are especially suited for the discovery of genes and gene promoters in crop plants [31, 32]. Many HSP70 genes from different species such as *Agave sisalana* [10]; *Solanum tuberosum* [33, 34]; common bean [35]; *A. thaliana* [36]; Populus [37] have been described at transcription that expresses under heat stress. Four transcriptional regulatory systems have also been reported, two of them are ABA-independent (Abscisic acid) whereas other two are ABA-dependent. Genetic and molecular analyses suggested cross talk between these regulatory systems. Genomic analyses of stress-inducible genes have recently revealed cross talk in stress-responsive gene expression [33, 34, 38, 39].

4. HSP70 gene family leading to improve abiotic stress tolerance in plant

Plants accumulate specific stress responsive proteins under harsh environmental conditions [40]. Heat-shock proteins (HSPs) and late embryogenesis abundant (LEA) proteins accumulate under salinity, extreme temperature and water stress. These proteins have been shown to be involved in cellular protection during the stress [41, 42]. Enzymes and proteins are not able to function during abiotic stresses. Therefore, it is necessary for cell survival to prevent them from aggregation and maintain their functional conformations [3]. HSP70s are synthesized when environmental changes disturb an organism's whole physiological system to such an extent that results in denaturation of proteins [43, 44]. Under such situation many stress associated proteins especially HSPs have been proven to act as molecular chaperones which play significant role in protein synthesis, maturation, degradation and targeting in an extensive array of ordinary mobile processes. Furthermore, molecular chaperones stabilize the proteins and membranes, in addition to assist in refolding of protein beneath stress conditions [45, 46]. They had been broadly documented in many plant species together with Arabidopsis, soybean, tobacco, rice, maize and wheat. They may be regularly expressed in plants constitutively but their expression is regulated by using various environmental conditions which include heat and salt [47]. Research have verified their presence in cytosol, mitochondria and chloroplast and play vital function in remodeling machines that participate in maintaining the integrity of the mobile proteome via facilitating protein reworking, disaggregation, reactivation or degradation of misfolded and inactive protein [13].

The cDNA coding HSP70 solubilizes and releases the aggregated protein in a kingdom that can be replenished with the assistance of small HSP gene circle of relatives as stated by way of Nillegoda et al., [48]. They're generally cytoprotective, presenting thermo-tolerance that is specifically crucial for plant life [49]. The HSP70 superfamily's genomic evaluation revealed an evolutionary history as phylogenetic tree of all HSP70 participants that cautioned the similarity of HSP70s in 12 subgroups, including the ones expressed formerly to the mammalian HSP110 and GRP170 in the identical sub-cell component. Growth in the expression of HSP70 in one of a kind plant species underneath heat stress situations has been studied appreciably by using proteomics and practical genomics [50]. A widespread osmoprotective effect changed into received in *Escherichia coli* transformed with the cytosolic chaperonin CCP-1a from *Bruguiera sexangula* [51].

HSPs have been studied extensively through plant transformation in response to heat stress [52]. Over expression of HSP101 from Arabidopsis in rice transgenic plants

Organism	Nomenclature	Length (bps)	Localization	Reference
Arabidopsis	AT 3G12580.1	650	wall/plasma membrane Mitochondrion /cell	[55]
	AT 1G11660.1	763	Nucleus	
	AT 1G16030.1	646	Nucleus	
	AT 1G56410.1	617	Chloroplast/cytoplasm	
	AT 1G79920.1	831	Nucleus/cell wall/plasma membrane	
	AT 1G79930.1	831	Plasma membrane	
	AT 2G32120.1	563	Cytoplasm	
	AT 3G09440.1	649	Golgi apparatus/cytoplasm/plasma membrane	
	AT 4G16660.1	867	Golgi apparatus/chloroplast/vacuole membrane/ER	
	AT 4G17750.1	495	Cytoplasm/nucleus	
	AT 4G37910.1	682	Mitochondrion/cell wall	
	AT 5G02490.1	653	Plasma membrane/cell wall/golgi apparatus/ nucleus	
	AT 3G17880.1	380	Cytoplasm	
Tomato	Solyc08g079260.2	422	Unpredicted	[33, 34]
	Solyc12g043110.1	852	Unpredicted	
	Solyc01g106210.2	681	Mitochondrion	
	Solyc06g005440.1	118	Nucleus	
	Solyc11g020040.1	692	Unpredicted	
	Solyc12g042560.1	210	Unpredicted	
	Solyc03g117630.1	654	Nucleus/cytoplasm	
	Solyc06g052050.2	619	Nucleus/cytoplasm	
	Solyc03g117620.2	186	Nucleus/cytoplasm	
	Solyc11g020300.1	443	Chloroplast	
	Solyc03g082920.2	667	Nucleus/cytoplasm	
	Solyc09g010630.2	669	Nucleus/cytoplasm	
	Solyc11g066100.1	654	Nucleus/cytoplasm	
	Solyc08g082820.2	666	Nucleus/cytoplasm	
	Solyc11g066060.1	698	Unpredicted	
	Solyc01g106260.2	670	Mitochondrion	
	Solyc08g079170.2	579	Nucleus/cytoplasm	
	Solyc12g043120.1	846	Nucleus/cytoplasm	
	Solyc01g060400.1	80	Nucleus/cytoplasm	
	Solyc03g117630.1	654	Nucleus/cytoplasm	
	Solyc06g052050.2	619	Nucleus/cytoplasm	
	Solyc03g117620.2	186	Nucleus/cytoplasm	

Organism	Nomenclature	Length (bps)	Localization	Reference
	Solyc11g020300.1	443	Chloroplast	
	Solyc03g082920.2	667	Nucleus/cytoplasm	
	Solyc09g010630.2	669	Nucleus/cytoplasm	
	Solyc11g066100.1	654	Nucleus/cytoplasm	
	Solyc08g082820.2	666	Nucleus/cytoplasm	
	Solyc11g066060.1	698	Unpredicted	
	Solyc03g117630.1	654	Nucleus/cytoplasm	
	Solyc06g052050.2	619	Nucleus/cytoplasm	
	Solyc03g117620.2	186	Nucleus/cytoplasm	
	Solyc11g020300.1	443	Chloroplast	
	Solyc03g082920.2	667	Nucleus/cytoplasm	
	Solyc09g010630.2	669	Nucleus/cytoplasm	
	Solyc11g066100.1	654	Nucleus/cytoplasm	
	Solyc08g082820.2	666	Nucleus/cytoplasm	
	Solyc11g066060.1	698	Unpredicted	
	Solyc01g106260.2	670	Mitochondrion	
	Solyc08g079170.2	579	Nucleus/cytoplasm	
	Solyc12g043120.1	846	Nucleus/cytoplasm	
	Solyc09g011030.	2397	Nucleus/cytoplasm	
	Solyc07g043560.2	890	Nucleus/cytoplasm	
	Solyc04g011440.2	651	Nucleus/cytoplasm	
	Solyc02g080470.2	753	Nucleus/cytoplasm	
	Solyc07g005820.2	654	Nucleus/cytoplasm	
Rice	Os01g62290	649	Nucleus/cytoplasm	
	Os03g16860	651	Nucleus/cytoplasm	
	Os03g16880	561	Nucleus/cytoplasm	
	Os03g16920	654	Nucleus/cytoplasm	
	Os03g60620	650	Nucleus/cytoplasm	
	Os05g38530	647	Nucleus/cytoplasm	[56, 57]
	Os11g47760	650	Nucleus/cytoplasm	
	Os11g08440	578	Nucleus/cytoplasm	
	Os11g08445	658	Nucleus/cytoplasm	
	Os01g62290	649	Nucleus/cytoplasm	
	Os03g16860	651	Nucleus/cytoplasm	
	Os03g16880	561	Nucleus/cytoplasm	
	Os11g08460	563	Nucleus/cytoplasm	
	Os11g08470	468	Nucleus/cytoplasm	

Organism	Nomenclature	Length (bps)	Localization	Reference
	Os12g38180	216	Cytoplasm/Nucleus	
	Os01g33360	609	ER	
	Os02g53420	680	Mitochondria	
	Os03g02260	677	Mitochondria	
	Os09g31486	685	Mitochondria	
	Os05g23740	690	Chloroplast	
	Os12g14070	699	Chloroplast	
	Os01g49430	540	Unpredicted	
	Os01g08560	846	Cytoplasm	
	Os02g48110	903	ER	
	Os03g11910	579	Cytoplasm	
	Os05g08840	854	Cytoplasm	
	Os05g51360	438	ER	
	Os06g10990	471	Cytoplasm	
	Os06g46600	754	Cytoplasm	
	Os12g05760	462	ER	
	Os02g43020	579	Nucleus/plasma membrane	
	Os03g16460.	385	Nucleus/Cytoplasm	
	Os03g60780	380	Nucleus/Cytoplasm	
	Os04g35900	430	Membrane	
	Os02g01030	410	Unpredicted	
	Os01g33360	609	ER	
	Os02g53420	680	Mitochondria	
	Os03g02260	677	Mitochondria	
	Os09g31486	685	Mitochondria	
	Os05g23740	690	Chloroplast	
	Os12g14070	699	Chloroplast	
	Os01g49430	540	Unpredicted	
	Os01g08560	846	Cytoplasm	
	Os02g48110	903	ER	
	Os03g11910	579	Cytoplasm	
	Os05g08840	854	Cytoplasm	
	Os05g51360	438	ER	
	Os06g10990	471	Cytoplasm	
	Os06g46600	754	Cytoplasm	
Agave sisalana	Os12g05760	462	ER	
	Os02g43020	579	Nucleus/plasma membrane	

Organism	Nomenclature	Length (bps)	Localization	Reference
	Os03g16460.	385	Nucleus/Cytoplasm	
	Os03g60780	380	Nucleus/Cytoplasm	
	Os04g35900	430	Membrane	
	Os02g01030	410	Unpredicted	
	MH555298.1	328	ER	[10]
	MH555299.1	933	Secretory pathway	
	MH555300.1	908	Chloroplast	
	MH555301.1	242	Unpredicted	
	MH555302.1	271	Mitochondria	
	MH555303.1	251	Secretory pathway	
	MH555304.1	251	Unpredicted	
	MH555305.1	229	Unpredicted	
	MH555306.1	228	Cytoplasm	
	MH555307.1	241	Unpredicted	
	MH555308.1	213	Chloroplast	
	MH555309.1	567	Unpredicted	
	MH555310.1	241	ER	
	MH555311.1	221	Unpredicted	
	MH555312.1	376	Secretory pathway	
	MH555313.1	2229	Secretory pathway	
	MH555314.1	311	Cytoplasm	
	MH555315.1	352	Chloroplast	
	MH555316.1	550	Secretory pathway	
	MH555317.1	215	Unpredicted	
	MH555318.1	239	Secretory pathway	
	MH555319.1	2043	Secretory pathway	
	MH555320.1	344	Secretory pathway	
	MH555321.1	2072	Unpredicted	
	MH555322.1	981	Unpredicted	
	MH555323.1	2117	Unpredicted	
	MH555324.1	358	Secretory pathway	
	MH555325.1	962	Unpredicted	
	MH555326.1	221	Chloroplast	
	MH555327.1	2029	Secretory pathway	
	MH555328.1	1495	Unpredicted	
	MH555329.1	893	Cytoplasm	
	MH555330.1	2229	Secretory pathway	

Organism	Nomenclature	Length (bps)	Localization	Reference
	MH555331.1	356	Unpredicted	
	MH555332.1	434	Secretory pathway	
	MH555333.1	1188	Mitochondria	
	MH555334.1	1154	Cytoplasm	
	MH555335.1	512	Chloroplast	
	MH555336.1	2477	Unpredicted	
	MH555337.1	372	Chloroplast	
	MH555338.1	1961	Cytoplasm	
	MH555339.1	2006	ER	
	MH555340.1	1660	Chloroplast	
	MH555341.1	2756	Mitochondria	
	MH555342.1	1790	Cytoplasm	
	MH555343.1	510	Chloroplast	
	MH555344.1	810	ER	
	MH555345.1	468	Secretory pathway	
	MH555346.1	432	Chloroplast	
	MH555347.1	405	Unpredicted	
	MH555348.1	303	Unpredicted	
	MH555349.1	903	Chloroplast	
	MH555350.1	1005	Secretory pathway	
	MH555351.1	257	Mitochondria	
	MH555352.1	550	Unpredicted	
	MH555353.1	376	Secretory pathway	
	MH555354.1	646	Unpredicted	
	MH555355.1	306	Mitochondria	
	MH555356.1	311	Cytoplasm	
	MK759669.1	2229	Secretory pathway	
	MK759670.1	2064	Chloroplast	
	MK759671.1	2477	Chloroplast	
	MK759672.1	2578	Unpredicted	

Table 1.
Sequence characteristic & subcellular localization of HSP70 gene family in different plants species.

improved the growth performance after recovery from heat stress [53]. Heikkila et al. [54] demonstrated that exposure of seedlings of corn to ABA, water stress, heat shock, and wounding increases the synthesis of HSP70. Study of HSP70 in Rice, Tomato and Arabidopsis genomes revealed that about 37 genes of HSP70 were identified in rice, 30 in tomato 13 in Arabidopsis and 68 in *A. sisalana* and has been documented to be present within every cellular compartment of plants (**Table 1**).

5. Structure and function of HSP70

The HSP70s have two huge functional domains, an ATPase domain within the N-terminal part of the protein and a peptide-binding domain within the C-terminal part of the protein [58]. The two domain names are more or less 40 & 25 kDa lengthy, respectively and are separated by means of a hinge location at risk of protease cleavage. Since the HSP70s are determined inside the cytosol, endoplasmic reticulum (ER), mitochondria and plastids, the N-terminal transit peptide of variable sequence is present in the precursor form of these individuals that make part for importation into the organelle [59]. For several HSP70s, a C-terminal subdomain of 5 kDa or much less is needed for plenty co-chaperone interactions (**Figure 1**) [36]. HSP70s exercising their role in several cell techniques by way of binding uncovered hydrophobic residues of non-native proteins during protein folding, stopping protein aggregation, selling the regeneration of combination proteins and keeping proteins in an importable translocation surroundings to subcellular cubicles [9]. HSP70s interact in protein folding by means of chaperone strategies that encompass repeated cycles of peptide binding, ATP hydrolysis and peptide release. Cytosolic Hsp70s are lively in cellular strategies such as protein folding, denatured protein folding, protein aggregation prevention and protein retention in an import-capable eukaryotic surroundings [60]. There are some medical studies of plant cytosolic Hsp70s, but they are acknowledged to behave like different prokaryotic and eukaryotic HSP70s. In vitro experiments have proven that plant cytosolic Hsp70s mixes mysterious precursor polypeptides with nascent. Further, whilst wheat germ extract become removed from cytosolic Hsp70, co-translocation and processing of precursor proteins have become inefficient and the

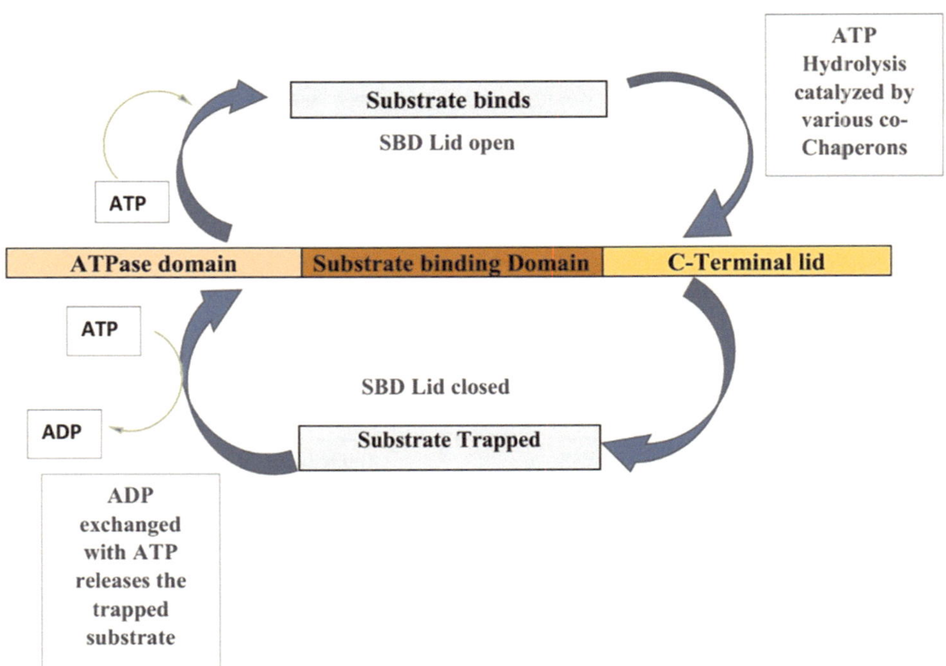

Figure 1.
Structure of HSP70: SBD (substrate binding domain), ATP (adenosine 5′-triphosphate), ADP (adenosine diphosphate).

incorporation of cytosolic Hsp70 restored the translocation and processing of precursor proteins [61]. It is proposed that the cytoplasmic Hsp70s are involved within the ER-translocation precursor protein.

Chloroplast Hsp70 homologs recognized to be energetic in import strategies are within the outer envelope membrane facing the cytoplasm and inter membrane space, the stroma and the thylacoid lumen. Hsp70 is concerned in early protein imports by using associating with chloroplast precursor proteins, based on move-linking and immune precipitation studies [62]. The precise characteristic of Hsp70 stays unclear, however it is proposed that precursor proteins are moved from the cytosolic HSP70s to the chloroplast HSP70s. This is confirmed through studies with Dnak as a model system and shows that HSP70s can join chloroplast precursor protein transit peptides [7]. Mitochondrial precursor protein translocation also occurs post-translationally and precursor proteins are once again preserved in an import-ready state with the useful resource of cytosolic HSP70s. Its homologs are positioned inside the outer mitochondrial membrane facing the cytosol and in the mitochondrial bean matrix [63]. An HSP70 homolog located in the outer membrane changed into suspected to be involved in protein translocation within the outer membrane with the aid of attaching the precursor proteins launched from the cytosolic Hsp70 and integrating them into the outer membrane. However, there may be no proof to guide this process. Matrix Hsp70, which has a strong homology to DnaK, is determined to be closely connected with the inner mitochondrial membrane of bean [36]. It is proposed that the HSP70 matrix may additionally result in internal membrane import by means of pulling precursor proteins into the matrix in an ATP-based method aided by using the GrpE mitochondrial co-chaperone homolog [64].

6. Mechanism of HSP70 in plants

The ability of plants to tolerate dangerous effects of intense high temperature without irreversible harm is heat stress tolerance [65]. Effect of temperature contributes to a number of bad changes in plants' life: extreme dehydration and dryness, chlorophyll burning and different physiological disorders. The cessation of protein synthesis improves the degradation and accumulation of ammonia poisonous substances [66]. However, the heat tolerance mechanisms in flora have been partly understood that HSP70 gene protect flowers from oxidative damage [67]. Additional mechanisms probably contributing to heat tolerance involve phytohormones, second messenger molecules which includes calcium (Ca^{++}) and an expansion of transcription factors [68]. The various downstream tactics, safety towards oxidative damage and protein aggregation at some stage in heat stress are critical for preserving mobile membrane integrity and photosynthesis. Consequently, over-expression of HSP70s initially regarded to be a promising technique for engineering to evaluate the heat tolerance in vegetation; but, best restricted success has been reported in past many years [69], and no field tests, heat tolerant and transgenic line has been stated. These observations advocate that a single stress tolerance mechanism might not be sufficient and additional mechanisms will be had to generate durable heat-tolerant cultivars. Highly conserved protein, HSP70s are omnipresent proteins first-class acknowledged for their susceptibility to numerous stresses consisting of heat stress [70]. HSP70 assist to place every protein inside the organelles of the cellular and interplay the mitochondrion and chloroplast proteins [11]. They've a link to proteasomal degradation pathway mediated through ubiquitin. In addition unfolded outer membrane proteins in the

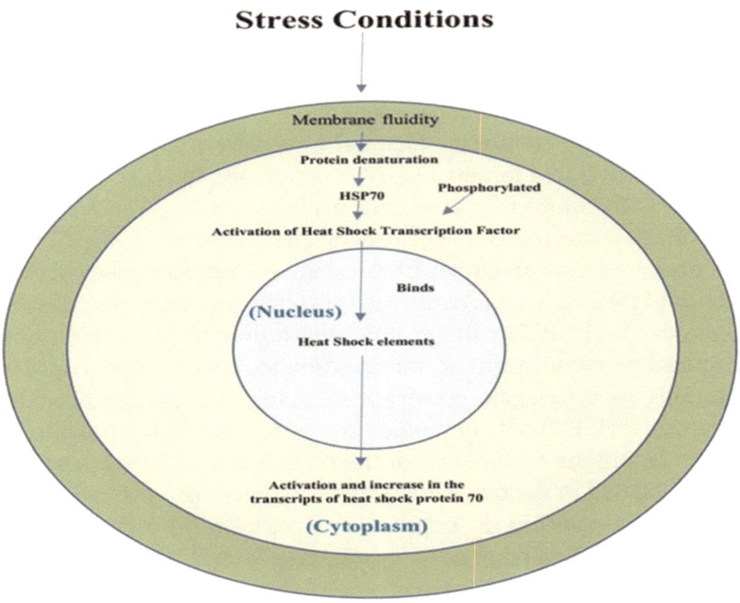

Figure 2.
Response of HSP70 against.

intercellular spaces transduce a signal to the inner membrane proteins under hot temperature inside the cytosol. This causes the heat shock transcription factors to be activated [71]. These heat shock transcription factors (HSFs) associated with HSP70 are one of the maximum reported protein families. They commonly hold collectively with heaat shock induced factors (HSEs) within the promoter areas to set off their expression, which transcribes the HSP70 (**Figure 2**).

7. HSP70 confers the tolerance to heat stress in plants

As the primary pigment of plants, chlorophyll (Chl) plays a crucial role in the mechanism of photosynthesis and its contents. Role of HSP70 in the prevention of heat, stress, chlorophyll and water breakdown was determined in transgenic tobacco and cotton seedlings. As seen in research conducted by Batcho et al., [52] & Wang et al., [72], overall output of chl, chl (a) and chl (b) content of the non-transgenic plants were decreased with the extension of treatment time after treatment with heat stress. However, the total Chl, Chla and Chlb content of HSP70 transgenic plants was higher and the reduction was slower when compared to the controls. Assay of soluble sugar content and comparative electrical conductivity of transgenic plants was improved during heat treatment when compared to control plants. This suggests that the relative electrolyte leakage of the control plants was evidently higher and the damage to the cell membrane was severe. This is consistent with the studies of [36, 73] indicating that HSP70 is involved in response to heat stress in plants.

Overexpression of HSP70 was found to decrease the malondialdehyde (MDA) content and increased production of superoxide dismutase (SOD) and peroxidase (POD) in transgenic plants when compared to control [74]. Some reactive oxygen species (ROS) such as superoxide, hydrogen peroxide and hydroxyl radicals will be synthesized and accumulated when plants are heat stressed. These ROS are cytotoxic through inactivating enzymes and killing essential cellular components such as cell membranes by oxidative processes' damage. MDA is the final product of peroxidation of the membrane. The higher the peroxidation, larger the amount of MDA produced. Plants have developed several defensive pathways to reduce oxidative damage and mitigate adverse effects. Transgenic tobacco plants demonstrated the higher overall activity of SOD and POD. This suggested that there would be less accumulation of ROS in transgenic and a better state of growth under heat stress. It has been found that overexpression of HSP70 increased the soluble sugar content and decreased the electrical conductivity in transgenic plants. The cell membrane also experiences primary physiological injuries which results the cell electrolyte leakage under heat stress [75].

8. Conclusion and future prospects

Heat Shock Protein Gene70 (HSP70) is one of the solutions to induce heat stress tolerance in agriculturally important crop plants. These genes identified, isolated from local environment/habitat and local plant species will be helpful to make the genetic transformation of local varieties of desirable plants. The modern genomic approaches will be helpful for the characterization of genes at transcriptional or promoter level to modify the gene and to enhance the gene expression in transgenic crops. Hence the HSP70 will be a suitable target to combat the crops against global warming threat to crops.

Author details

Fatima Batool[1*†], Batcho Anicet Agossa[1,2†], Zainab Y. Sandhu[3], Muhammad Bilal Sarwar[1], Sameera Hassan[1] and Bushra Rashid[1]

1 University of the Punjab Lahore, Centre of Excellence in Molecular Biology, Pakistan

2 Faculty of Agriculture and Environmental Sciences, Catholic University of the West Africa Cotonou, Benin

3 Montclair State University, Montclair, New Jersey, USA

*Address all correspondence to: fatima.batool@cemb.edu.pk

† These authors contributed equally.

References

[1] Khan Z, Shahwar D. Role of heat shock proteins (HSPs) and heat stress tolerance in crop plants. In: Roychowdhury R, Choudhury S, Hasanuzzaman M, Srivastava S, editors. Sustainable Agriculture in the Era of Climate Change. Switzerland: Springer, Cham; 2020. pp. 211-234

[2] Barna J, Csermely P, Vellai T. Roles of heat shock factor 1 beyond the heat shock response. Cellular and Molecular Life Sciences. 2018;**75**(16): 2897-2916

[3] Yadav A, Singh J, Ranjan K, Kumar P, Khanna S, Gupta M, et al. Heat shock proteins: Master players for heat-stress tolerance in plants during climate change. In: Wani SH, Kumar V, editors. Heat Stress Tolerance in Plants: Physiological, Molecular and Genetic Perspectives. London: John Willey and Sons Press; 2020. pp. 189-211

[4] Trivedi DK, Huda KMK, Gill SS, Tuteja N. Molecular chaperone: Structure, function, and role in plant abiotic stress tolerance. In: Tuteja N, Gill SS, editors. Abiotic Stress Response in Plants. London: Willey Press; 2016. DOI: 10.1002/9783527694570.ch7

[5] Khan A, Ali M, Khattak AM, Gai WX, Zhang HX, Wei AM, et al. Heat shock proteins: Dynamic biomolecules to counter plant biotic and abiotic stresses. International Journal of Molecular Sciences. 2019;**20**(21):5321

[6] Wruck F, Avellaneda MJ, Koers EJ, Minde DP, Mayer MP, Kramer G, et al. Protein folding mediated by trigger factor and Hsp70: New insights from single-molecule approaches. Journal of Molecular Biology. 2018;**430**(4): 438-449

[7] Rosenzweig R, Nillegoda NB, Mayer MP, Bukau B. The Hsp70 chaperone network. Nature Reviews. Molecular Cell Biology. 2019;**20**(11): 665-680

[8] Kandasamy G, Andréasson C. Hsp70–Hsp110 chaperones deliver ubiquitin-dependent and-independent substrates to the 26S proteasome for proteolysis in yeast. Journal of Cell Science. 2018;**131**(6):jcs210948

[9] Dahiya V, Buchner J. Functional principles and regulation of molecular chaperones. In: Donev R, editor. Advances in Protein Chemistry and Structural Biology. Netherlands: Elsevier Science Direct Publishres; 2019. pp. 1-60

[10] Batcho AA, Sarwar MB, Tariq L, Rashid B, Hassan S, Husnain T. Identification and characterisation of heat shock protein gene (HSP70) family and its expression in Agave sisalana under heat stress. The Journal of Horticultural Science and Biotechnology. 2020;**95**(4):470-482

[11] Usman MG, Rafii MY, Martini MY, Yusuff OA, Ismail MR, Miah G. Molecular analysis of Hsp70 mechanisms in plants and their function in response to stress. Biotechnology and Genetic Engineering Reviews. 2017;**33**(1):26-39

[12] Hasan M, Cheng Y, Kanwar MK, Chu XY, Ahammed GJ, Qi ZY. Responses of plant proteins to heavy metal stress—A review. Frontiers in Plant Science. 2017;**8**:1492

[13] Chatterjee BK, Puri S, Sharma A, Pastor A, Chaudhuri TK. Molecular chaperones: Structure-function relationship and their role in protein folding. In: Asea A, Kaur P, editors.

Regulation of Heat Shock Protein Responses. Switzerland: Springer International Publishing; 2018. pp. 181-218

[14] Guo YM, Samans B, Chen S, Kibret KB, Hatzig S, Turner NE, et al. Drought-tolerant Brassica rapa shows rapid expression of gene networks for general stress responses and programmed cell death under simulated drought stress. Plant Molecular Biology Reporter. 2017;**35**(4):416-430

[15] Leng P, Zhao J. Transcription factors as molecular switches to regulate drought adaptation in maize. Theoretical and Applied Genetics. 2020;**133**(5): 1455-1465

[16] Sarazin V, Duclercq J, Guillot X, Sangwan B, Sangwan RS. Water-stressed sunflower transcriptome analysis revealed important molecular markers involved in drought stress response and tolerance. Environmental and Experimental Botany. 2017;**142**: 45-53

[17] Singh B, Salaria N, Thakur K, Kukreja S, Gautam S, Goutam U. Functional genomic approaches to improve crop plant heat stress tolerance. F1000 Research. 2019;**8**:1721

[18] Singh V, Ali MN. Analysis of differential gene expression under salinity through differential display reverse transcription polymerase chain reaction (DDRTPCR) technique: A review. Electronic Journal of Biology. 2016;**12**:394-401

[19] Pallavi V, Abida PS. Identification of differentially expressed MRNA in black pepper (Piper nigrum L.) genotype for drought tolerance. Journal of Pharmacognosy and Phytochemistry. 2018;**7**(2):2953-2956

[20] Røsok Ø, Sioud M. Discovery of differentially expressed genes. In: Mouldy S, editor. Target Discovery and Validation Reviews and Protocols: Emerging Strategies for Targets and Biomarker Discovery. Vol. 1. Springer Nature Switzerland: Humana Press; 2007. pp. 115-129

[21] Assmann EM, Ottoboni LMM, Ferraz A, Rodríguez J, De Mello MP. Iron-responsive genes of Phanerochaete chrysosporium isolated by differential display reverse transcription polymerase chain reaction. Environmental Microbiology. 2003;**5**(9):777-786

[22] Carginale V, Maria G, Capasso C, Ionata E, La Cara F, Pastore M, et al. Identification of genes expressed in response to phytoplasma infection in leaves of Prunus armeniaca by messenger RNA differential display. Gene. 2004;**332**:29-34

[23] Visioli G, Maestri E, Marmiroli N. Differential display-mediated isolation of a genomic sequence for a putative mitochondrial LMW HSP specifically expressed in condition of induced thermotolerance in Arabidopsis thaliana (L.) Heynh. Plant Molecular Biology. 1997;**34**(3):517-527

[24] Jamil A, Riaz S, Ashraf M, Foolad MR. Gene expression profiling of plants under salt stress. Critical Reviews in Plant Sciences. 2011;**30**(5):435-458

[25] Kavar T, Maras M, Kidrič M, Šuštar-Vozlič J, Meglič V. Identification of genes involved in the response of leaves of Phaseolus vulgaris to drought stress. Molecular Breeding. 2008;**21**(2):159-172

[26] Chikara SK, Pandey M, Pandey S, Vaidya K, Chaudhary S. Next generation sequencing: A revolutionary tool for plant variety improvement. AJSIH Special Issue. 2014:137-154

[27] Mccouch SR, Mcnally KI, Wang W, Sackville HR. Genomics of gene banks: A case study in rice. American Journal of Botany. 2012;**99**(2):407-423

[28] Goff SA, Ricke D, Lan TH, Presting G, Wang R, Dunn M, et al. A draft sequence of the rice genome (Oryza sativa L. ssp. japonica). Science. 2002;**296**(5565):92-100

[29] Matsumoto E, Hayashida N, Sakamoto K, Ohi M. Behavior of DNA markers linked to a clubroot resistance gene in segregating populations of Chinese cabbage (Brassica rapa ssp. pekinensis). Journal of the Japanese Society for Horticultural Science. 2005; **74**(5):367-373

[30] Yu XQ, Zhu YF, Ma C, Fabrick JA, Kanost MR. 2002. Pattern recognition proteins in Manduca sexta plasma. Insect Biochemistry and Molecular Biology. 2002;**32**(10):1287-1293

[31] Casaretto JA, El-Kereamy A, Zeng B, Stiegelmeyer SM, Chen X, Bi YM, et al. Expression of OsMYB55 in maize activates stress-responsive genes and enhances heat and drought tolerance. BMC Genomics. 2016;**17**(1):1-15

[32] Qi Y, Liu Y, Zhang Z, Gao J, Guan Z, Fang W, et al. The over-expression of a chrysanthemum gene encoding an RNA polymerase II CTD phosphatase-like 1 enzyme enhances tolerance to heat stress. Horticulture Research. 2018;**5**(1): 1-10

[33] Liu S, Lv Z, Liu Y, Li L, Zhang L. Network analysis of ABA-dependent and ABA-independent drought responsive genes in Arabidopsis thaliana. Genetics and Molecular Biology. 2018a;**41**(3): 624-637

[34] Liu J, Pang X, Cheng Y, Yin Y, Zhang Q, Su W, et al. The Hsp70 gene family in Solanum tuberosum: Genome-wide identification, phylogeny, and expression patterns. Scientific Reports. 2018b;**8**(1):1-11

[35] Büyük İ, Inal B, Ilhan E, Tanriseven M, Aras S, Erayman M. Genome-wide identification of salinity responsive HSP70s in common bean. Molecular Biology Reports. 2016;**43**(11): 1251-1266

[36] Sung DY, Kaplan F, Guy CL. Plant Hsp70 molecular chaperones: Protein structure, gene family, expression and function. Physiologia Plantarum. 2001; **113**(4):443-451

[37] Yer EN, Baloglu ME, Ziplar UT, Ayan S, Unver T. Drought-responsive Hsp70 gene analysis in populus at genome-wide level. Plant Molecular Biology Reports. 2016;**34**(2):483-500

[38] Banerjee A, Roychoudhury A. Abscisic-acid-dependent basic leucine zipper (bZIP) transcription factors in plant abiotic stress. Protoplasma. 2017; **254**(1):3-16

[39] Nejat N, Mantri N. Plant immune system: Crosstalk between responses to biotic and abiotic stresses the missing link in understanding plant defence. Current Issues in Molecular Biology. 2017;**23**:1-16

[40] Sharma B, Deswal R. Ecophysiolomic analysis of stress tolerant Himalayan shrub Hipppophae rhamnoides shows multifactorial acclimation strategies induced by diverse environmental conditions. Physiologia Plantarum. 2020;**168**(1):58-76

[41] Parankusam S, Bhatnagar-Mathur P, Sharma KK. Heat responsive proteome changes reveal molecular mechanisms underlying heat tolerance in chickpea.

Environmental and Experimental Botany. 2017;**141**:132-144

[42] Saux M, Ponnaiah M, Langlade N, Zanchetta C, Balliau T, El-Maarouf-Bouteau H, et al. A multiscale approach reveals regulatory players of water stress responses in seeds during germination. Plant, Cell & Environment. 2020;**43**(5): 1300-1313

[43] Archana PR, Aleena J, Pragna P, Vidya MK, Niyas APA, Bagath M, et al. Role of heat shock proteins in livestock adaptation to heat stress. Journal of Dairy Veterinary Animal and Research. 2017;**5**(1):00127

[44] Penke B, Bogár F, Crul T, Sántha M, Tóth ME, Vígh L. Heat shock proteins and autophagy pathways in neuroprotection: From molecular bases to pharmacological interventions. International Journal of Molecular Sciences. 2018;**19**(1):325

[45] Dubrez L, Causse S, Nb B, Dumetier B, Garrido C. Heat-shock proteins: Chaperoning DNA repair. Oncogene. 2020;**39**:516-529

[46] Kumar N, Suyal DC, Sharma IP, Verma A, Singh H. Elucidating stress proteins in rice (Oryza sativa L.) genotype under elevated temperature: A proteomic approach to understand heat stress response. 3 Biotech. 2017; **7**(3):205

[47] Sánchez-Martín J, Keller B. Contribution of recent technological advances to future resistance breeding. Theoretical and Applied Genetics. 2019; **132**(3):713-732

[48] Nillegoda NB, Wentink AS, Bukau B. Protein disaggregation in multicellular organisms. Trends in Biochemical Sciences. 2018;**43**(4): 285-300

[49] Hassan FU, Nawaz A, Rehman MS, Ali MA, Dilshad SM, Yang C. Prospects of HSP70 as a genetic marker for thermo-tolerance and immuno-modulation in animals under climate change scenario. Animal Nutrition. 2019;**5**(4):340-350

[50] Wang D, Luthe DS. Heat sensitivity in a bentgrass variant. Failure to accumulate a chloroplast heat shock protein isoform implicated in heat tolerance. Plant Physiology. 2003;**133**(1): 319-327

[51] Yamada A, Saitoh T, Mimura T, Ozeki Y. Expression of mangrove allene oxide cyclase enhances salt tolerance in Escherichia coli, yeast, and tobacco cells. Plant and Cell Physiology. 2002;**43**(8): 903-910

[52] Batcho AA, Sarwar MB, Rashid B, Hassan S, Husnain T. Heat Shock Protein Gene Identified from Agave sisalana (AsHSP70) Confers Heat Stress Tolerance in Transgenic Cotton (Gossypium Hirsutum). Theoretical and Experimental Plant Physiology. 2021;**33** (2):141-156

[53] Katiyar-Agarwal S, Agarwal M, Grover A. Heat-tolerant basmati rice engineered by over-expression of hsp101. Plant Molecular Biology. 2003; **51**(5):677-686

[54] Heikkila JJ, Papp JT, Schultz GA, Bewley JD. Induction of heat shock protein messenger RNA in maize mesocotyls by water stress, abscisic acid, and wounding. Plant Physiology. 1984; **76**(1):270-274

[55] Lin BI, Wang JS, Liu HC, Chen RW, Meyer Y, Barakat A, et al. Genomic analysis of the Hsp70 superfamily in Arabidopsis thaliana. Cell Stress & Chaperones. 2001;**6**(3): 201-208

[56] Jung KH, Gho HJ, Nguyen MX, Kim SR, An G. Genome-wide expression analysis of HSP70 family genes in rice and identification of a cytosolic HSP70 gene highly induced under heat stress. Functional & Integrative Genomics. 2013;**13**(3):391-402

[57] Sarkar NK, Kundnani P, Grover A. Functional analysis of Hsp70 superfamily proteins of rice (Oryza sativa). Cell Stress & Chaperones. 2013; **18**(4):427-437

[58] Takayama S, Xie Z, Jc R. An evolutionarily conserved family of Hsp70/Hsc70 molecular chaperone regulators. The Journal of Biological Chemistry. 1999;**274**(2):781-786

[59] Agranovsky AA, Folimonova SY, Folimonov AS, Denisenko ON, Zinovkin RA. The beet yellows closterovirus p65 homologue of HSP70 chaperones has ATPase activity associated with its conserved N-terminal domain but does not interact with unfolded protein chains. The Journal of General Virology. 1997;**78**(3): 535-542

[60] Mayer MP, Gierasch LM. Recent advances in structural and mechanistic aspects of Hsp70 molecular chaperones. The Journal of Biological Chemistry. 2019;**294**(6):2085-2097

[61] Horwich AL, Fenton WA. Chaperonin-assisted protein folding: A chronologue. Quarterly Reviews of Biophysics. 2020;**53**:E4. DOI: 10.1017/S0033583519000143

[62] Jarvis P. Targeting of nucleus-encoded proteins to chloroplasts in plants. The New Phytologist. 2008; **179**(2):257-285

[63] Lee DW, Hwang I. Evolution and design principles of the diverse chloroplast transit peptides. Molecules and Cells. 2018;**41**(3):161-167

[64] Flores-Pérez Ú, Jarvis P. Molecular chaperone involvement in chloroplast protein import. Biochimica et Biophysica Acta (BBA)-Molecular Cell Research. 2013;**1833**(2):332-340

[65] Levitt J. Water, Radiation, Salt, and Other Stresses. New York: Elsevier Academic Press Inc. A subsidiary of Harcourt Brace Jovanovich, Publishers; 2015

[66] Uprety DC, Reddy VR. Crop Responses to Global Warming. Singapore: Springer Singapore; 2016. Sep 13

[67] Ohama N, Sato H, Shinozaki K, Yamaguchi-Shinozaki K. Transcriptional regulatory network of plant heat stress response. Trends in Plant Science. 2017; **22**(1):53-65

[68] Wani SH, Kumar V, Shriram V, Sah SK. Phytohormones and their metabolic engineering for abiotic stress tolerance in crop plants. The Crop Journal. 2016;**4**(3):162-176

[69] Semwal VK, Khanna-Chopra R. Enhanced oxidative stress, damage and inadequate antioxidant defense contributes towards insufficient recovery in water deficit stress and heat stress combination compared to either stresses alone in Chenopodium album (Bathua). Physiology and Molecular Biology of Plants. 2020;**26**(7):1331-1339

[70] Jacob P, Hirt H, Bendahmane A. The heat shock protein/chaperone network and multiple stress resistance. Plant Biotechnology Journal, 2017;**15**(4): 405-414

[71] Quirós PM, Mottis A, Auwerx J. Mitonuclear communication in

homeostasis and stress. Nature Reviews Molecular Cell Biology. 2016;**17**(4): 213-226

[72] Wang X, Yan B, Shi M, Zhou W, Zekria D, Wang H, et al. Overexpression of a Brassica campestris HSP70 in tobacco confers enhanced tolerance to heat stress. Protoplasma. 2016;**253**(3): 637-645

[73] Sung DY, Vierling E, Guy CL. Comprehensive expression profile analysis of the Arabidopsis Hsp70 gene family. Plant Physiology. 2001a;**126**(2): 789-800

[74] Gosavi G, As J, Aa K, Sr G, Pawar B, Vp C. Effect of heat stress on proline, chlorophyll content, heat shock proteins and antioxidant enzyme activity in sorghum (Sorghum bicolor) at seedlings stage. Indian Journal of Biotechnology. 2014;**13**(3):356-363

[75] Xu H, Liu G, Liu G, Yan B, Duan W, Wang L, et al. Comparison of investigation methods of heat injury in grapevine (Vitis) and assessment to heat tolerance in different cultivars and species. BMC Plant Biology. 2014;**14**(1): 156

Chapter 8

Abiotic Stress in Plants

Shubham Dey and Ayan Raichaudhuri

Abstract

Stress in plants refers to external conditions, which drastically affect the growth, development, or productivity of plants. Stress triggers a wide range of plant responses, such as altered gene expression, cellular metabolism, changes in growth rates, and crop yields. Some abiotic stresses, such as low or high temperature, deficient water, and ultraviolet radiation, make plant growth and development unfavorable, leading to a fall in crop yield worldwide. The following writeup incorporated the abiotic stress factors related to the growth and development of plants, such as temperature, drought, heat, cold, and many more. Abiotic stress factors are the nonliving factors influencing the metabolism, growth, and development of the plant tissues at that particular time when such abiotic stress affects them. As a result of such abiotic stresses, the plants have generated many stress tolerance factors. Various stress-responsive genes are thus being formulated in response to the abiotic stresses, so the plants can survive even in such extreme conditions as well. Henceforth, it can be concluded that the abiotic stress factors imposed on the plants adversely impact their growth and developmental procedures, and at the same time, they also produce some stress tolerance factors to minimize the damage.

Keywords: biotic stress, abiotic stress, temperature, salt, stress regulators

1. Introduction

Abiotic stress can be defined as the adverse impacts created by the abiotic factors on the plant tissues [1]. Abiotic stress is caused by nonliving factors that are in contrast to biotic stress, which is caused by living organisms. The various factors impacting the plant tissues interrupt their normal metabolism. In response to this stress, the plants adapt newer metabolic reactions to resist the stress. The majority of such reactions aid them to regulate and sustain themselves against various environmental factors [1].

In **Figure 1**, stresses influencing the growth and developmental patterns of plants are shown. As stated by the figure, plant stress depends upon the stress factors being living or nonliving and thus the stress gets segmented as biotic (living) stress and abiotic (nonliving) stress in plants [2].

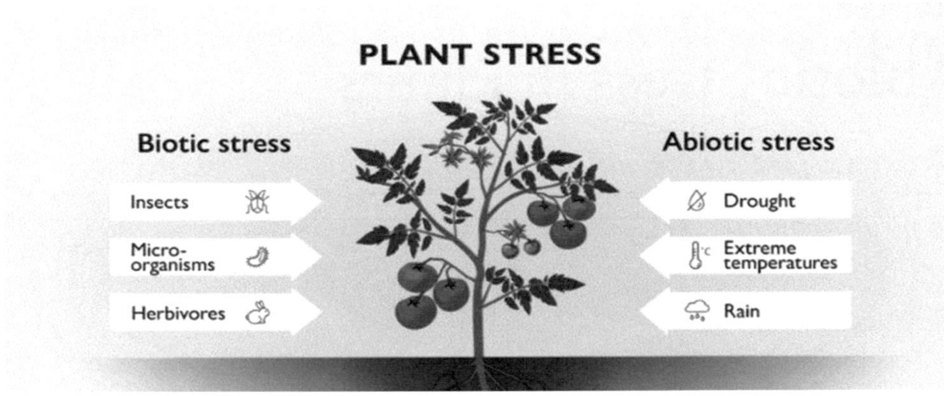

Figure 1.
Some of the common plant stresses: both biotic and abiotic stresses have effects on plants.

2. Stress impacts on plants

The consequence of the stress factors on the plant tissue is their influence on their growth and development pattern [2]. As a result of the stresses, various types of plant metabolism get triggered, such as the altered expression of the inherited genes, metabolism of the cells of plants, changed patterns of growth types, crop yields, and much more [2]. However, as stated by Zhang et al. [3], there are two types of stresses—biotic and abiotic that are observed among the plant tissues.

2.1 Biotic stress

Biotic stress is caused by living organisms, such as viruses, bacteria, fungi, nematodes, insects, weeds, and many others [4]. Such stressors deprive the host plants of the growth factors and nutrients within them and eventually the plants die. Thus, biotic stress factors become the major reason for the plants pre-and post-harvest losses.

In **Figure 2**, the influence of the endophytic fungi on desert plants is. The figure shows how fungi grow with the association of the plants and thus take in salts, water, and other nutrients from the plant roots. In such cases, the growing plants become deprived of salt and water and continuous salt stress and water deficiency are encountered [5]. As a result of all such negative impacts, the deterioration of the inherent metabolism of the plant parts occur. Additionally, in the case of the desert plants, already deserts are known for lower availability of water, thus any such further disturbances by the endophytic fungi have adversely impacted the plant growth [5].

2.2 Abiotic stress

Abiotic stress factors are the nonliving factors influencing the present metabolism, growth, and development of the plant tissues [6]. As stated by Sharma et al. [7], abiotic stress factors impacting plants are excessive hot temperature, extreme cold temperature, salinity, drought, mineral availability or toxicity, and much more. Such abiotic stress factors have, thus, negatively impacted the overall crop yields, and,

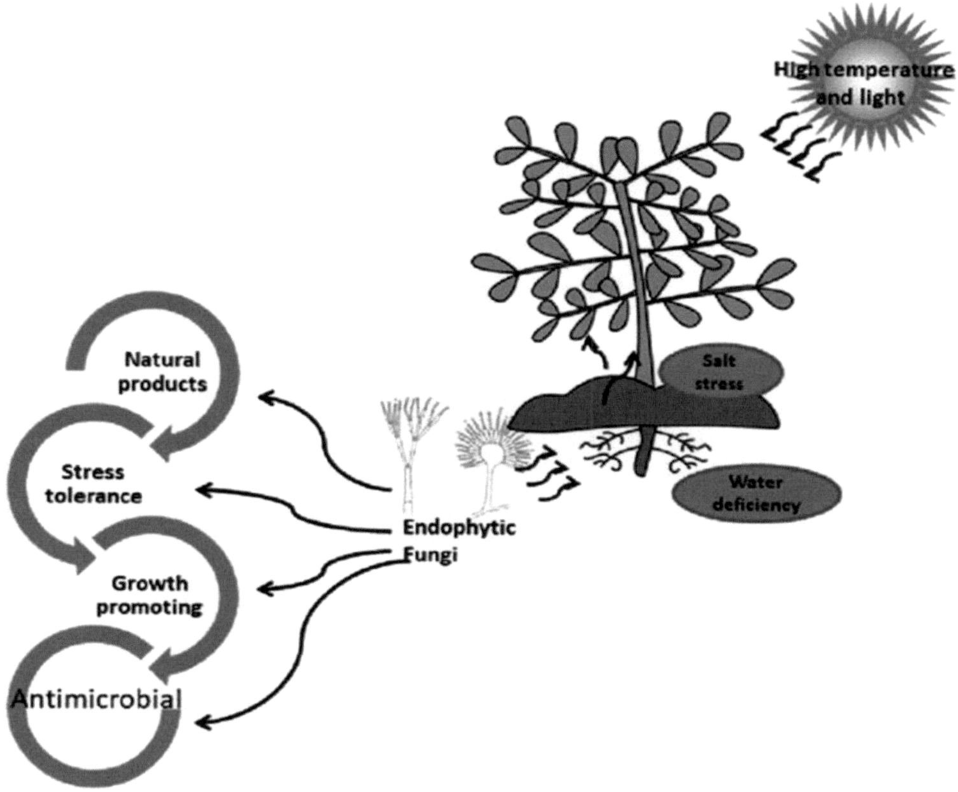

Figure 2.
Biotic stress on the desert plant: the endophytic fungi are taking the natural product of the soil meant for the plant and creating biotic stress.

thus, there is a need of generating resistant plant varieties that can sustain against abiotic stress factors [6].

Figure 3 showcases the various abiotic stress impacting plant parts and their relative outcomes. As stated in the figure, sunlight, cold weather, salinity, and mineral availability or deficiency are the major abiotic stress factors for a plant [8]. As a result of various stress factors, multiple responses can be seen. For example—the amount of ROS (reactive oxygen species) can be increased, plant growth and yield reduced, and simultaneously photosynthetic activity can be reduced. The adverse environmental temperatures deteriorated the plants' growth and developmental patterns [8].

3. Mechanism of abiotic stress

The abiotic stress is known to impact the internal metabolism of the plant parts, and, thus, the overall productivity of the plants gets reduced [9]. This could be the major adverse impact as the abiotic stress factors are widely known to cause mostly negative impacts only [9]. The several abiotic stress factors and their simultaneous mechanisms are being discussed as follows:

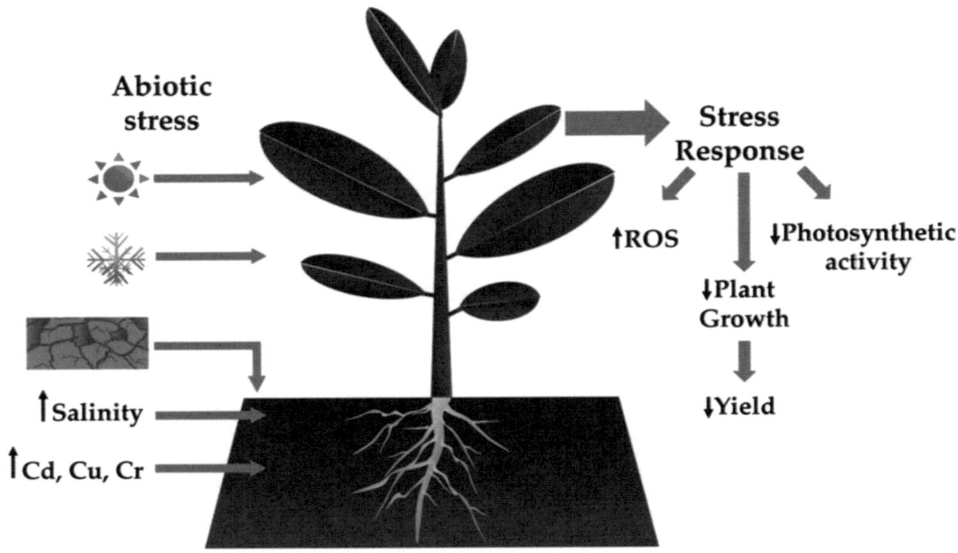

Figure 3.
Abiotic stress factors in plants: major abiotic stress factors for a plant deficiency are sunlight, cold weather, salinity, and mineral availability or deficiency. They are the major abiotic stress factors for plant deficiency.

3.1 Cold

Every plant is known to survive at a particular temperature only. The alteration of the required temperature changed the overall sustenance patterns of the individual plants [10]. In case of plants growing in cold temperatures, cold temperature would result in disruption of the plant tissues, as a result, it would lead to deterioration of the life cycle of plants.

Figure 4 explains the various mechanisms adopted by the plants to mitigate the cold stress factors against the sustenance of plants [10]. As seen from the figure, the incorporation of such stress factors aids the plants to generate signals and further transcriptional control so that the genes of stress signaling could be simultaneously activated. This will ultimately lead to re-establishment of the cellular homeostasis and functional and structural protection of protein and membranes [10]. All such optimistic sequential steps impacted positively internal cellular membranes and, thus, lead to stress tolerance or resistance against abiotic stress factors by plants [10].

3.2 Salt

The salt concentration is one of the major factors impacting plants' growth and development [11]. Higher amounts of salt lead to the re-release of genes for minimizing stresses against salt concentration and thus optimizing the plants to survive in such hazardous situations as well.

Figure 5 illustrates the fact that accumulation of solute concentration leads to mineral absorption in an excessive amount and simultaneously it leads to cell wall modification and incorporation of transporters that lead to re-transportation of salts to mesophyll, homeostasis of potassium, and nitrate ions and thus generating optimistic stress responses [12].

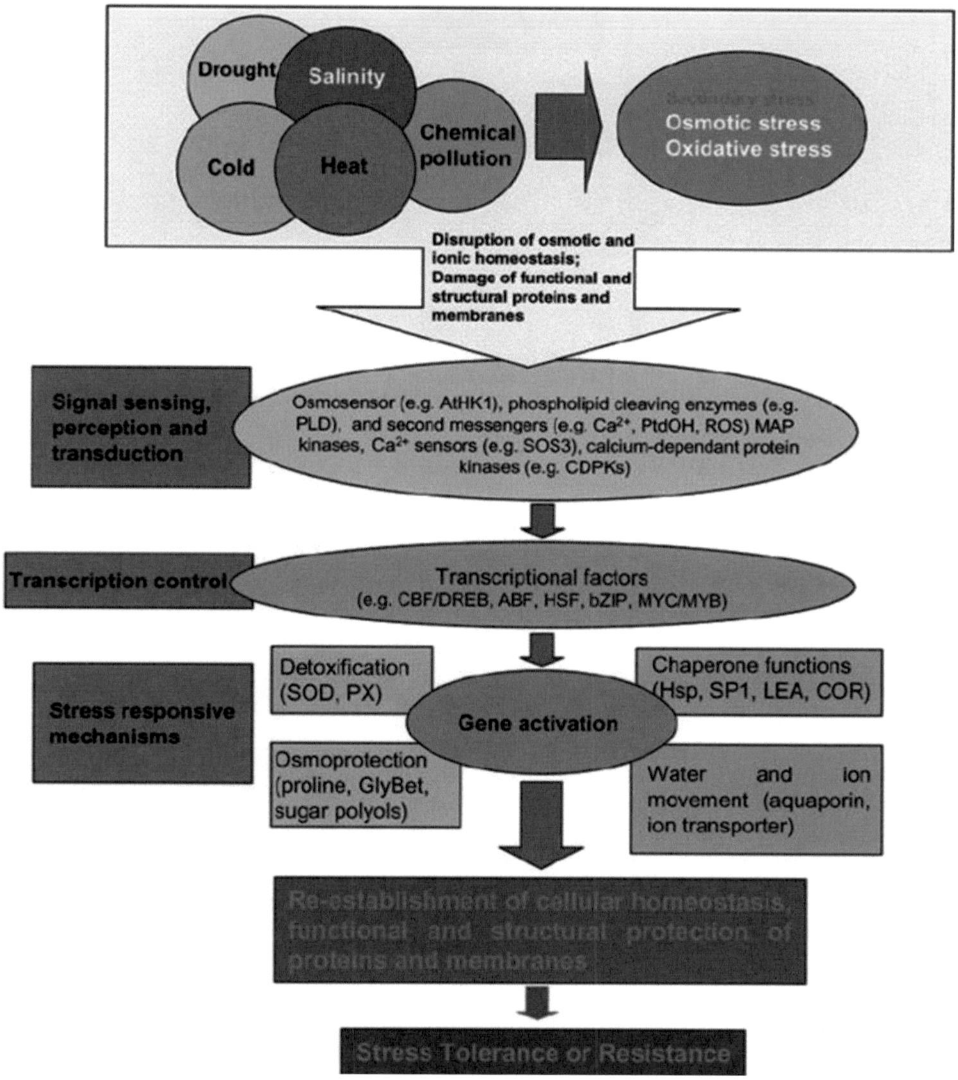

Figure 4.
Mechanism of abiotic stress factors on plants: genes of abiotic stress signaling could be simultaneously activated. This will ultimately lead to re-establishment of the cellular homeostasis, functional and structural stress tolerance, or resistance protection of proteins and membranes.

3.3 Toxin

Toxins are chemicals released by the plant tissues in response to several abiotic stress factors [13]. Toxins are also considered to be the surrounding stress factors adversely impacted by the environment, and, thus, the plants are getting negatively impacted by such abiotic stress factors.

Figure 6 illustrates the toxins released by the cell wall, cell membranes, cytoplasm, chloroplast, mitochondrion, endoplasmic reticulum, peroxisome, and nucleus [14]. Signal integration of the stress factors occurs now and the stress response genes are activated and released. These responses lead to the sustenance of the overall growth and development of the plant parts [14].

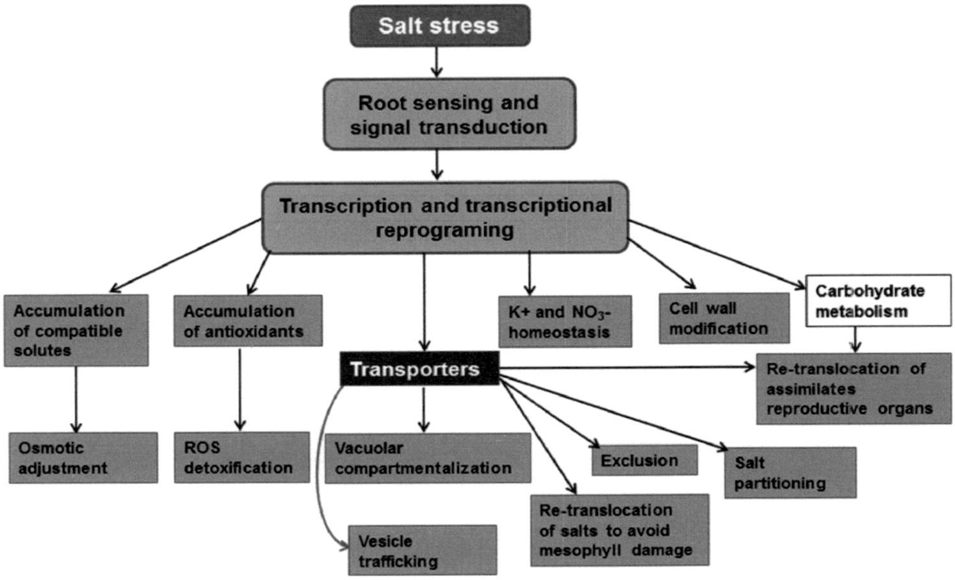

Figure 5.
Factors impacting salt stress in plants: excessive amount of mineral absorption leads to cell wall modification and incorporation of transporters that lead to re-translocation of salts to mesophyll, homeostasis of potassium and nitrate ions, and thus generating optimistic stress responses.

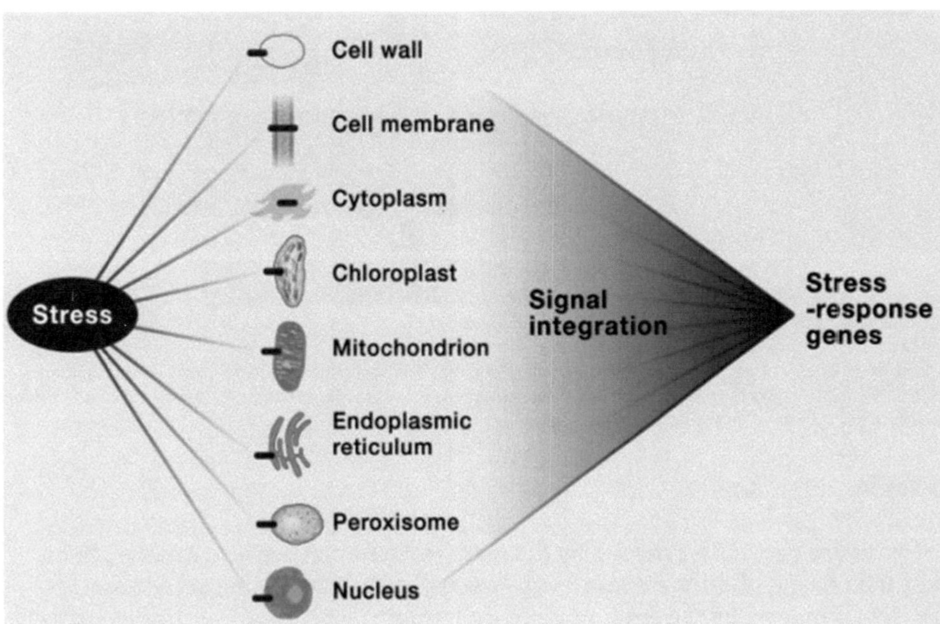

Figure 6.
Mitigation strategies against toxins: signal integration of the stress factors occurs and the stress response genes are activated and released. These responses lead to the sustenance of the overall growth and development of the plant parts.

4. Mitigation strategies adopted by plants for overcoming abiotic stress factors

The abiotic stress factors are the ones that cannot be sustained and mitigated by the plants externally, henceforth plants are known for developing fresh mechanisms within their inner metabolism to balance the excessive adverse impacts created by the outside environment [7]. Many such mitigation strategies are adopted by plants to overcome such abiotic stress factors [7].

In **Figure 7**, the sequential steps adopted by plants to overcome the abiotic stress impacts are shown. The figure states that with abiotic stresses, the plants tend to develop excessive ROS. Such an excessive production leads to the further incorporation of 3 steps—activation of oxygen antioxidants, up-regulation of osmolytes, and activation of stress-responsive genes [15]. The activation of the stress-responsive genes makes the plants much tolerant and thus they can survive against such hazardous temperatures as well. So identification of targeted genes is necessary as the overall mechanism depends upon such gene regulations only [3]. Henceforth it can be concluded that oxidative stress reduction results in an increase in the stress tolerance factors and long-term sustainability of the plants in such adverse conditions as well [3].

5. Conclusion

The research on abiotic stress factors on plant growth and development reveals that they are the major factors that influence and lead to the deterioration of the plant

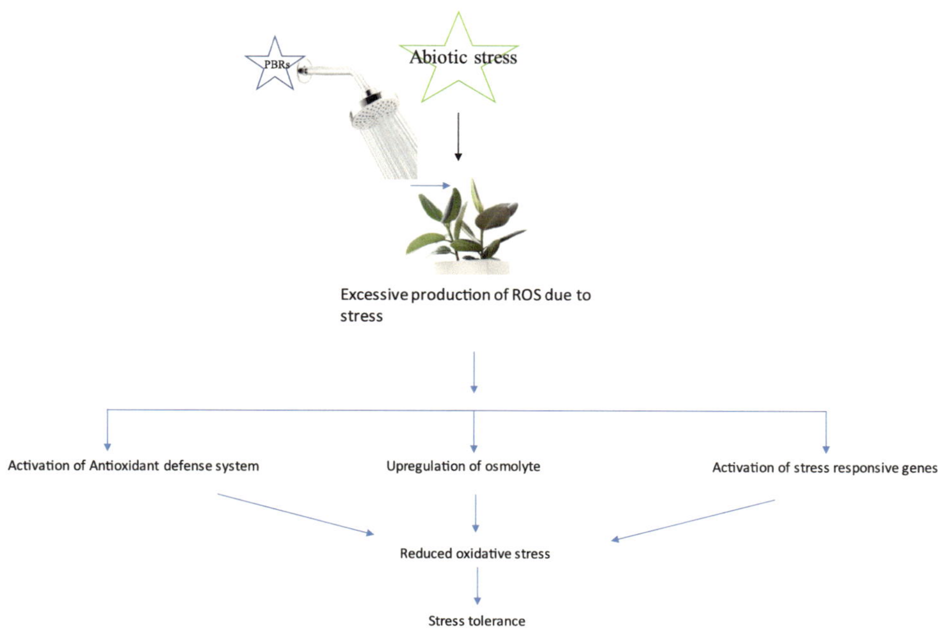

Figure 7.
Steps for overcoming abiotic stress by plants: in abiotic stresses, the plants tend to develop excessive ROS in oxidative stress reduction, which results in increase of stress tolerance factors and long-term sustainability of the plants.

species. Heat, cold, drought, salinity, and toxins are various abiotic stressors impacting adversely the overall development of plants. Various stress-responsive genes are formulated in response to abiotic stresses so that the plants can survive extreme conditions as well. Rapid population growth, economic development, and international economic integration have intensified resource use in every sector of the world. The human population is expected to increase to a total of 9 billion by 2050. So production of more food from the same area of land will be needed and this can happen only by reducing the adverse environmental impacts on plants. This is what has been called sustainable intensification, for feeding, clothing, and providing energy to such a large population. Transgenic approaches have been proven to show as powerful tools to help understand and manipulate the responses of plants to stress. Global research analyses indicate that transcription to proteins and metabolites occurs during abiotic stress. These findings will advance our understanding of major metabolic pathways and provide direction for achieving abiotic stress-tolerant plants. The viable evaluation of transgenes that enhance crop performance under both stress and optimal conditions is a prolonged, tedious, and expensive process. It is being proposed that the current stance on plant stress tolerance can be significantly polished by thorough characterization of individual genes and evaluating their contribution to stress tolerance.

The molecular mechanisms of plants to create stress tolerance against salt, drought, and temperature involve a number of regulatory proteins, such as transcription factors. The study of such mechanisms enabled us to increase our knowledge of enhanced plant survival and increased crop yields in spite of abiotic stresses. Further research is needed for accurate evaluation in the field of genotypes for abiotic stress resistance, a deeper understanding of the transcription factors that regulate major stress-responsive genes, and cross-talks between divergent signaling components. We are to advance our knowledge on traits that are associated with root architecture and plasticity, especially in agronomically superior genotypes under abiotic stress conditions. Crop tolerance to various abiotic stresses is a matter of continued research to increase our knowledge further and to help plants from deterioration and extinction. The stress biotechnology research in the recent future will emphasize on strength and stress-induced expression of the transgenes, combined with the regulatory machinery involving transcription factors as a new genetic manipulation tool for controlling the expression of many different stress-responsive genes.

In conclusion, plant sciences currently achieve good models of how model plants react to environmental factors by transcriptional and metabolic reprogramming. However, especially molecular research efforts in crops have to be strengthened considerably. Plant stress physiology is a very complex matter and needs future biocomputational integration of multiple omics and meta-omics to understand it properly. This needs further effort in developing innovative research tools and fundamental resources for crop plant research, such as reference genomes, proteomes, and metabolomes with comprehensive annotations and structure-function relationships, respectively. Even for the model *Arabidopsis*, these resources are not fully available. Nevertheless, in several cases, *Arabidopsis* and other model plants have already been proved suitable for the translation of fundamental research into agronomically relevant crop traits. This is encouraging but requires further and significant investment into translational research. Besides this, it remains indispensable to investigate abiotic stress resistance mechanisms directly in elite crop plants and in the genetic resources available for breeding.

DOI: http://dx.doi.org/10.5772/intechopen.105944

Author details

Shubham Dey and Ayan Raichaudhuri*
Amity Institute of Biotechnology, Amity University, Kolkata, India

*Address all correspondence to: araichaudhuri@kol.amity.edu

References

[1] Andjelkovic V. Introductory chapter: Climate changes and abiotic stress in plants. In: Plant, Abiotic Stress and Responses to Climate Change. IntechOpen; 2018. p. 1

[2] Bechtold U, Field B. Molecular mechanisms controlling plant growth during abiotic stress. Journal of Experimental Botany. 2018;**69**(11):2753-2758

[3] Zhang H, Zhu J, Gong Z, Zhu JK. Abiotic stress responses in plants. Nature Reviews Genetics. 2021:1-16

[4] Chen D, Shao Q, Yin L, Younis A, Zheng B. Polyamine function in plants: Metabolism, regulation on development, and roles in abiotic stress responses. Frontiers in Plant Science. 2019;**9**:1945

[5] Corpas FJ. Hydrogen sulfide: A new warrior against abiotic stress. Trends in Plant Science. 2019;**24**(11):983-988

[6] Dresselhaus T, Hückelhoven R. Biotic and abiotic stress responses in crop plants. Agronomy. 2018;**8**(11):267

[7] Sharma A, Shahzad B, Rehman A, Bhardwaj R, Landi M, Zheng B. Response of phenylpropanoid pathway and the role of polyphenols in plants under abiotic stress. Molecules. 2019;**24**(13):2452

[8] Gull A, Lone AA, Wani NUI. Biotic and abiotic stresses in plants. In: Abiotic and Biotic Stress in Plants. 2019. pp. 1-19

[9] Hasanuzzaman M, Bhuyan MHM, Zulfiqar F, Raza A, Mohsin SM, Mahmud JA, et al. Reactive oxygen species and antioxidant defense in plants under abiotic stress: Revisiting the crucial role of a universal defense regulator. Antioxidants. 2020;**9**(8):681

[10] Kosová K, Vítámvás P, Urban MO, Prášil IT, Renaut J. Plant abiotic stress proteomics: The major factors determining alterations in cellular proteome. Frontiers in Plant Science. 2018;**9**:122

[11] Lata R, Chowdhury S, Gond SK, White JF Jr. Induction of abiotic stress tolerance in plants by endophytic microbes. Letters in Applied Microbiology. 2018;**66**(4):268-276

[12] Martínez-Noël GM, Tognetti JA. Sugar signaling under abiotic stress in plants. In: Plant Metabolites and Regulation under Environmental Stress. Elsevier Inc.; 2018. pp. 397-406

[13] Nadarajah KK. ROS homeostasis in abiotic stress tolerance in plants. International Journal of Molecular Sciences. 2020;**21**(15):5208

[14] Park CJ, Park JM. Endoplasmic reticulum plays a critical role in integrating signals generated by both biotic and abiotic stress in plants. Frontiers in Plant Science. 2019;**10**:399

[15] Zandalinas SI, Fichman Y, Devireddy AR, Sengupta S, Azad RK, Mittler R. Systemic signaling during abiotic stress combination in plants. Proceedings of the National Academy of Sciences. 2020;**117**(24):13810-13820

Chapter 9

Heterologous Expression of Genes in Plants for Abiotic Stresses

Shahzad Ali, Nadir Zaman, Waqar Ali, Majid Khan, Muhammad Aasim, Asmat Ali and Muhammad Usman

Abstract

Abiotic stresses are considered to be the major factors causing a decrease in crop yield globally, these stresses include high and low temperature, salinity, drought, and light stress etc. To overcome the consistent food demand for the ever-growing population, various genes from micro-organisms and non-plant sources have been expressed in transgenic plants to improve their tolerance against abiotic stresses. Gene expression in transgenic plants through conventional methods are time-consuming and laborious that's why advanced genetic engineering methods for example *Agrobacterium*-mediated transformation and biolistic methods are more accurate, useful, and less time-consuming. This review provides an insight into various bacterial genes for example *mtlD*, *codA*, *betA*, *ADH*, *IPT*, *DRNF1* and *ggpPS*, etc. that have been successfully expressed in transgenic plants against various abiotic stress for stress tolerance enhancement and crop yield improvement which exhibited good encouraging results. Genes from yeast (*Saccharomyces cerevisiae*) have been introduced in transgenic plants against drought and salinity stress. All these genes expressed from non-plant sources in plants can be very helpful to enhance crops for better yield productivity in the future to meet the demands of the consistently rising population of the world.

Keywords: abiotic stresses, heterologous expression, bacterial, yeast, fish and insect genes

1. Introduction

Plant stress is a condition in which plant growing in an unfavorable condition that mainly causes growth problems, deficiencies in crop yields, and even death when the stress-causing factors cross the limit that plants can tolerate [1]. It refers to external environmental conditions that adversely affect the overall growth, progress, or production of plants [2].

There are two types of stresses to which plants are subjected that is abiotic stress and biotic stress. The crop loss worldwide is mainly due to abiotic stress which consists of drought, cold, salinity, high environmental temperature, and radiation, etc. [3]. While biotic stresses are the attack of various pathogens on plants including bacteria, fungi, herbivores, and nematodes) etc. [4]. Due to the sessile nature

IntechOpen

of plants they cannot avoid these environmental factors but develop several mechanisms to tackle these abiotic and biotic stresses for their survival and environmental adaptation.

1.1 Abiotic stress mechanism in plants

Plants usually sense the environmental stress and then stimulate appropriate suitable response takes place, cell surface receives the stimuli and the transformation to the transcriptional system in the nucleus takes place via various pathways that help in transduction, make plants resistant to various environmental stress by the activation of molecular, biochemical, and physiological suitable response [5]. The first line of defense of plants is situated in roots to overcome abiotic stress. If the plant growing in the soil is healthy and there is biological diversity the chances of survival against the abiotic stress of the plant will be high. High salinity affects the growth and development of plants. The disruption of (Na^+) and (K^+) ratio in the cytoplasm is mainly the primary response shown by the plants against stress. Living microorganisms need to ensure effective growth and generate an effective environmental response, this especially very important in plants because of their immobility and encountering large changes/alterations in temperature, humidity, light, and availability of nutrients in the environment. Massive agricultural losses happen due to environmental stresses [6, 7] and the improvement of crop resistance is a major goal for crop programs.

A genetic locus that keeps productivity maintained even in serious conditions are situated within the germplasm of existing crops, their relative species that are earlier adopted to severe environments. Selective breeding in combination with other loci has improved crops yield in extremely challenging environmental conditions throughout agricultural history. An efficient advanced paradigm is the precise selection of genetic factors of stress adaptation that have been in nature for years and passes on by plants to their higher verities [8]. Abiotic stress causes biosynthetic capacity and nutrient decrease which leads to inhibition in plant growth and has been further elaborated by various researchers in their work by knowing the response to abiotic stress through various signaling pathways involving several genes, mechanism of post-transcriptional modification, and proteins. Those pathways are MAPK, ABF/bZIP, Ca2+-CBL-CIPK, and CBF/DREB which enables much stress responding transcription factors to initiate downstream signals needed for abiotic stress defense [9]. These signaling pathways can predict the effects generated by abiotic stress to control growth and plant adaptation. Recently genes have been identified which control plant growth during stress conditions for example molecular mechanism which controls leaf progress and growth under drought conditions relates both transcriptional signals to the circadian clock. Importantly (ERFs), ERF2 and ERF8 related to ethylene response factors showed to affect leaf in drought and wet conditions [10].

Abscisic acid plays a huge role in helping plants for their environmental adaptation against cold, drought, alteration in temperature, salinity, and wounding [11]. During extreme environmental conditions, the level of Abscisic acid goes up through the ABA biosynthesis process. High-level ABA combines with receptor for the initiation of signal transduction which leads to the cellular response to stress [12]. Various mechanisms that help in the protection of plant survival against abiotic stress are very much important, yet they are activated at the cost of plant growth and its productivity which is essential for agriculture. Recent studies in molecular genetics help us to understand the basis of abiotic stress tolerance [13]. **Figure 1** illustrates the various signaling pathways involved in abiotic stress mechanisms in plants [9].

1.2 Abiotic stresses (factors) that affect plants

1.2.1 Temperature

Temperature is a very important abiotic stress factor that affects plant from seed germination to reproduction [14]. Significant temperature changes can lead to permanent disturbance in the plant cycle which even leads to death. It causes plant stress by two means; extremely cold and hot temperature, severe cold conditions below the optimum temperature can cause physical and mechanical changes to the plant and leads to severe cell disruption [15]. In various areas extremely low temperature causes agricultural crop productivity and affects the cultivation process [16]. While due to uncontrolled rise in temperature affects the rate of photosynthesis, water availability to plants, and fruit ripening. Due to climatic changes an appreciable rise in temperature in the coming times will cause rainfall reduction, alteration in wind speed, and snow leads to less growing plant season and eventually will harm crop production and quality [17]. The effects of verglas/frost and high temperature have been evaluated recently on the production of Wheat (*Triticum aestivum L*), fruitless plants and

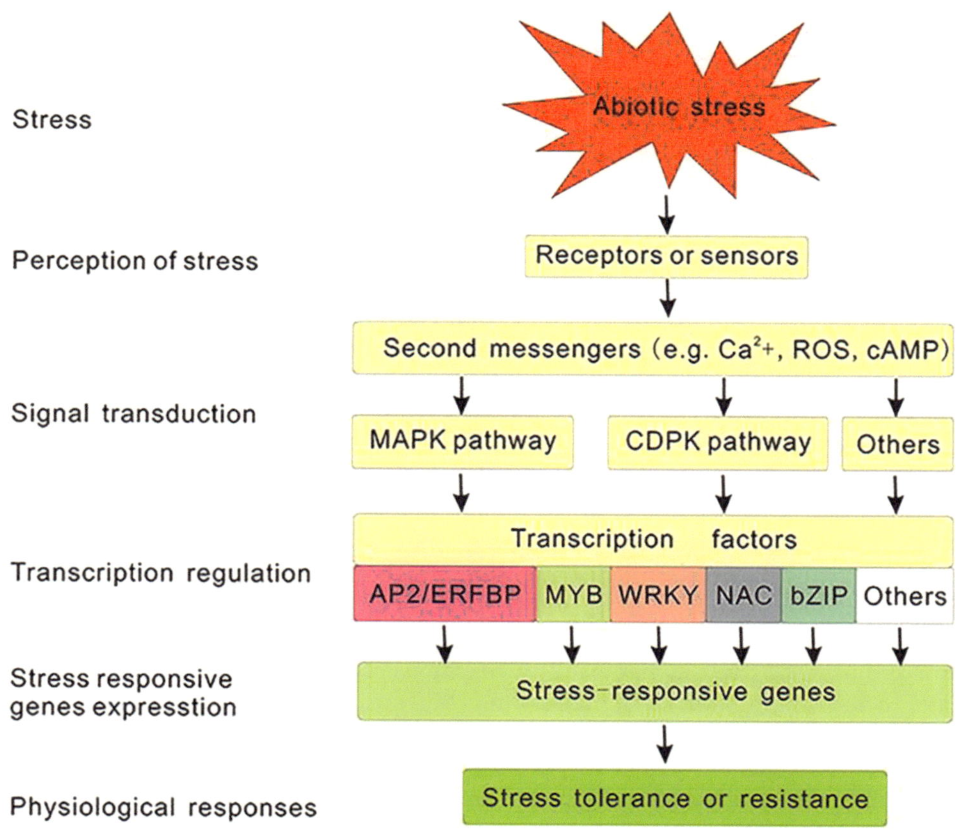

Figure 1.
Illustration of various pathways involved in abiotic stress mechanism in plants [9].

termination of matured grains was due to frost while extreme temperature caused a decrease in grains number during the filling grain period [18]. These noteworthy effects due to extreme climatic changes in crop production will result in food insecurity and crop production trends in the future [19].

1.2.2 Drought/water stress

To obtain maximum crops yield globally drought or water stress is a very important factor it affects plants in many ways; during the growth phase, water stress decreases leaf expansion development, photosynthetic process, the height of the plant, and the overall area of leaf. The early symptoms caused by drought stress are leaf rolling and dryness of leaf tip, cell elongation is seriously affected by drought stress water scarcity blocks stomata, and reduces transpiration [20]. It has a huge negative impact on plant growth and the potential quality of yield in the agricultural system. *Miscanthus* has very good potential for the production of biofuel it was observed after an experiment that drought the weight of the plant significantly by about 45% and cell wall composition and biomass were affected by drought stress during the plant growth phase [21]. The availability of water to plants is very necessary however waterlogging in the area surrounding the roots can be very damaging it can cause lack of oxygen and even death of the plant due to its lethality [22]. The transfer of free oxygen exchange between the soil and atmosphere is caused by water stress suffocation [23]. Waterlogging is often caused by floods, heavy rain, and snow in winter, such soils have limited or lack of oxygen due to less gas exchange [24].

1.2.3 Light stress

For plants, the energy production process through photosynthesis sunlight plays an important role. Plants adapt themselves to change in light which alter considerably at various times. That is why plants can develop certain mechanisms that help in maximum use of existing light during irradiance state while other mechanisms to escape the long-term sunlight exposure [25]. As a result of low light or reduction in solar energy significant decrease happens in metabolic rate which leads to a reduction in crop yields and lower growth rates. An increase in reactive oxygen species (ROS) and photo-damage is caused by prolonged exposure of plants to sunlight [26].

1.2.4 Salinity stress on plants

Soil salinity is considered as one of the major abiotic stress affecting the performance of crop plants adversely around the globe, it can create a cluster of diverse interactions that harms the nutrition uptake, metabolic process, and plant vulnerability to various biotic stresses as well [27]. Minerals and nutrients present in the soil have valuable importance but the unwanted existence of salts results in extreme ionic and osmotic stress in plants [22]. The cations present in inorganic soils or water includes potassium (K^+), magnesium (Mg^+), calcium (Ca^+), and sodium (Na^+) while the important anions are NO_3^-, HCO_3^-, SO_4^{2-}, Cl^-, and CO_2^{-3} other components include SiO_2, Al^{3+}, Sr^{2+}, B, Mo, and Ba^{2+} [28]. Enzymes inactivation, cell death, and subsequently whole plant can diminish due to high salinity [29]. Salinity stress in plants leads to a huge decrease in dry and fresh weight obtained from stem, roots, and leaves [30]. An excessive amount of salt increases osmotic pressure in plants which reduces the chances of minerals like (K^+ Ca^{2+}) and nutrient uptake for survival, such primary effects leads to secondary effects as a non-proper expansion of cell, decrease in membrane function and a significant decrease in cytosol metabolic activity [5]. According to FAO world's 6% of the land is affected by salt. **Table 1** shows the distribution of salt-affected land around the world.

2. Heterologous expression of genes in plants for abiotic stresses tolerance

During the past two decades the use of recombinant DNA technologies, the methods of gene transfer, and tissue culture techniques have improved the transformation and transgenics in many varieties of crop production in agriculture. Transformation techniques provide larger accessibility to the pool of genes as compared to conventional methods because the genes are inserted from bacteria, animals, viruses, yeast, fungi, and even from various synthetic chemicals prepared in the laboratory (Chahal and Gosal 2002). Various methods are used for genetic transformation of crop plants, Biolistic bombardment, and *Agrobacterium*-mediated are the most common methods used for gene transfer in plants [31].

2.1 Genetic engineering through bacterial genes in plants against abiotic stresses

Cloned genes insertion has produced transgenics against abiotic stresses in plants [32]. Many bacterial genes have been expressed in plants to confer abiotic stresses like salinity, drought, temperature, cold, and light stress. Those bacterial genes include *mtlD* which is expressed in several transgenic crops like tomato (*Lycopersicon esculentum* L. *var. Pusa Uphar*) was transformed against salinity, high and cold temperature and drought stress by the insertion of *mtlD* gene [33] Wheat (*Triticum aestivum* L.) was transformed by the expression of this gene against salinity and waterlogging stress, as a result, the transgenic plants showed good resistance than WT non-transformed plants [34]. Tobacco plants were transformed by the expression of this gene against various abiotic stresses and the results were very improved in comparison to wild type plants [35]. Finger millet (*Eleusine coracana*) is a major food crop consumed and cultivated around the globe has been genetically transformed by the expression of this bacterial gene against drought and salinity [36]. Peanut (*Arachis hypogaea* L.) has been genetically transformed by the insertion of bacterial *mtlD* gene against salinity and drought stress [37]. Moreover Indica rice

Regions	Total area (Mha)	Saline soil (Mha)	Percent %	Sodic soil (Mha)	Percent%
Africa	1899	39	2	34	1.3
Asia, the Pacific and Australia	3107	195	6.3	249	8
Europe	2011	7	0.3	73	3.6
Latin America	2039	61	3	51	2.5
Near East	1802	92	5.1	14	0.8
North America	1929	5	0.2	15	0.8
Total	**12,781**	**397**	**3.1**	**434**	**3.4**

Table 1.
Various parts of the world affected by salinity stress [5].

was transformed through the above-mentioned gene to improve its productivity in drought and saline environment [37]. Another Bacterial gene codA which has been isolated from (*Arthrobacter globiformis)* coding for choline oxidase and expressed Arabidopsis thaliana to improve its resistance to salinity, freezing, and high temperature [38]. The same bacterial *codA gene* was expressed in tomato (*Lycopersicon escuelentum*) to enhance its tolerance against high temperature, chilling, and drought stress [39]. Other Bacterial gene *IPT i*solated from *Agrobacterium tumefaciens* and expressed for the enhancement of various crop plants to improve their tolerance against various abiotic stresses like sugarcane (*Saccharum* spp.) cv. enhancement against cold stress [40] the same gene was inserted in rice for its enhancement against drought stress and increase in crop yield as well [41]. ADH *sysr1* gene of *cyanobacteria* was expressed in tobacco plants for salt tolerance improvement and that showed encouraging results in comparison to wild type plants [42]. *IPT* gene was also introduced in canola (*Brassica napus* L.) plants for delayed leaf senescence leading to crop improvement in drought stress [43]. The expression of Bacterial *ggpPS* gene isolated from (*Azotobacter vinelandii*) for glucosyl glycerol biosynthesis confers salt and drought stress tolerance in *Arabidopsis thaliana* [44]. Gene encoding for bacterial chaperons have been inserted in various transgenic plants for the improvement of tolerance to abiotic stresses successfully [45].

Abiotic stress tolerance related genes from micro-organisms are considered to be very valuable for the production of transgenic plants. A cyanobacterium (*Nostoc flagelliforme*) that can tolerate water deficit conditions is proved to be very useful prokaryotic organism for gene isolation. Salt tolerant gene *DRNF1* having P-loop NTPase (nucleoside-triphosphatase) domain, has been expressed in *Arabidopsis thaliana* the results indicated improvement in the growth of shoots and seed germination in saline conditions [46].

2.1.1 The expression of mtID bacterial gene for the improvement of various abiotic stresses in plants

To improve abiotic stress tolerance in transformed tomato plants a bacterial mannitol-1-phosphate dehydrogenase (*mtID*) gene was inserted through Agrobacterium-mediated method supported by CaMV35S promoter, Rt PCR, and southern blotting analysis was used for the confirmation of transient integration, reverse transcription (RT)-PCR and direct activity of mtID gene was analyzed for the confirmation of transgene expression [33]. Transgenic tomato plants upon exposure to low temperature round about 4°C in a cold chamber survived for almost 2 days as compared to untransformed plants that were not able to survive and their death accrued slowly. During the exposure of transgenic tomato plants to chilling effect showed a significant decrease in electrolyte leakage in the plant membrane, when they are exposed to stress the leakage starts to the surrounding environment, and the damage caused to the cell with hardiness can be identified through the leaked contents conductivity in water by the comparison of injured and non-injured plants [47] while an increase to lipid peroxidation [48], antioxidant enzymes [49] and relative water content [50] as compared to the non-transformed plants. Drought stress was tested through polyethylene glycol in the medium, and salinity by the content of sodium chloride (NaCl) showed a greater response to these stress than non-transformed plants. By the indication of all these observations, it was clear that the introduction of bacterial (mtID) gene in tomato showed a good response to abiotic stress than transformed plants. As **Figure 2** shows the various transformation phases of tomato [33].

Figure 2.
*Different tomato (*Lycopersicon esculentum L. *var.* Pusa Uphar*) transformation phases [33].*

The above work has shown that the accumulation of mannitol in several transgenic plants can improve plant tolerance against abiotic stresses. At the cellular level due to *mtlD* gene insertion from *E. coli* [51] has been reported for the protection of Wheat (*Triticum aestivum* L.) against the adverse impact of waterlogging and soil salinity. Through the exposure of calli to polyethylene glycol ranging from 8000 mm of NaCl. The stress in the T2 plant was caused by the addition of a large amount of water and 150 mm of NaCl to the nutrition medium, the fresh weight of *mtlD* calli was decreased by about 40% in the existence of PEG and 37% during salt stress, no effect was observed in the growth of +*mtlD* callus, while in the plants of -*mtlD* the content of fresh and dry weight, height of the plant, and leaf areas was decreased by about 70%, 56%, 40%, and 45% in a comparison with 40%, 8%, 18% and 29% comprehensively with that of +*mtlD* plants. Salinity stress decreased shoots of fresh and dry weight, changes in the height of plants were observed, and the leaf area was reduced by 77%, 73%, 25%, and 36% in -*mtlD* plants as compared to 50%, 30%, 12%, and 20% in plants having +*mtlD*. As a result, no effect was seen on the growth of *mtlD* callus and transgenic wheat plants showed significant tolerance against salinity and waterlogging stress due to the insertion of this bacterial gene [34].

For expression in higher plants against abiotic stress a bacterial gene that codes for mannitol-l phosphate dehydrogenase, *mtlD* was inserted stably in tobacco plants which translated in tobacco through a functional enzyme, led to the accumulation of mannitol, which was identified and detected through NMR and mass spectrometry, the concentration of mannitol increased by d 6 jumol/g (fresh weight) in roots and leaves of the transgenic tobacco while these sugar were not detected in wild or untransformed tobacco plants that were passed through the same treatment. This study could help us in understanding sugar role of alcohol in the enhancement of plant tolerance against abiotic stresses in higher plants comprehensively [35].

In Asia, Nepal, India, and almost 25 countries of Africa, finger millet (*Eleusine coracana*) is cultivated and consumed as a major crop for food, it covers more than 12% of the world's millet cultivating area [52, 53]. It also has better nutritional properties and ingredients than wheat and other major crops [54, 55]. It is very vulnerable to various abiotic stresses like drought and salinity in fields during the early stages of seed germination and the development of seedlings, therefore it is very important to make finger millet plants resistant to drought, salinity, and oxidative stress. Proper radical scavenging capability and cell protection by osmotic modification during several abiotic stresses are major mechanisms in plants, mannitol is an osmolyte [56] that helps in the neutralization of various free hydroxyl radicals produced due to abiotic stresses and decreases stress disruption in various plant species. Through *Agrobacterium*-mediated transformation, the biosynthetic pathway gene from bacteria mannitol-1-phosphate dehydrogenase (*mtlD*) was expressed in finger millet plants to understand the performance of transgenic plant upon their exposure to drought and salinity stress simultaneously. The results obtained through these experiments showed that transgenic finger millet had better performance in saline and drought stresses as compared to wild type plants [36].

Peanut (*Arachis hypogaea L.*) is an important crop grain cultivated in tropical and sub-tropical zones in about 21–24 M ha production areas, it is generally grown in rain-fed areas where drought is a major crop decreasing factor occurring in semiarid lands that cover 70% of the peanut cultivation areas [57] one of the many strategies is to genetically transform peanut plants that can resist in drought conditions [34]. Plant breeding through conventional methods is time-consuming with lesser success and far more laborious, while genetic engineering techniques show great potential to develop peanuts plants that can tolerate drought conditions. Plants have developed several mechanisms against drought stress for their survival [58]. To overcome drought stress in peanuts bacterial *mtlD* genes were expressed through CaMV35S promotor using *Agrobacterium tumefaciens*-mediated transformation. The transformed plants showed significant resistance to water deficit conditions as a result of mannitol accumulation during these experiments, **Figure 3** illustrates the whole process [37].

The accumulation of mannitol an osmolyte plays an important role in abiotic stress. So, through the insertion of *mtlD* gene from *E. coli* for the improvement of *Indica* Basmati rice against salinity and drought stress, by agrobacterium-mediated transformation, many putative transformed plants were generated. Transgene existence was confirmed in early transformed plants by PCR through *mtlD* and hygromycin phosphotransferase, the transgenic lines showed better performance against salinity and drought stress as compared to wild type plants [59].

2.1.2 RNA chaperones genes of bacteria confer abiotic stresses in transgenic plants

With a consistent increase in the world's population, constant supply of food demands, and a decrease in water shortage alongside cultivating land, it is necessary

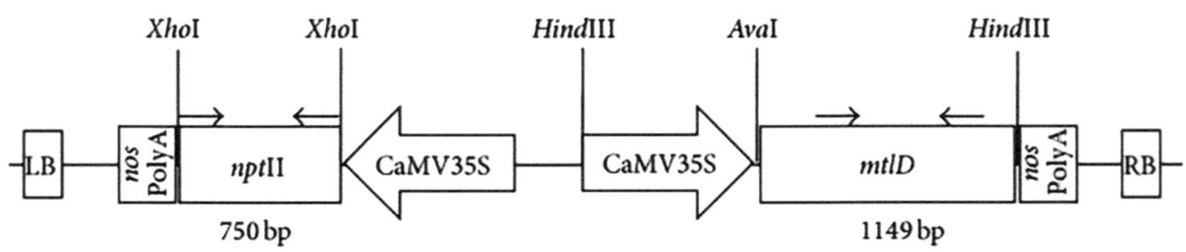

Figure 3.
Schematic illustration of the T-DNA section of pCAMBIA 1380 binary plasmid used for transformation of deembryonated cotyledons with Agrobacterium tumefaciens strain LBA 4404. The location of the primers used in PCR assays is shown by arrows on the top of the mtlD gene. LB, left T-DNA border sequence; RB, right border sequence; 35S, CaMV35S promoter; and mtlD, mannitol-1-phosphate dehydrogenase [37].

to transform crops like rice that can grow in salt-affected areas [60]. High saline condition seriously affects the growth of rice like leaf expansion ability, root, and shoot formation [61]. The decrease in leaf expansion occurs in rice due to the low rate of osmotic turgor pressure under saline and cold conditions [62]. Results obtained from various research studies have shown that chloroplast and mitochondria in rice plants are seriously affected by salt and chilling stress [63, 64]. During abiotic stress conditions, bacterial RNA chaperones play a major role in stable messenger RNA expression, in salinity stress these bacterial genes develop transgenic rice plants that can tolerate even cold stress apart from salinity [60].

Drought is the major factor that causes crop yield reduction globally leading to socioeconomic complications. During an estimation, it was observed that a 40% loss in Maize crop is caused by drought stress alone in North America annually [65]. Maize crops are vulnerable to drought stress through-out their growing stages, effects of stresses that initiate during the flower development phases either before the start of floral events or post pollination results in a significant reduction of crop yields at the end of the season [66, 67]. In 2013 the first drought-resistant maize crop was genetically transformed by the expression of bacterial genes that codes for chaperonin showed significant improvement in resistance to water deficit stress [45].

The expression of bacterial CSPs (cold shock proteins) exhibited improvement against cold stress in transgenic *Arabidopsis thaliana* seedlings cultivated at very low temperatures on standard agar media in Petri dishes as illustrated in **Figure 4**. The tests were conducted of the transformed Arabidopsis seedlings having *CspA* and *CspB* to check the improvement for cold stress using non-transformed seedling as controls. The seedlings were exposed to low temperatures for 6 weeks and the results suggested improve tolerance against cold stress in comparison to non-transformed wild type plants [45].

2.1.3 *The expression of* ADH (alcohol dehydrogenase gene) *isolated from* cyanobacteria Synechocystis *sp. improves salt tolerance in tobacco plants*

A gene PCC 6906 (*sysr1*) from *synechocystis* that shows a good response to salt was engineered and stably inserted in higher developed tobacco plants. The tolerance response of the gene *sysr1* (An ADH superfamily member) was investigated through quantitative real-time PCR, gas chromatography-mass spectrometry, and bioassays. The tobacco plants having *ADH* showed considerably improved tolerance to salt stress, besides that the activity of many

Figure 4.
Demonstration of Arabidopsis thaliana seedlings exposed to cold temperatures for 6 weeks [45].

stress-responsive genes was up-regulated and enhanced due to the expression of (*sysr1*). The results suggested that the expression of *ADH* genes could significantly improve transgenic tobacco plants against salt stress through genetic engineering techniques in the future. **Figure 5** shows the identification of *sysr1* gene plants in three transgenic lines (1, 4 and 7) [42].

2.1.4 Bacterial codA *gene enhances tolerance against various abiotic stresses in plants*

Arabidopsis thaliana was genetically transformed by a gene isolated from *Arthrobacter globiformis* that codes for choline oxidase, an enzyme used for the synthesis of glycine betaine from choline, which remarkably improved cold or freezing stress in plants. Moreover, the photosynthesis machinery was more resistant to freezing/cold stress than non-transformed plants. These results also indicated the accumulation of glycine betaine in transformed plants, enhanced their ability to extremely cold temperatures [68].

In the control of RNA CaMV35S promotor, *A. thaliana* was genetically transformed by *codA* gene of bacterial *Arthrobacter globiform* coding for choline oxidase. Subsequently, the accumulation of increased glycinebetaine occurred in the seeds of transformed plants. The transformation of *codA* gene significantly boosted

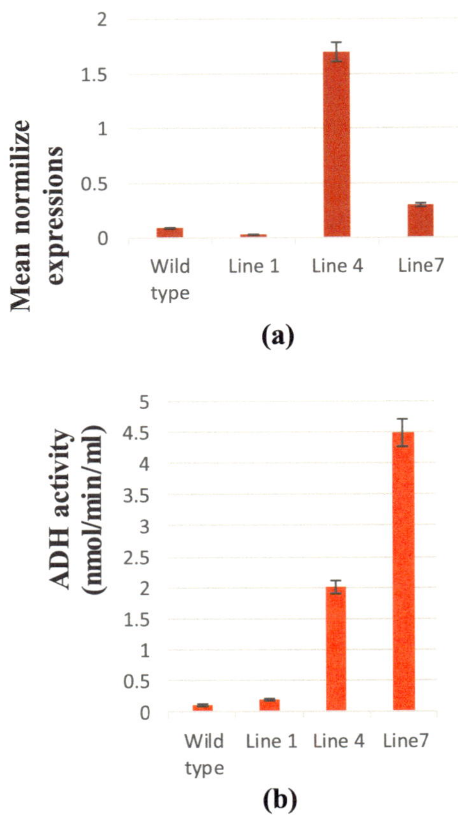

Figure 5.
Identification of transgenic plants with sysr1 gene. (a) Results obtained from the analysis of quantitative real-time PCR in three transformed lines (lines 1, 4 and 7) used for the assays of salt tolerance. (b) the ADH activity of leaf of transgenic tobacco plants with the control [42].

the ability of plants to high-temperature stress in the period of seed germination and the growth phase of young seedlings. The level of improvement of the resistance to high temperature was evaluated with the extent of the expression of choline oxidase and the accumulation of glycinebetaine in transgenic plants [69].

Tomato (*Lycopersicon escuelentum*) was genetically transformed by the introduction of *codA* gene from *Arthrobacter globiform* bacterium for choline oxidase that had been allowed to target both mitochondria and cytosol. The accumulation of glycinbetaine was detected in the seeds of transformed plants by about 1 μmol g^{-1} dry weight while no accumulation of glycinbetaine was seen in wild type/non-transformed plants. The transformed codA seeds germinated at fast speed during high temperatures. After heat stress, the content of small mitochondrial heat shock proteins, 70 heat shock proteins, and cognate 70 were much increased in transformed seeds during the heat stress phase than non-transformed seeds. Cognat 70 (HSP70) accumulation was more obvious in codA transgenic seeds than non-transgenic seeds. The results suggested that the transformation of tomato seeds with codA gene showed improved tolerance to a higher temperature in tomato plants [70].

Genetically transformed tomato (*Lycopersicon escuelentum*) plants which can synthesize glycinbetain was produced by the introduction of the bacterial *codA* gene. The expression of the gene was examined through RT-PCR analysis and in combination with RNA blotting hybridization. During the seed germination phase, the transformed plants exhibited greater tolerance to salt stress following the growth of young seedlings as well. The insertion of *codA* gene resulted in high-stress resistance ability in leaves and overall plants. Results from the experiments revealed that the developed leaves of *codA gene* transformed plants showed more water content, chlorophyll content, and enhanced proline levels in comparison with non-transformed plants during salinity and water stress [39].

They are vulnerable to chilling stress because of lower glycinbetaine synthesis ability. The cold temperature lower than 10°C causes severe injuries to tomato plants leading to lesser yield production. Bacterial *codA* gene has been introduced into the genome of the tomato by targeting chloroplast. The transformed plants expressed this gene and synthesize choline, by the accumulation of glycinbetain in leaves and the formation of shoots up to 0.3 and 0.2 μm/g fresh weight. The chloroplast of transgenic plants contained 86% of glycinbatain, in different developmental stages during the seed sprouting and fruit production process, the glycinbetain containing plants were more resistant to chilling stress than their wild types, 10–30% increase was seen in fruit production on average during abiotic stress, thus the introduction of GB biosynthesis pathways is an important strategy against chilling stress in tomato plants [71].

A. thaliana was genetically transformed through the bacterial *codA* gene which encodes for choline oxidase. The photosynthetic activity examined for chlorophyll fluorescence of transgenic plants were more resistant to light stress than non-transformed wild type plants. This improvement in resistance to light stress was eventually because of the high speed of the recovery process of the photosystem II complex from the photo-inactive stage. It showed that in vivo production of glycinbetain and no changes in the lipids membrane or H_2O_2 level, it ensured the protection of photosystem II complex in transgenic plants from the possible damage due to light stress [38].

2.1.5 The expression of IPT *gene against various abiotic stresses in plants*

To increase the cold stress tolerance, *IPT* gene was introduced in sugarcane (*Saccharum* spp.) cv. RB855536, in the control of a promoter (AtCOR15a), through biolistic, non-biological transformation method. The leaves extracted from genetically transformed plants showed good resistance and decrease leaf senescence upon their exposure to low temperature as compared to wild-type control plants. Improved enhancement against cold stress was seen due to the expression of this gene in non-acclimatized plants when the transgenic plants were exposed to extremely cold temperatures. The chlorophyll content of leaf was 31% more than non-transformed plants. A decrease in malondialdehyde level and the leakage of electrolyte showed lesser damage caused by chilling stress in transgenic plants. So, stress-inducible promoter *COR15a* used in the insertion of the *IPT* gene in transgenic plants shown no adverse effect while improving them against cold stress [40].

To delay the process of leaf senescence would allow capturing sunlight for longer periods, which leads to photosynthetic improvement and its contribution to plant growth and enhanced seed yield. Moreover, delayed senescence would allow the slow degeneration of source tissues so that the metabolites, proteins, nutrients could be slowly and gradually released to the sink tissues. Increase in plant potential biomass, maintenance of photosynthetic process, the higher influx of nitrate, increase in the life of flowers after harvesting, improved drought resistance, and greater seeds yield are the benefits of delayed leaf senescence [72, 73]. Cytokinin; a plant hormone that plays an important role in the process of cell division, cell growth, and differentiation, and it influences various developmental and phycological characteristics in plants ranging from seed germination, the flowering period of the plant, apical dominance, developmental process of flowering, fruits and leaf senescence [74, 75]. In various plants the role of a plant hormone cytokinin in delaying leaf senescence has been reported by [73, 76–78]. A gene *IPT* isolated from *Agrobacterium tumefaciens* has been inserted in several plants to enhance cytokinin level as this gene synthesize the rate-limiting step in cytokinin. Canola plants were transformed by *IPT* gene in combination with *AtMYB32* promoter, the insertion of *IPT* gene in transformants caused delayed leaf senescence cultivated under control condition and various field experiments at two separate geographical areas for one season. As a result, the transformed Canola (*Brassica napus* L.) plants maintained high chlorophyll content for a longer period and an increase in seed yield under drought and irrigated conditions was observed as compared to wild type non-transformed plants. In comparison to control plants, all of the seed quality parameters and oleic acid content in transformed plants were exactly similar, as a result of the experiments, it was concluded that the introduction of bacterial *IPT* gene can significantly improve crop yield and seed quality under irrigated and drought stress conditions in various plants [43].

In rain-fed areas drought is a major hindrance to rice crop productivity [79, 80]. To fulfill the constant demand of rice by 2030 a remarkable increase by almost 35% in yield is necessary [81] that is why the development of transgenic rice to drought stress and improved productivity is an important challenge, various studies have indicated that the expression of bacterial *IPT* gene using different promotors could help in delaying leaf senescence to improve crop productivity under drought stress conditions [82]. In tobacco plants *IPT* gene was introduced under the control of senescence-related receptor [83] and a promotor to induce stress exhibited enhancement in photosynthetic capacity leading to improvement in drought tolerance in tobacco plants [73]. Moreover, transgenic rice plants resistant to drought stress were produced by the insertion of *IPT* gene under the control of P_{SARK} a stress-inducible promoter. The plants were tested against drought stress tolerance at two yield sensitive developmental phases; pre and post-anthesis. During both treatments, the transformed rice plants showed remarkable resistance to drought stress and an increase in yield grain as compared to non-transformed wild type control plants [41].

2.1.6 The expression of bacterial ggpPS *gene isolated from* Azotobacter vinelandii *for glucosyl glycerol biosynthesis confers salt and drought stress tolerance in transgenic plant*

Various organisms generally accumulate compatible solutes to show response against salt and drought stress, which includes heterotrophic and cyanobacteria which shows resistance to salty environment and produces glucosyl glycerol as their major compound for protection. To know the potential of glucosyl glycerol to enhance salt resistance in higher plants, a gene *ggpPS* that codes combinedly for GG-phosphate synthase/phosphatase was isolated through PCR from the chromosomal DNA of the cells treated with lysozyme from a heterotrophic (*proteobacterium A. vinelandii*) and introduced into model plant *A. thaliana*. The high accumulation of glucosyl glycerol was observed due to the expression of this gene. In various growth experiments, three separate Arabidopsis lines were tested that showed varied glucosyl glycerol levels. Plants having a low level of glucosyl glycerol within leaves showed no changes in growth development in the control condition, rather an improvement to salt tolerance. While plants having a very low or higher glucosyl glycerol content exhibited growth delay and no enhancement of salt resistance was observed, the results suggested that the suitable solute synthesis has a positive impact on the stress tolerance of plants as long as the accumulation extent does not interfere adversely with the metabolic process of plants [44].

2.1.7 The expression of bacterial betA *gene confers abiotic stress tolerance in transgenic plants*

Drought stress exists in most of the areas where sugarcane is grown and cultivated, which has no support of irrigation system and has lower rainfall. To know psychological and biochemical mechanisms better, underlying plants response to water deficit stress, have been overcome by the development of drought-resistant plants through biotechnological techniques. To tackle water stress plants use various strategies like variations in gene expression and the accumulation of compatible solutes for survival and growth. A bacterial gene *betA* that codes for *CDH* choline dehydrogenase has been effectively expressed in sugarcane to produce drought resistant plants. The function of *CDH* is the conversion of choline in betaine aldehyde that is then transformed into glycinebetaine GB, the expression of *betA* gene improves the level of glycinbetaine that act as an osmoprotectant and help in the acclimatization of sugarcane in water deficit stress, the drought-resistant sugarcane was first developed by Ajinomoto Company in Tokyo [84].

Transgenic cotton (*Gossypium hirsutum* L.) was genetically transformed by the expression of a bacterial *betA* gene from *E. coli* for the enhancement of glycinbetaine, its accumulation was identified at three stages. Five lines expressing this gene showed significant improvement to drought stress than wild type non-transformed plants from seedlings to flowering plants. The five transgenic lines showed better relative water content, a decrease in the leakage of ions, and less malondialdehyde content in comparison to wild-type plants. The glyycinbetaine content was positively related with water deficit tolerance in water stress, the results indicated that the expression of the *betA* gene not only provide protection to cell membrane against drought stress but also act in the osmotic adjustment in transgenic cotton plants, more importantly, line 4 among five lines showed a significant increase in cotton seed yield after exposure to drought stress which will help a great deal in cotton production in future [85].

The similar *betA* gene of *E. coli* was expressed through *Agrobacterium*-mediated transformation in maize to improve its tolerance against cold or chilling stress, five transgenic lines were tested in which four lines exhibited a higher level of glycinebetaine than WT plants. At lower temperatures 10 and 15°C three transformed lines showed an increase in germination stages, as identified through the progress of germination and presented lower inhibition in the speed of shoot growth in seedlings than non-transformed lines. Upon exposure to chilling stress the tolerance of transgenic plants was significantly improved in cell membrane injury, the level of damage caused by cold stress, survival rate, and photosynthetic capacity in transgenic lines than WT plants [86].

Tobacco plants were also genetically transformed by the expression of this gene from *E. coli* and improvement in glycinbetaine was observed leading to improvement in the resistance of transgenic plants to chilling and salinity stress than wild-type plants [87, 88].

3. Gene expression from yeast (*Saccharomyces cerevisiae*) in plants against abiotic stresses tolerance

Just like the above bacterial genes expression in plants to improve their tolerance against abiotic stresses, yeast genes have also been introduced in transgenic plants to enhance their tolerance, *TPSI* gene of *Saccharomyces cerevisiae* has been expressed in transgenic tobacco against salt and drought stress and the results were very much better than WT plants [89] *HAL1* gene was expressed in tomato against salt tolerance and the results showed better improvement in comparison to wild type plants [90] similarly *HAL1* and *HAL3* genes were introduced in *A. thaliana* for its enhancement against saline stress and the transgenic plants exhibited much better tolerance than wild type non transformed lines [90], the procedures of the expression of these genes have been discussed below.

3.1 Insertion of a yeast gene *TPSI* in transgenic tobacco plants against drought and salt stress

A gene trehalose-6-phosphate synthase from yeast was introduced in tobacco plants by the control of Cauliflower mosaic virus (CaMV35S) regulation sequence. *Agrobacterium*-mediated transformation method was used for the introduction of a gene into the genomic DNA of tobacco (*Nicotiana tabacum* L) plants. The accumulation of trehalose was found in transgenic plants through ion-exchange chromatography in combination with ampometry detection procedure. The disaccharide that was non-reducing accumulated almost 0.17 per gram of fresh weight in leaf extracts of the transformants. The plants with trehalose accumulation had various changes in phenotypes like dwarfness, pointed or lancet leaves pattern, and decrease in sucrose level. Moreover, the expression of *TPS1* gene in tobacco plants showed significant tolerance to drought and salt stress as illustrated in **Figure 6** [91].

3.2 The role of yeast *HAL1*, and *HAL3* genes against salt tolerance in plants

To overcome salinity stress in *Arabidopsis thaliana*, the yeast genes *HAL1* and *HAL3* were introduced under the control of 35S promoter via the

Figure 6.
Drought tolerance in transgenic tobacco plants by the overexpression of the TPSI gene from yeast. The left 2 rows consist of non-transformed control plants while the right two rows contain the transgenic homozygous plants. No water has been given to all plants for almost 15 days. The results obtained are similar by exposure to drought stress with 400 mM NaCl. The visible better changes can be seen in transgenic plants with TPSI gene [91].

Agrobacterium-mediated method. Almost 33 plants showing resistance to kanamycin were obtained from 70,000 plus seeds. Southern blotting analysis showed that *HAL1* and *HAL3* genes were introduced into all the genomes of the transgenic plants. The copy number of the yeast gene in all plants was in the range of 1–3 by the confirmation of southern blotting analysis, there was no difference in the phenotype of the transgenic plants compared to wild ones. Most of the transformants were self-pollinated, the progenies of transformants and non-transform *A. thaliana* plants were observed through different experiments for gene expression to know the salt resistance. The measurement of (K^+) and (Na^+) showed that the transgenic plants accumulated fewer (Na^+) as compared to the control lines. In light of several tests, it was observed that the introduction of yeast *HAL1* gene exhibited more resistance to saline soil in comparison to non-transform plants [92].

In past, remarkable advancements have been made in the identification and isolation of various genes which could be used in the process of abiotic stress protection in plants. It is hard to believe that a single gene insertion would make a dramatic improvement to salt stress directly producing a fresh salt-resistant transgenic plant that could be enough for breeding purpose point of view. Yeast *HAL1* gene was introduced in tomato (*Lycopersicon esculentum*) through a well-modified plasmid containing the elements of enhancer and salt resistance was evaluated in transgenic plants from progenies. The result showed that transgenic lines having one copy of the *HAL1* gene had higher salt tolerance than non-transformed plants [90].

For the production of transformed watermelon plants, and adjusted *agrobacterium* mediated protocols were maintained. The efficient transformation rate was 2.8–5.3% in the cultivars. Yeast *HAL1* gene under the control of 35S cauliflower mosaic virus having a double sequenced enhancer was cloned in pBiN19 plasmid. RNa4 from Alfalfa mosaic virus was used alongside 35S. The vector was introduced in the LBA4404 strain of *agrobacterium tumefaciens* for the inoculation of watermelon cotyledon explants. PCR and Southern hybridization analysis were used for the assessment of the *HAL1* gene in new transformants. Improved elongation of leaves and new roots emergence was seen in plantlets in culture media having NaCl. It was observed that the *HAL1* gene as a molecular tool for genetic engineering could be very useful to protect crop plants in the future [93].

3.3 *HAL1* gene mode of action in (*Saccharomyces cerevisiae*)

Yeast (*S. cerevisiae*) *HAL1* gene was initially found in the screening process for various genes that could be expressed in various plasmid copies that improve saline resistance in yeast (*S. cerevisiae*). It codes for a soluble protein in the cytoplasm, even though there is no significant information available about this gene, still it is a major affective ions regulator during the homeostatic process, slight expression of its promotor generally have an impact on the potassium levels inside the cells [94]. However, a significant expression by strong promotor had an impact on (K^+) and (Na^+) homeostasis [95]. The expression of the *HAL1* gene decreases the loss of (K^+) from cells affected by salt stress a phenomenon initiated through an unknown K^+ efflux system. The cells with *HAL1* contain a high level of potassium in cells, and a low level of sodium within the cells, and an increased K^+/Na^+ ratio as compared to control cells the last one indicating the enhancement in salt tolerance [96]. Currently, it is not known how a protein product from the cytoplasm of the *HAL1* gene can control the transportation of sodium and potassium efflux. Besides the lack of information available about this process *HAL1* gene possess a high capability to improve salt tolerance of various plants, and it was selected in the first trials for expression of genes in transgenic plants [89].

4. Anti-freeze proteins

During the study on fishes in the waters of temperate oceans proteins that act as antifreeze elements were found, in winter the temperature of these waters reaches (−1.9°C) but fishes under these waters still survive. NaCl is the most common electrolyte in blood serum of most species, but to inhibit freezing environment it only helps in 40–50% of the examined freezing point depression [97] the other substances due to which freezing point depression occurred were marked as proteins and glycoproteins [98–100] the molecular masses of antifreeze-glycoproteins ranges from 2.6 to 34 kD. They consist of tripeptide repeats (A l a-A l a-T h r) along with the moiety of disaccharide (-Naga-Gal) having the residue of threonyl [101].

4.1 AFP gene mechanism of action

Many researchers have studied the ant-freeze protein from winter flounder because of their small size and are very effective for structural mechanism requirements, there are some changes in the size and AFP amino acid composition which depends on the isolation technique from the serum of the fish [102]. Through southern blot and restriction maps of genomic clones analysis, the pattern of antifreeze protein multigene family was observed in winter flounder [103]. Most of them are equal in number to 40 AFP genes in this fish are present in 7–8 kbp DNA elements which act like tandem repeats, in every repeat, there is 1 kbp long AFP gene having same transcription shape and orientation, they also have some restriction site polymorphism ability even though the repeats are homologous. When winter flounder genomic DNA goes through the digestion phase mainly by Restriction endonuclease which normally does not cut inside the repeats, many of the AFP genes goes to 40 kbp long fragments that represent five or more repeats in tandem as clusters. After the digestion of genomic DNA, these genes reside in the fragments of extremely high mol. Weight indicating the groups of clusters in the genome [104]. By the combination of protein and DNA sequencing methods, the precursor of amino acid in the second AFP protein B gene has been observed in winter flounder. The precursor containing 82 amino acid residues is only different in three main sites to AFP, A gene that acts in the process of substitution, various other changes, all are grouped inside the DNA that codes for the mature portion of protein. In the process of post-transcriptional modification, the c-terminal glycine residue removal takes place [105].

4.2 The introduction of fish antifreeze AFP gene in transgenic plants

The quality of fruits and vegetables can be compromised by adverse effects due to the formation of ice crystals inside the frozen tissues. At lower concentrations, some proteins from the blood of fishes have shown the ability to help in the inhibition of ice crystals formation. To know whether the expression of certain genes improves freezing properties of the plant tissues, the transgenic tomato and tobacco have been produced by the expression of anti-freeze gene *AFA3* were introduced at higher steady mRNA levels in the leaves of transgenic plants but no crystals inhibition was observed in tissues extracts. As a result of these experiments, ice crystal inhibition was seen in transformed tomato and tobacco plant tissues [106]. The freezing rate and temperature of the storage site are the two factors that influence ice formation in frozen fruits and vegetables, when plants tissues gradually freeze the large and randomly distributed extracellular gaps are filled by ice crystals in comparison to small intracellular and extracellular gaps which freezes rapidly during storage temperature changes lead to large shaped ice crystals and reorganization of ice in food [107].

AFP genes isolated from fish and insects are more useful in the inhibition of frost or crystal formation in several crop plants. AFPs isolated from insects and then their expression in plants against freezing stress are much better than those of fish because of their survival ability in freezing temperatures. AFPs can decrease water freezing level (thermal hysteresis) has generated the phenomenon that the damage could be avoided by those plants which are much more sensitive to frost at the end of autumn and the start of spring due to the expression of higher activity genes coding antifreeze proteins allowing them to be unfrozen in extremely cold and freezing temperatures. During the last two decades, the effectiveness of this idea has been conducted in several different research studies that produce transgenic plants by the expression of various AFPs. Earlier the anti-freezing proteins isolated from fish were used in these studies but later on, as AFPs of insects with high levels of anti-freezing activity were discovered and now being used for plant transformation studies as a choice. A chemically synthesized antifreeze gene from winter flounder fish was introduced through the *Agrobacterium*-mediated transformation method in potato *solanum Tuberosum L. cv.* which decreased the electrolyte leakage from the leaves at freezing temperature [108].

Spring wheat which is vulnerable to the damage caused by frost can also be transformed to show tolerance to frost by the expression of winter flounder gene AFPs in the cytoplasm and apoplast of the plant where ice formation leads to damage at the cellular level. The transformed wheat lines which were targeted by apoplast antifreeze proteins showed the highest anti-freezing activity and exhibited remarkable protection against frost at very lower temperatures [109].

Various marine species survive in extremely cold seawater below the freezing point temperature of their non-protected blood serum by producing anti-freezing proteins and glycoproteins [110, 111]. These proteins and glycoproteins have subsequently been considered for the neutralization of ice nucleator agents [112] to protect the cell from ice crystallization potential damage by hypothermic temperatures [113]. The introduction of these proteins in transgenic plants has been a very important tool for increasing their cold stress tolerance against freezing temperatures. In early work, an AFP gene that codes for alanine-rich, α-helical Type I AFP from winter flounder fish was introduced into tobacco plants through the *Agrobacterium*-mediated transformation method. The transformed plants produced antifreeze proteins mRNA and upon exposure to cold showed the accumulation of AFP to a detectable extent. The observation from the results was that fish antifreeze gene could be very useful in protecting plants from cold and freezing stress [114].

An anti-freezing gene (IIA7 cDNA) was isolated from a fish winter flounder *Pseudopleuronectes americanus* which can survive below the freezing temperature point under cold sea waters, which encodes for 91 amino acid and then proceeded to a mature protein of 53 amino acids. Only mature antifreeze proteins are encoded by this gene, a start methionine was also cloned alongside a plasmid that allowed improved expression from a double cauliflower mosaic virus CaMV 35S promotor. A binary vector pMON200 and intermediate vector pBI121 was used for the subcloning of anti-freezing protein. Various Kanamycin resistant seedlings were tested against the frost tolerance more than 30% of plants survived as compared to the control wild type plants these results confirmed that these genes can help in the resistance to frost in tobacco plants [115].

4.3 Transformation of plants with insects AFPs

The first transgenic plants were produced by the expression of insect AFPs [116], a chemically synthesized gene based on the anti-freezing proteins from an insect *Choristoneura fumiferana*, was introduced into (*Nicotiana Tubaccum*) tobacco plants through Cauliflower mosaic virus 35S promoter. The transformation success was determined by the levels of properly shaped transcripts through real-time PCR, in crude leaf homogenates the recrystallization inhibition activity and the apoplast plant extracts, and the most importantly the degree of water freezing point 0.37% in the apoplast liquid [117].

Transgenic *A. thaliana* was produced by the gene isolated from an insect (*Dendroides canadensis*), AFPs were introduced by *agrobacterium* mediated transformation. The AFP genes simultaneously with and without peptide signals sequence were expressed in transgenic plants. The thermal hysteresis activity showed the existence of active AFPs in proteins isolated from plants that expressed both proteins and were found in fluids of leaf apoplast of plants expressing AFPs alongside signal peptide. The transformed lines did not show any enhancement to survive in freezing temperatures in comparison to wild type plants, however, when cooled under four different stages the transformed lines containing active AFPs apoplast fluid froze at significantly low temperatures in comparison to wild type, especially when there was no intrinsic nucleation [118].

To illustrate the activity of AFPs from beetle (*Microdera punctipennis*) from the deserts of Xinjiang *China,* for freezing stress resistance in plants the *MpAFP149* gene, alongside the signal peptide sequence used for the secretion of *MpAFP149* into the apoplast gaps in the control of cauliflower virus 35S was expressed in tobacco plants through *Agrobacterium tumefaciens* transformation method. The transformants were determined by reverse transcription-polymerase chain reaction analysis of leaf fragments, and those plants having higher transcripts contents were identified for further experiments and analysis. The introduced AFPs were restricted to cell walls of transformants by the use of immune-gold label procedure, and the existence of AFPs in apoplast liquid was indicated by western blot. The inhibition of crystal formation and thermal hysteresis tests to observe the expressed AFPs active state were not done. However, it could be expected that a slight extent of activity existed, the resistance to freez stress of transformed plants near to non-transformed plants was identified through ion-exchange chromatography technique, ion leakage, and malondialdehyde (membrane lipid peroxidase product) measurement release leading to the plant exposure to −1°C for varied periods for 72 h. Upon exposure to −1°C for 2 and 3 days, non-transformed plants were observed to be more adversely affected than the transformed, as by the assessment of more wilted leaves in them. The transgenic plants seemed fully recovered after 1 day at 25–28°C while the non-transformed plants appeared stressed by the indication of wilted leaves, which was obvious even after 5 days of recovery time. After 1 day at −1°C lower ion leakage and malondialdehyde level was observed in transformed and wild type plants but the level increased significantly after 2–3 days at the same temperature. Therefore the AFPs genes protected transgenic tobacco plants from frost stress [119].

The AFPs synthesized from Spruce budworm (*C. fumiferana*) an insect in the *Choristoneura* genus and introduced into *A. thaliana* by plant codon and a peptide with PR-signal of tobacco, the expression vector in plants had a synthesize gene of AFP with double 35S promotor. The transgenic lines showing the high content of anti-freezing protein transcript were selected based on RT-PCR of total RNA from *Arabidopsis* leaves. After 3 weeks growth progress was determined at 23°C under the condition of extended photoperiod, wild and transformed plants were moved to 4°C for 48 h (at long and short photoperiods simultaneously) and further exposed to a very low temperature of −20°C for 30 min. The plants were then maintained at 4°C at night and transferred back to the facility of growth chamber having 23°C temperature. Through visual inspection, the death of most wild-type plants in comparison to the survival of most transformed plants was observed, although the exact numbers were not determined. The transformed lines having a high level of *AFP* transcript showed better survival ability in comparison with wild type plants that exhibited very poor survival capabilities. The rise in electrolyte leakage and malondialdehyde content was observed in all plants upon their exposure to cold treatment, but the levels were much higher in wild type than transgenic plants. The results showed that the expression of AFPs gene from Spruce budworm (*C. fumiferana*) in transgenic *A. thaliana* plants increased their tolerance to freezing temperatures and helped in the removal of injuries [120].

The lists of genes that have been expressed in plants for abiotic stresses tolerance improvement are shown in **Tables 2** and **3**.

Several genes have been expressed in transgenic plants from bacteria for abiotic stresses tolerance that exhibited good results in many transgenic plants for example tomato, tobacco, finger millet, peanut, potato, *A. thaliana*, wheat etc. are shown in the following **Table 2**.

Other genes from insects, fish, and yeast have been introduced in transgenic plants that exhibited better tolerance against various abiotic stresses are shown in **Table 3**.

Gene	Origin	Plant	Abiotic stress	Reference
mtID	*Escherichia coli*	Tomato	Cold and drought	[33]
mtID	*E. coli*	Wheat	Salinity and flooding	[34]
mtID	*E. coli*	Finger millet	Drought, salinity	[36]
ADH	*Cyanobacteria Synechocystis* sp	Tobacco	Salinity	[42]
CodA	*Arthrobacter globiformis*	*Arabidopsis thaliana*	Cold, light stress, and high temperatures	[68]
CodA	*A. globiformis*	Tomato	Chilling and high temperatures	[70]
IPT	*Agrobacterium tumefaciens*	sugarcane	Cold	[40]
ggpPS	*Azotobacter vinelandii*	*A. thaliana*	Salinity and drought	[44]
BetA	*E. coli*	Sugarcane	Drought	[84]
BetA	*E. coli*	Maize	Chilling	[86]
DRNF1	*Nostoc flagelliforme*	*Arabidopsis Thaliana*	Salinity	[46]
IPT	*A. tumefaciens*	Canola	Drought	[43]

Table 2.
Bacterial genes expressed in plants for abiotic stresses tolerance.

Genes	Origin	Plants	Abiotic stress	Reference
TPS1	*Saccharomes cerevisiae*	Tobacco	Drought and salinity	[89]
HAL1, HAL3	*Saccharomyces cerevisiae*	*Arabidopsis thaliana*	Salinity	[92]
HAL1	Yeast	Tomato	Salinity	[90]
AFA3	Winter flounder	Tomato	Freezing stress	[106]
AFA5	Winter flounder	Potato	Frost	[108]
IIA7	*Pseudopleuronectes americanus*	Tobacco	Freezing stress	[115]
MpAFP149	*Microdera punctipennis*	Tobacco	Frost	[119]

Table 3.
List of heterologous expression of genes in transgenic plants for abiotic stresses tolerance from yeast, fish and insects.

5. Conclusion

In this study, the use of various genes isolated from non-plant sources have been expressed in plants for improving their tolerance against abiotic stresses that adversely affect plant growth, and crop yield productivity are reviewed comprehensively. Gene expression in transgenic plants through conventional methods are time consuming and laborious that is why advanced genetic engineering methods for example *Agrobacterium*-mediated transformation and biolistic methods are more accurate, useful, and less time consuming. This review of the chapter provides an extensive insight into various bacterial genes for example *mtID*, *codA*, *betA*, *ADH*, *IPT*, *DRNF1* and *ggpPS*, etc. that have been successfully expressed

DOI: http://dx.doi.org/10.5772/intechopen.105171

in transgenic plants against various abiotic stresses for stress tolerance enhancement and crop yield improvement which exhibited good encouraging results. Genes from yeast (*Saccharomyces cerevisiae*) have been introduced in transgenic plants against drought and salinity stress, other genes isolated from fish for example *AFA3* and *AFA5* which codes for anti-freezing proteins improve transgenic plants against frost stress. Genes from insects have also been inserted in plants to improve their resistance. According to the available literature, several genes isolated from bacteria, yeast, fish, and insets have been expressed in transgenic plants for their enhancement against high and low temperatures, drought, light, and salinity stress. Various research studies have been conducted to improve transgenic plants for the fulfillment of the constant demands of the ever-increasing population. Further work can be done in the future to enhance crop and transgenic plants through new sophisticated technologies. The above mentioned genes can be tested on various other crops to improve their resistance, better yield productivity, longevity in shelf life and enhanced resistance against abiotic and biotic stresses.

Acknowledgements

First of all, I am extremely thankful to ALMIGHTY ALLAH, the ever-magnificent, greatest, merciful, gracious, and pervasive, who provided me the audacity and knowledge to commence and complete this task.

I am also extremely thankful to our holy Prophet Mohammad (SAW), who is a light for humanity and who began his preaching from learning.

I convey my bottomless sense of gratitude to my supervisor Dr. Nadir Zaman associate professor, Department of Biotechnology, University of Malakand, for his welcoming support, and guidance from the beginning till the completion of the present work, who has always been kind in all phases.

I would like to express my profound sense of admiration to all teachers especially Dr. Fazal Hadi chairman department of Biotechnology for his able guidance and support throughout the B.S program, Dr. Syed Muhammad Jamal, Dr. Aftab Ali Shah, Dr. Waqar Ali, Dr. Ayaz Ali Khan, Dr. Muhammad Aasim, Dr. Tariq khan, Dr. Alam Zeb ex-chairman Department of Biotechnology, and Mr. Taqweem Ul Haq for sharing expertise and providing a friendly learning environment.

I am cordially thankful to my family for the generous support they provided me throughout my entire life and particularly through the process of pursuing the BS (Hons.) degree. Because of their unconditional love and prayers, I have the chance to complete this thesis.

I also want to express my profound gratitude to all my friends for their company and continue support.

Abbreviations

Ala	alanine
ABA	abscisic acid
ADH	alcohol dehydrogenase
AFPs	anti-freeze proteins
Al_3^+	aluminum ion

B	boron
Ba_2^+	barium
Ca^+	calcium
CaMV	cauliflower mosaic virus
CaMV35S	cauliflower mosaic virus regulatory sequence
Cl	chlorine
CO_2^{-3}	carbonate
CodA	choline oxidase gene
CSPs	cold shock proteins
DNA	deoxyribonucleic acid
E. coli	*Escherichia coli*
FAO	food and agriculture organization
GG	glycinbetaine
ggpPS	geranylgeranyl diphosphate synthase
HAL1	yeast gene
HCO_3	bicarbonate
IPT gene	isopentenyltransferase gene

Author details

Shahzad Ali[1*], Nadir Zaman[2], Waqar Ali[1], Majid Khan[1], Muhammad Aasim[3], Asmat Ali[3] and Muhammad Usman[3]

1 Institute of Biotechnology and Genetic Engineering (IBGE), The University of Agriculture Peshawar, Pakistan

2 Department of Biotechnology, University of Malakand, Pakistan

3 University of Malakand, Pakistan

*Address all correspondence to: alis60597@gmail.com

References

[1] Mosa KA, Ismail A, Helmy M. Omics and System Biology Approaches in Plant Stress Research. Plant Stress Tolerance. Cham: Springer; 2017. pp. 21-34

[2] Parween A, Singh V, Bajpai M. Abiotic stress and red clover: A less explored area of research. Advances in Research. 2020;**21**(6):1-5

[3] Kanojia A, Dijkwel PP. Abiotic Stress Responses are Governed by Reactive Oxygen Species and Age. 2018. pp. 295-326

[4] Pedrol N, González L, Reigosa MJ. Allelopathy and abiotic stress. Allelopathy. Dordrecht: Springer; 2006. pp. 171-209

[5] Gull A, Lone AA, Wani NUI. Biotic and Abiotic Stresses in Plants. Abiotic and Biotic Stress in Plants. London, IntechOpen; 2019

[6] Charles H, Godfray J, Beddington JR, Crute IR, Haddad L, Lawrence D, et al. Food security: The challenge of feeding 9 billion people. Science. 2010;**327**(5967):812-818

[7] Cramer GR, Urano K, Delrot S, Pezzotti M, Shinozaki K. Effects of abiotic stress on plants: A systems biology perspective. BMC Plant Biology Journal. 2011;**11**(1):163

[8] Mickelbart MV, Hasegawa PM, Bailey-Serres J. Genetic mechanisms of abiotic stress tolerance that translate to crop yield stability. Nature Reviews Genetics. 2015;**16**(4):237-251

[9] Wang H, Wang H, Shao H, Tang X. Recent advances in utilizing transcription factors to improve plant abiotic stress tolerance by transgenic technology. Frontiers in Plant Science. 2016;**7**:67

[10] Dubois M, Claeys H, Van den Broeck L, Inzé D. Time of day determines Arabidopsis transcriptome and growth dynamics under mild drought. Plant, Cell & Environment. 2017;**40**(2):180-189

[11] Sah SK, Reddy KR, Li J. Abscisic acid and abiotic stress tolerance in crop plants. Frontiers in Plant Science. 2016;**7**:571

[12] Ng LM, Melcher K, Teh BT, Xu HE. Abscisic acid perception and signaling: Structural mechanisms and applications. Acta Pharmacologica Sinica. 2014;**35**(5):567-584

[13] Bechtold U, Field B. Molecular Mechanisms Controlling Plant Growth during Abiotic Stress. Oxford University: Press UK; 2018

[14] Kranner I, Minibayeva FV, Beckett RP, Seal CE. What is stress? Concepts, definitions and applications in seed science. New Phytologist. 2010;**188**(3):655-673

[15] Olien C, Smith MN. Ice adhesions in relation to freeze stress. Plant Physiology. 1977;**60**(4):499-503

[16] Grover M, Ali SZ, Sandhya V, Rasul A, Venkateswarlu B. Role of microorganisms in adaptation of agriculture crops to abiotic stresses. World Journal of Microbiology and Biotechnology. 2011;**27**(5):1231-1240

[17] Naglaa A, Moses V, Prakash C. Climate change effects on abiotic stress in plants. GM Crop Food. 2014;**5**:77-80

[18] Barlow K, Christy B, O'leary G, Riffkin P, Nuttall J. Simulating the impact of extreme heat and frost events on

wheat crop production: A review. Field Crops Research. 2015;**171**:109-119

[19] Djanaguiraman M, Perumal R, Ciampitti I, Gupta S, Prasad P. Quantifying pearl millet response to high-temperature stress: Thresholds, sensitive stages, genetic variability and relative sensitivity of pollen and pistil. Plant, Cell & Environment. 2018;**41**(5):993-1007

[20] Kallarackal J, Milburn J, Baker D. Water relations of the Banana. III. Effects of controlled water stress on water potential, transpiration, photosynthesis and leaf growth. Functional Plant Biology. 1990;**17**(1):79-90

[21] Van der Weijde T, Huxley LM, Hawkins S, Sembiring EH, Farrar K, Dolstra O, et al. Impact of drought stress on growth and quality of miscanthus for biofuel production. GCB Bioenergy. 2017;**9**(4):770-782

[22] Sazzad K. Exploring Plant Tolerance to Biotic and Abiotic Stresses. Uppsala: Swedish University of Agricultural Sciences; 2007

[23] Drew MC. Oxygen deficiency and root metabolism: Injury and acclimation under hypoxia and anoxia. Annual Review of Plant Biology. 1997;**48**(1):223-250

[24] Jackson M, Colmer T. Response and adaptation by plants to flooding stress. Annals of Botany. 2005;**96**(4):501-505

[25] Adamiec M, Drath M, Jackowski G. Redox state of plastoquinone pool regulates expression of Arabidopsis thaliana genes in response to elevated irradiance. Acta Biochemical Polonica. 2008;**55**(1):161-174

[26] Barta C, Kálai T, Hideg K, Vass I, Hideg É. Differences in the ROS-generating efficacy of various ultraviolet wavelengths in detached spinach leaves. Functional Plant Biology. 2004;**31**(1):23-28

[27] Safdar H, Amin A, Shafiq Y, Ali A, Yasin R, Shoukat A, et al. A review: Impact of salinity on plant growth. Natural Science. 2019;**17**(1):34-40

[28] Manchanda G, Garg N. Salinity and its effects on the functional biology of legumes. Acta Physiologiae Plantarum. 2008;**30**(5):595-618

[29] HanumanthaRao B, Nair RM, Nayyar H. Salinity and high temperature tolerance in mungbean [Vigna radiata (L.) Wilczek] from a physiological perspective. Frontiers in Plant Science. 2016;7:957

[30] Qados AMA. Effect of salt stress on plant growth and metabolism of bean plant *Vicia faba* (L.). Journal of the Saudi Society of Agricultural Sciences. 2011;**10**(1):7-15

[31] Mubeen H, Naqvi R, Masood A, Shoaib M, Raza S. Gene transformation: Methods, uses and applications. Journal of Pharmaceutical and Biological Sciences. 2016;**4**(2):54

[32] Altieri MA. The ecological impacts of transgenic crops on agroecosystem health. Ecosystem Health. 2000;**6**(1):13-23

[33] Khare N, Goyary D, Singh NK, Shah P, Rathore M, Anandhan S, et al. Transgenic tomato cv. Pusa Uphar expressing a bacterial mannitol-1-phosphate dehydrogenase gene confers abiotic stress tolerance. Plant Cell, Tissue and Organ Culture (PCTOC). 2010;**103**(2):267-277

[34] Abebe T, Guenzi AC, Martin B, Cushman JC. Tolerance of mannitol-accumulating transgenic wheat to water

stress and salinity. Plant Physiology. 2003;**131**(4):1748-1755

[35] Tarczynski MC, Jensen RG, Bohnert HJ. Expression of a bacterial *mtlD* gene in transgenic tobacco leads to production and accumulation of mannitol. Proceedings of the National Academy of Sciences of the United States of America. 1992;**89**(7):2600-2604

[36] Hema R, Vemanna RS, Sreeramulu S, Reddy CP, Senthil-Kumar M, Udayakumar M. Stable expression of mtlD gene imparts multiple stress tolerance in finger millet. PLoS One. 2014;**9**(6):e99110

[37] Bhauso TD, Radhakrishnan T, Kumar A, Mishra GP, Dobaria JR, Patel K, et al. Overexpression of bacterial mtlD gene in peanut improves drought tolerance through accumulation of mannitol. The Scientific World Journal. 2014;**2014**

[38] Kondo Y, Sakamoto A, Nonaka H, Hayashi H, Saradhi PP, Chen TH, et al. Enhanced tolerance to light stress of transgenic Arabidopsis plants that express the codA gene for a bacterial choline oxidase. Plant Molecular Biology. 1999;**40**(2):279-288

[39] Goel D, Singh AK, Yadav V, Babbar SB, Murata N, Bansal KC. Transformation of tomato with a bacterial *codA* gene enhances tolerance to salt and water stresses. Journal of Plant Physiology. 2011;**168**(11):1286-1294

[40] Belintani N, Guerzoni J, Moreira R, Vieira L. Improving low-temperature tolerance in sugarcane by expressing the *IPT* gene under a cold-inducible promoter. Biologia Plantarum. 2012;**56**(1):71-77

[41] Peleg Z, Reguera M, Tumimbang E, Walia H, Blumwald E. Cytokinin-mediated source/sink modifications improve drought tolerance and increase grain yield in rice under water-stress. Plant Biotechnology Journal. 2011;**9**(7):747-758

[42] Yi SY, Ku SS, Sim H-J, Kim S-K, Park JH, Lyu JI, et al. An alcohol dehydrogenase gene from Synechocystis sp. confers salt tolerance in transgenic tobacco. Frontiers in Plant Science. 2017;**8**:1965

[43] Kant S, Burch D, Badenhorst P, Palanisamy R, Mason J, Spangenberg G. Regulated expression of a cytokinin biosynthesis gene *IPT* delays leaf senescence and improves yield under rainfed and irrigated conditions in canola (*Brassica napus* L.). PLoS One. 2015;**10**(1):e0116349

[44] Klähn S, Marquardt DM, Rollwitz I, Hagemann M. Expression of the *ggpPS* gene for glucosylglycerol biosynthesis from Azotobacter vinelandii improves the salt tolerance of *Arabidopsis thaliana*. Journal of Experimental Botany. 2009;**60**(6):1679-1689

[45] Castiglioni P, Warner D, Bensen RJ, Anstrom DC, Harrison J, Stoecker M, et al. Bacterial RNA chaperones confer abiotic stress tolerance in plants and improved grain yield in maize under water-limited conditions. Plant Physiology. 2008;**147**(2):446-455

[46] Cui L, Liu Y, Yang Y, Ye S, Luo H, Qiu B, et al. The drnf1 gene from the drought-adapted cyanobacterium Nostoc flagelliforme improved salt tolerance in transgenic Synechocystis and Arabidopsis plant. Genes. 2018;**9**(9):441

[47] Mattsson A. Predicting field performance using seedling quality assessment. New Forests. 1997;**13**(1-3):227-252

[48] Cheng L, Zou Y, Ding S, Zhang J, Yu X, Cao J, et al. Polyamine accumulation in transgenic tomato enhances the tolerance to high temperature stress. Journal of Integrative Plant Biology. 2009;**51**(5):489-499

[49] Almeselmani M, Deshmukh P, Sairam R, Kushwaha S, Singh T. Protective role of antioxidant enzymes under high temperature stress. Plant Science. 2006;**171**(3):382-388

[50] Gaxiola RA, Li J, Undurraga S, Dang LM, Allen GJ, Alper SL, et al. Drought-and salt-tolerant plants result from overexpression of the AVP1 H+-pump. Proceedings of the National Academy of Sciences. 2001;**98**(20):11444-11449

[51] Davis T, Yamada M, Elgort M, Saier MH Jr. Nucleotide sequence of the mannitol (mtl) operon in *Escherichia coli*. Molecular Microbiology. 1988;**2**(3):405-412

[52] Gowda B, Halaswamy B, Seetharam A, Virk D, Witcombe J. Participatory approach in varietal improvement: A case study in finger millet in India. Current Science. 2000:366-368

[53] Hamza NB. Progress and prospects of millet research in Sudan. In: Tadele Z, editor. 2009. pp. 260-260

[54] Mbithi-Mwikya S, Van Camp J, Mamiro PR, Ooghe W, Kolsteren P, Huyghebaert A. Evaluation of the nutritional characteristics of a finger millet based complementary food. Journal of Agricultural Food Chemistry. 2002;**50**(10):3030-3036

[55] Oghbaei M, Prakash J. Bioaccessible nutrients and bioactive components from fortified products prepared using finger millet (Eleusine coracana). Journal of Science Food Agriculture. 2012;**92**(11):2281-2290

[56] Bhatt D, Negi M, Sharma P, Saxena SC, Dobriyal AK, Arora S. Responses to drought-induced oxidative stress in five finger millet varieties differing in their geographical distribution. Physiology and Molecular Biology of Plants. 2011;**17**(4):347-353

[57] Reddy T, Reddy V, Anbumozhi V. Physiological responses of groundnut (*Arachis hypogea* L.) to drought stress and its amelioration: A critical review. Plant Growth Regulation. 2003;**41**(1):75-88

[58] Ravi K, Vadez V, Isobe S, Mir RR, Guo Y, Nigam SN, et al. Identification of several small main-effect QTLs and a large number of epistatic QTLs for drought tolerance related traits in groundnut (*Arachis hypogaea* L.). Theoritical Application of Genetics. 2011;**122**(6):1119-1132

[59] Pujni D, Chaudhary A, Rajam M. Increased tolerance to salinity and drought in transgenic Indica rice by mannitol accumulation. Journal of Plant Biochemistry and Biotechnology. 2013;**16**

[60] Ganie SA, Ahammed GJ, Wani SH. Vascular plant one zinc-finger (VOZ) transcription factors: Novel regulators of abiotic stress tolerance in rice (*Oryza sativa* L.). Genetic Resources and Crop Evolution. 2020;**67**(4):799-807

[61] Hussain S, Zhang J-h, Zhong C, Zhu L-f, Cao X-c, Yu S-m, et al. Effects of salt stress on rice growth, development characteristics, and the regulating ways: A review. Journal of Integrative Agriculture. 2017;**16**(11):2357-2374

[62] Paparella S, Araújo SS, Rossi G, Wijayasinghe M, Carbonera D,

Balestrazzi A. Seed priming: State of the art and new perspectives. Plant Cell Reports. 2015;**34**(8):1281-1293

[63] Matsumuro T, Miyake H, Takeoka Y. Salinity-induced ultrastructural alterations in leaf cells of Rice (*Oryza sativa* L.). Plant Production Science. 2000;**3**:422-429

[64] Sun J, Zheng T, Yu J, Wu T, Wang X, Chen G, et al. TSV, a putative plastidic oxidoreductase, protects rice chloroplasts from cold stress during development by interacting with plastidic thioredoxin Z. New Phytologist. 2017;**215**(1):240-255

[65] Boyer JS. Plant productivity and the environment. Science. 1982;**218**(4571):443-448

[66] Boyer J, Westgate M. Grain yields with limited water. Journal of Experimental Botany. 2004;**55**(407):2385-2394

[67] Claassen M, Shaw RH. Water deficit effects on corn. II. Grain components 1. Agronomy Journal. 1970;**62**(5):652-655

[68] Sakamoto A, Valverde R, Alia T, Chen H, Murata N. Transformation of Arabidopsis with the codA gene for choline oxidase enhances freezing tolerance of plants. Plant Journal. 2000;**22**(5):449-453

[69] Alia H, Hayashi AS, Murata N. Enhancement of the tolerance of Arabidopsis to high temperatures by genetic engineering of the synthesis of glycinebetaine. Plant Journal. 1998;**16**(2):155-161

[70] Li S, Li F, Wang J, Zhang W, Meng Q, Chen TH, et al. Glycinebetaine enhances the tolerance of tomato plants to high temperature during germination of seeds and growth of seedlings. Plant Cell Environmental Biology. 2011;**34**(11):1931-1943

[71] Park EJ, Jeknić Z, Sakamoto A, DeNoma J, Yuwansiri R, Murata N, et al. Genetic engineering of glycinebetaine synthesis in tomato protects seeds, plants, and flowers from chilling damage. Plant Journal. 2004;**40**(4):474-487

[72] Chang H, Jones ML, Banowetz GM, Clark DG. Overproduction of cytokinins in petunia flowers transformed with PSAG12-IPT delays corolla senescence and decreases sensitivity to ethylene. Plant Physiology. 2003;**132**(4):2174-2183

[73] Rivero RM, Kojima M, Gepstein A, Sakakibara H, Mittler R, Gepstein S, et al. Delayed leaf senescence induces extreme drought tolerance in a flowering plant. Proceedings of the National Academy of Sciences. 2007;**104**(49):19631-19636

[74] Haberer G, Kieber JJ. Cytokinins. New insights into a classic phytohormone. Plant Physiology. 2002;**128**(2):354-362

[75] Miller CO, Skoog F, Okumura F, Von Saltza M, Strong F. Isolation, structure and synthesis of kinetin, a substance promoting cell division1, 2. Journal of the American Chemical Society. 1956;**78**(7):1375-1380

[76] Gan S, Amasino RM. Inhibition of leaf senescence by autoregulated production of cytokinin. Science. 1995;**270**(5244):1986-1988

[77] Qin H, Gu Q, Zhang J, Sun L, Kuppu S, Zhang Y, et al. Regulated expression of an isopentenyltransferase gene (IPT) in peanut significantly improves drought tolerance and increases yield under field conditions. Plant and Cell Physiology. 2011;**52**(11):1904-1914

[78] Robson PR, Donnison IS, Wang K, Frame B, Pegg SE, Thomas A, et al. Leaf senescence is delayed in maize expressing the agrobacterium IPT gene under the

control of a novel maize senescence-enhanced promoter. Plant Biotechnology Journal. 2004;**2**(2):101-112

[79] Farooq M, Wahid A, Basra S. Improving water relations and gas exchange with brassinosteroids in rice under drought stress. Journal of Agronomy and Crop Science. 2009;**195**(4):262-269

[80] Serraj R, Kumar A, McNally K, Slamet-Loedin I, Bruskiewich R, Mauleon R, et al. Improvement of drought resistance in rice. Advances in Agronomy. 2009;**103**:41-99

[81] Bouman B, Humphreys E, Tuong T, Barker R. Rice and water. Advances in Agronomy. 2007;**92**:187-237

[82] Ma Q-H. Genetic engineering of cytokinins and their application to agriculture. Critical Reviews in Biotechnology. 2008;**28**(3):213-232

[83] Hajouj T, Michelis R, Gepstein S. Cloning and characterization of a receptor-like protein kinase gene associated with senescence. Plant Physiology. 2000;**124**(3):1305-1314

[84] Sugiharto B. Biotechnology of drought-tolerant sugarcane. Sugarcane: Technology and Research. 16 May 2018;**138**

[85] Lv S, Yang A, Zhang K, Wang L, Zhang J. Increase of glycinebetaine synthesis improves drought tolerance in cotton. Molecular Breeding. 2007;**20**(3):233-248

[86] Quan R, Shang M, Zhang H, Zhao Y, Zhang J. Improved chilling tolerance by transformation with betA gene for the enhancement of glycinebetaine synthesis in maize. Plant Science. 2004;**166**(1):141-149

[87] Holmström KO, Somersalo S, Mandal A, Palva TE, Welin B. Improved tolerance to salinity and low temperature in transgenic tobacco producing glycine betaine. Journal of Experimental Botany. 2000;**51**(343):177-185

[88] Yilmaz JL, Bülow L. Enhanced stress tolerance in *Escherichia coli* and *Nicotiana tabacum* expressing a betaine aldehyde dehydrogenase/choline dehydrogenase fusion protein. Biotechnology Progress. 2002;**18**(6):1176-1182

[89] Serrano R, Culiañz-Maciá FA, Moreno V. Genetic engineering of salt and drought tolerance with yeast regulatory genes. Scientia Horticulturae. 1998;**78**(1-4):261-269

[90] Gisbert C, Rus AM, Bolarín MC, López-Coronado JM, Arrillaga I, Montesinos C, et al. The yeast *HAL1* gene improves salt tolerance of transgenic tomato. Plant Physiology. 2000;**123**(1):393-402

[91] Romero C, Bellés JM, Vayá JL, Serrano R, Culiáñez-Macià FA. Expression of the yeast trehalose-6-phosphate synthase gene in transgenic tobacco plants: Pleiotropic phenotypes include drought tolerance. Planta. 1997;**201**(3):293-297

[92] Yang SX, Zhao YX, Zhang Q, He YK, Zhang H, Luo D. HAL1 mediate salt adaptation in *Arabidopsis thaliana*. Cell Research. 2001;**11**(2):142-148

[93] Ellul P, Ríos G, Atarés A, Roig LA, Serrano R, Moreno V. The expression of the Saccharomyces cerevisiae *HAL1* gene increases salt tolerance in transgenic watermelon [Citrullus lanatus (Thunb.) Matsun. & Nakai.]. Theoretical and Applied Genetics. 2003;**107**(3):462-469

[94] Gaxiola R, De Larrinoa IF, Villalba JM, Serrano R. A novel and conserved salt-induced protein is an important determinant of salt

tolerance in yeast. The EMBO Journal. 1992;**11**(9):3157-3164

[95] Rios G, Ferrando A, Serrano R. Mechanisms of salt tolerance conferred by overexpression of the HAL1 gene in Saccharomyces cerevisiae. Yeast. 1997;**13**(6):515-528

[96] Serrano R. Salt tolerance in plants and microorganisms: Toxicity targets and defense responses. International Review of Cytology. 1996;**165**:1-52

[97] DeVries A. Role of glycopeptides and peptides in inhibition of crystallization of water in polar fishes. Philosophical Transactions of the Royal Society of London B, Biological Sciences. 1984;**304**(1121):575-588

[98] Burcham TS, Osuga DT, Chino H, Feeney RE. Analysis of antifreeze glycoproteins in fish serum. Analytical Biochemistry. 1984;**139**(1):197-204

[99] DeVries AL. [21] antifreeze glycopeptides and peptides: Interactions with ice and water. Methods in Enzymology. 1986;**127**:293-303

[100] Mantel LH, Farmer LL. Osmotic and ionic regulation. Internal Anatomy and Physiological Regulation. 1983;**5**:53-161

[101] DeVries AL, Komatsu SK, Feeney RE. Chemical and physical properties of freezing point-depressing glycoproteins from Antarctic fishes. Journal of Biological Chemistry. 1970;**245**(11):2901-2908

[102] Scott GK, Davies PL, Shears MA, Fletcher GL. Structural variations in the alanine-rich antifreeze proteins of the pleuronectinae. European Journal of Biochemistry. 1987;**168**(3):629-633

[103] Gourlie B, Lin Y, Price J, DeVries A, Powers D, Huang R. Winter flounder antifreeze proteins: A multigene family. Journal of Biological Chemistry. 1984;**259**(23):14960-14965

[104] Scott GK, Hew CL, Davies PL. Antifreeze protein genes are tandemly linked and clustered in the genome of the winter flounder. Proceedings of the National Academy of Sciences. 1985;**82**(9):2613-2617

[105] Pickett M, Scott G, Davies P, Wang N, Joshi S, Hew C. Sequence of an antifreeze protein precursor. European Journal of Biochemistry. 1984;**143**(1):35-38

[106] Hightower R, Baden C, Penzes E, Lund P, Dunsmuir P. Expression of antifreeze proteins in transgenic plants. Plant Molecular Biology. 1991;**17**(5):1013-1021

[107] Joslyn M. The freezing of fruits and vegetables. In: Meryman HT, editor. Cryobiology. 1966:565-607

[108] Wallis JG, Wang H, Guerra DJ. Expression of a synthetic antifreeze protein in potato reduces electrolyte release at freezing temperatures. Plant Molecular Biology. 1997;**35**(3):323-330

[109] Khanna HK, Daggard GE. Targeted expression of redesigned and codon optimised synthetic gene leads to recrystallisation inhibition and reduced electrolyte leakage in spring wheat at sub-zero temperatures. Plant Cell Reports. 2006;**25**(12):1336-1346

[110] Davies PL, Hew CL, Fletcher GL. Fish antifreeze proteins: Physiology and evolutionary biology. Canadian Journal of Zoology. 1988;**66**(12):2611-2617

[111] DeVries A. Antifreeze peptides and glycopeptides in cold-water fishes. Annual Review of Physiology. 1983;**45**(1):245-260

[112] Parody-Morreale A, Murphy KP, Di Cera E, Fall R, DeVries AL, Gill SJ. Inhibition of bacterial ice nucleators by fish antifreeze glycoproteins. Nature. 1988;**333**(6175):782-783

[113] Rubinsky B, Arav A, Fletcher G. Hypothermic protection—A fundamental property of "antifreeze" proteins. Biochemical and Biophysical Research Communications. 1991;**180**(2):566-571

[114] Kenward KD, Altschuler M, Hildebrand D, Davies PL. Accumulation of type I fish antifreeze protein in transgenic tobacco is cold-specific. Plant Molecular Biology. 1993;**23**(2):377-385

[115] Lee J-S. The Frost Tolerance of Tobacco Plants Transformed with the Gene Encoding the Antifreeze Protein from Winter Flounder. Louisiana State University and Agricultural & Mechanical College; 1991

[116] Holmberg N, Farrés J, Bailey JE, Kallio PT. Targeted expression of a synthetic codon optimized gene, encoding the spruce budworm antifreeze protein, leads to accumulation of antifreeze activity in the apoplasts of transgenic tobacco. Gene. 2001;**275**(1):115-124

[117] Tyshenko MG, Doucet D, Davies PL, Walker VK. The antifreeze potential of the spruce budworm thermal hysteresis protein. Nature Biotechnology. 1997;**15**(9):887-890

[118] Huang T, Nicodemus J, Zarka DG, Thomashow MF, Wisniewski M, Duman JG. Expression of an insect (*Dendroides canadensis*) antifreeze protein in *Arabidopsis thaliana* results in a decrease in plant freezing temperature. Plant Molecular Biology. 2002;**50**(3):333-344

[119] Wang Y, Qiu L, Dai C, Wang J, Luo J, Zhang F, et al. Expression of insect (Microdera puntipennis dzungarica) antifreeze protein MpAFP149 confers the cold tolerance to transgenic tobacco. Plant Cell Reports. 2008;**27**(8):1349

[120] Zhu B, Xiong A-S, Peng R-H, Xu J, Jin X-F, Meng X-R, et al. Over-expression of ThpI from *Choristoneura fumiferana* enhances tolerance to cold in Arabidopsis. Molecular Biology Reports. 2010;**37**(2): 961

Chapter 10

Reactive Oxygen Species, Oxidative Damage and Their Production, Detection in Common Bean (*Phaseolus vulgaris* L.) under Water Stress Conditions

Asmat Ara, Mahroofa Jan, Parvaze A. Sofi, Munezeh Rashid, Ajaz Ahmad Lone, Zahoor Ahmad Dar, Mohd. Ashraf Rather and Musharib Gull

Abstract

Reactive oxygen species (ROS) being small and highly reactive oxygen containing molecules play significant role in intracellular signaling and regulation. Various environmental stresses lead to excessive production of ROS causing progressive oxidative damage and ultimately cell death. This increased ROS production is, however, tightly controlled by a versatile and cooperative antioxidant system that modulates intracellular ROS concentration and controls the cell's redox status. Furthermore, ROS enhancement under stress serves as an alarm signal, triggering acclimatory/defense responses via specific signal transduction pathways involving H_2O_2 as a secondary messenger. Nevertheless, if water stress is prolonged over to a certain extent, ROS production will overwhelm the scavenging action of the anti-oxidant system resulting in extensive cellular damage and death. DAB (3,3′-diaminobenzidine) test serves as an effective assessment of oxidative damage under stress. It clearly differentiates the lines on the basis of darker staining of leaves under water stress. The lines showing greater per cent reduction in yield parameters show greater staining in DAB assay underlining the reliability of using this assay as a reliable supplement to phenotyping protocols for characterizing large germplasm sets.

Keywords: ROS, cell death, oxidative stress, DAB

1. Introduction

Abiotic stresses such as drought and high temperature invariably cause unfavorable changes in water status of plant cells as well as evolution of reactive oxygen species in cellular compartments resulting in acceleration of leaf senescence through lipid peroxidation and other oxidative damage [1]. Omae et al. [2] discovered a link between genotypic differences in bean leaf water status and crop productivity under drought conditions.

This implies that there are differences in leaf water status among bean cultivars, which could be related to drought tolerance mechanisms. Among the reactive oxygen species hydrogen peroxide (H_2O_2) is a non-radical reactive oxygen species (ROS) produced in a two- electron reduction of molecular oxygen. H_2O_2 being a strong oxidant, it can initiate localized oxidative damage in leaf cells leading to disruption of metabolic function and loss of cellular integrity, actions that result in senescence promotion.

In plants, reactive oxygen species (ROS) are formed by the leakage of electrons from the electron transport activities of mitochondria, chloroplasts and plasma membranes or as a byproduct of various biotic and abiotic stresses due to disruption of cellular homeostasis [3–5]. A cell is said to be under oxidative stress when the level of ROS exceeds the defense mechanism. Increased ROS production during various stresses endangers cells, causing lipid peroxidation, protein oxidation, enzyme inhibition, nucleic acid damage, activation of the programmed cell death pathway, and ultimately cell death [6–8]. The overproduction of H_2O_2 has been observed in plants exposed to a number of stress conditions and is considered as one of the factors causing oxidative stress [9].

2. ROS, sites of production and their effects

Reactive oxygen species are a group of free radicals, reactive molecules and ions that are derived from o_2. ROS are known for playing role as both deleterious and beneficial species depending on their concentration in plants. They are produced at several locations within the cell in both stressed and unstressed cells (**Figure 1**).

Production and removal of ROS needs to be controlled to avoid oxidative stress. When this level exceeds the defense mechanisms, a cell is said to be in a state of "oxidative stress". Increased level of ROS can cause damage to biomolecules like lipids, proteins and DNA (**Figure 2**). These reactions can alter intrinsic membrane properties like fluidity, loss of enzyme activity, ion transport, protein cross-linking, DNA damage, inhibition of protein synthesis ultimately resulting in cell death.

Under water stress, ROS production is enhanced in various ways. Inhibition of carbon dioxide assimilation coupled with changes in photosystem activities and photosynthetic

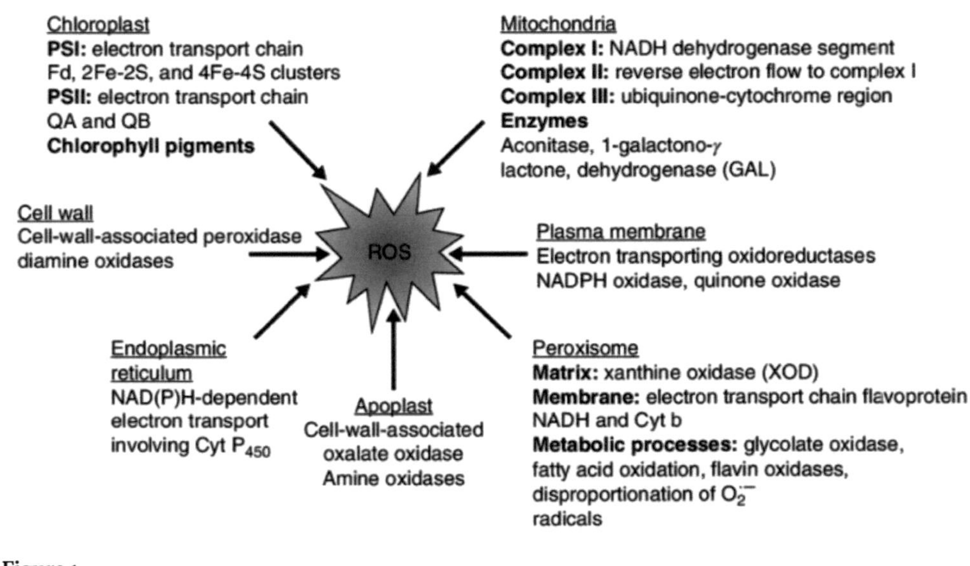

Figure 1.
Sites of production of reactive oxygen species (ROS) in plants.

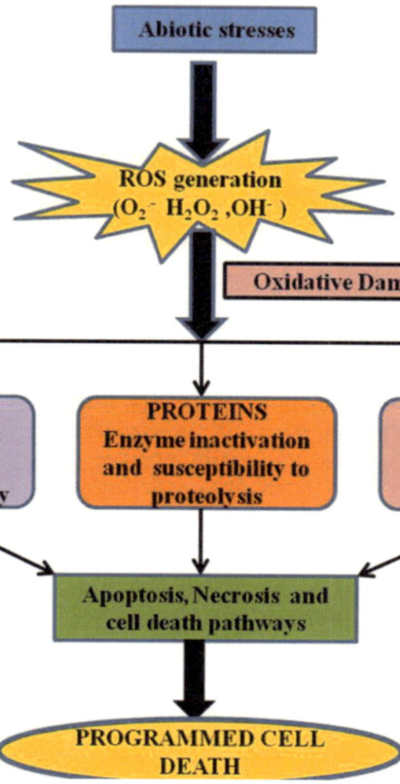

Figure 2.
Reactive oxygen species (ROS) induces oxidative damage to lipids, proteins and DNA.

transport capacity results in increased production of ROS [10]. Excess light energy dissipation in the PSII core and antenna generates ROS, which are potentially dangerous under water stress conditions [11]. The photorespiratory pathway is also increased, especially when RUBP oxygenation is optimum due to CO2 fixation limitation.

DAB (3,3′-diaminobenzidine) assay has been suggested as an effective qualitative assessment of plant response to biotic and abiotic stress and measures the intensity of oxidative burst under stress. Since the oxidative burst is an early response to stress, in terms of production of reactive oxygen species (ROS) including hydrogen peroxide through either NADPH oxidases or peroxidises (Bindschedler et al., 2006) that may exist singly or in combination in different plant species have been proposed for the generation of ROS. The qualitative evolution can be differentially tracked in different parts of plant under stress to assess the most vulnerable part under stress. It is done by staining with 3,3′-diaminobenzidine (DAB) which is oxidized by hydrogen peroxide and generates a dark brown precipitate.

3. Experimental method

The present study was conducted during 2016-2018 at the Division of Genetics & Plant Breeding, Faculty of Agriculture Wadura, SKUAST-K, Sopore. In the current study, fifty genotypes of common bean were evaluated under controlled conditions.

The genotypes used were chosen based on their yield screening trial performance and represented a wide range of market classes in terms of use category, growth habits, and seed characteristics. The material included 47 breeding lines as well as three released varieties, SR-1, SFB-1, and Arka Anoop. The experiment was designed in a completely randomized design.

The DAB assay was performed in accordance with Daudi & O'Brien [12]. In this protocol, hydrogen peroxide (one of several reactive oxygen species) is detected in situ by staining with 3,3′-diaminobenzidine (DAB). In the presence of some haeme-containing proteins, such as peroxidases, DAB is oxidized by hydrogen peroxide to produce a dark brown precipitate. This precipitate is used as a stain in plant cells to detect the presence and distribution of hydrogen peroxide. DAB staining solution was prepared by adding 50 mg DAB and 45 ml sterile H_2O for a final 1 mg ml^{-1} DAB solution in a 50 ml falcon tube. The tube was covered with aluminum foil as DAB is light-sensitive. About 25 µl Tween 20 (0.05% v/v) and 2.5 ml 200 mM Na_2HPO_4 to the DAB solution to produce 10 mM Na_2HPO_4 DAB staining solution. Similar, fully opened leaves were selected from each treatment and incubated for one hour in falcon tubes with 2 ml of the DAB staining solution with the volume being adjusted to ensure that leaves were immersed. The leaves from irrigated treatment were incubated with 2 ml of 10 mM Na_2HPO_4. All the falcon tubes from both drought and irrigated treatments were shaken for 4-5 h at 80-100 rpm. Following the incubation, the aluminum foil was replaced and the DAB staining solution replaced with bleaching solution (ethanol: acetic acid: glycerol in ratio of 3:1:1). For 15 minutes, the falcon tubes were immersed in a boiling water bath (90-95°C). The chlorophyll will be bleached out, but the brown precipitate formed by the DAB reacting with the hydrogen peroxide will remain. The time should be adjusted (5 minutes) depending on how the leaves look (they should be completely devoid of chlorophyll). After 15 minutes of boiling,

(a) (b)

Figure 3.
*DAB staining of common bean (*Phaseolus vulgaris *L.) genotypes under water stress conditions. Largely stained genotypes (3b) show higher production of hydrogen peroxide under stress conditions causing oxidative damage to cell structure.*

the bleaching solution was replaced with fresh bleaching solution and left to stand for 30 minutes. DAB staining was visualized directly on leaves.

4. Results and discussion

The DAB staining of common bean genotypes under irrigated and water stress conditions clearly differentiates the lines on the basis of darker staining of leaves under drought (**Figure 3**). The lines showing greater per cent reductions in yield parameters show greater staining in DAB assay underlining the reliability of using this assay as a reliable supplement to phenotyping protocols for characterizing large germplasm sets. However, DAB is only a qualitative test for evolution of reactive oxygen species such as H_2O_2 and the genotypes showing greater staining under drought can be further analyzed for the amount of H_2O_2 through various analytical methods.

The DAB staining of common bean genotypes clearly differentiates the genotypes on the basis of darker staining of leaves under water stress. The lines showing greater per cent reductions in yield parameters show greater staining in DAB assay underlining the reliability of using this assay as a reliable supplement to phenotyping protocols for characterizing large germplasm sets (**Table 1**). Reactive oxygen species play an important role as signaling molecules that initiate stress responses in plants. Environmental stresses are known to induce the production of H_2O_2 and other toxic oxygen species in cellular compartments, resulting in the acceleration of leaf senescence via lipid peroxidation and other oxidative damage, according to Upadhyaya et al. [1]. Because H_2O_2 is a strong oxidant, it can cause localized oxidative damage in leaf cells, disrupting metabolic function and causing cellular integrity loss, both of which promote senescence. Overproduction of H_2O_2 has been observed in plants subjected to a variety of stress conditions and is thought to be one of the causes of oxidative stress [9]. According to Foyer and Noctor [13], among the various forms of ROS, the central role in plant signaling, regulating plant development, and adaptation to abiotic and biotic stresses is played by hydroxyl radicals. ROS also act as signaling molecules to regulate development and stress responses [14]. Increased availability of H_2O_2 is commonly observed feature of plant stress response signature. The physiological context involves a continuous supply of environmental stimuli that can trigger intracellular H_2O_2 accumulation or modulate the response to such accumulation.

The detection of cellular levels of H_2O_2 was done by DAB staining method and our results shows a clear difference in the degree of staining achieved in the stressed plant. Under water stress, there was significant variation in staining in different genotypes indicating differential oxidative damage on account of production of H_2O_2. The lines which showed fair amount of tolerance to water stress in terms of higher yield and lower reduction had almost negligible staining while as the genotypes which showed lower yield showed higher reduction, distinctly darker staining. Less tolerant cultivars accumulated more H_2O_2 than more tolerant ones, and vulnerable variety showed noticeably greater increases in lipid peroxidation. Similar findings were reported by a number of prior studies (Chai et al., 2005; Zlatev et al., 2006).

Plants accumulate reactive oxygen species during drought stress (Verslues et al., 2006). ROS can cause cell death by destroying DNA, proteins, and carbohydrates through partially reduced or activated oxygen derivatives [3]. DAB staining investigations can efficiently differentiate ROS levels in transgenic lines of rice, with reduced staining in transgenic lines compared to control plants after drought-stress treatment, according to Jiang et al. (2016). DAB assay results were consistent with those

Genotype	Pods per plant			Pod length (cm)			Seeds per pod			100-seed weight (g)			Seed yield per plant (g)		
	S	NS	PC	S	NS	PC	S	NS	PC	S	NS	PC	S	NS	PC
WB-6	7.22	12.77	−43.48	10.73	13.71	−21.76	3.28	3.81	−14.02	39.34	47.66	−17.45	8.78	19.55	−55.09
WB-22	8.64	14.11	−38.76	9.63	12.14	−20.67	3.86	4.26	−9.27	44.39	50.57	−12.21	14.35	29.32	−51.04
WB-83	14.33	23.74	−39.62	8.60	11.11	−22.58	4.51	4.95	−8.98	24.72	28.20	−12.34	16.33	24.67	−33.81
WB-112	10.35	15.86	−34.74	11.78	13.08	−9.94	3.49	4.09	−14.65	41.27	44.94	−8.16	14.97	26.06	−42.54
WB-185	14.34	20.24	−29.14	11.90	12.98	−8.35	3.61	4.52	−20.22	40.31	44.36	−9.14	21.74	34.44	−36.88
WB-216	7.07	10.46	−32.39	11.02	13.96	−21.08	3.19	3.68	−13.29	40.93	47.22	−13.32	12.17	16.28	−25.27
WB-222	18.82	22.13	−14.95	9.10	10.09	−9.80	4.27	4.60	−7.27	24.99	28.28	−11.61	21.23	31.10	−31.73
WB-257	10.70	15.33	−30.09	11.30	12.97	−12.91	3.24	3.93	−17.43	41.87	48.72	−14.05	17.41	25.38	−31.41
WB-341	21.06	23.05	−8.61	9.61	10.47	−8.17	4.50	4.98	−9.64	26.68	29.00	−8.00	27.24	33.18	−17.88
WB-401	14.57	17.97	−18.91	9.06	10.36	−12.55	3.76	4.14	−9.18	25.36	27.65	−8.29	11.68	15.10	−22.61
WB-451	17.00	27.28	−37.67	8.47	10.35	−18.20	3.63	4.14	−12.30	25.98	29.57	−12.12	23.17	33.72	−31.28
WB-956	9.55	15.55	−38.57	11.56	12.93	−10.59	3.68	4.57	−19.47	38.21	42.84	−10.80	18.61	25.41	−26.74
WB-1446	11.47	14.11	−18.74	10.37	13.23	−21.62	3.52	4.06	−13.30	35.31	39.03	−9.53	17.84	24.96	−28.53
WB-1492	8.22	12.39	−33.68	9.60	10.84	−11.39	3.61	4.05	−10.97	29.42	32.13	−8.43	13.16	27.35	−51.86
WB-1587	4.16	9.07	−54.13	7.34	9.96	−26.34	3.12	3.77	−17.35	25.17	28.95	−13.07	8.26	13.50	−38.77
WB-1634	21.22	24.90	−14.77	10.62	11.67	−8.99	4.74	5.16	−8.13	25.18	26.71	−5.74	31.96	36.05	−11.33
WB-1643	16.05	19.27	−16.70	11.96	13.11	−8.77	3.79	4.61	−17.68	27.73	31.94	−13.18	20.69	25.52	−18.90
SR-1	6.84	10.61	−35.56	8.20	14.79	−44.52	3.32	4.44	−25.31	35.27	39.51	−10.73	13.43	20.18	−33.44
SFB-1	11.72	25.55	−54.12	13.92	16.90	−17.65	3.91	4.83	−19.13	23.15	26.53	−12.75	18.57	22.11	−16.01
Arka Anoop	10.71	17.17	−37.61	−12.73	15.78	−19.35	3.23	4.07	−20.64	24.00	29.60	−18.93	12.50	17.69	−29.33
Mean	12.26	17.73	30.88	10.38	**12.52**	**17.09**	**3.71**	**4.34**	**14.52**	**31.97**	**36.17**	**11.61**	**17.21**	**25.08**	**31.38**

Genotype	Pods per plant			Pod length (cm)			Seeds per pod			100-seed weight (g)			Seed yield per plant (g)		
	S	NS	PC	S	NS	PC	S	NS	PC	S	NS	PC	S	NS	PC
C.D ($p \leq 0.05$)		Genotypes = 2.882 Water regime = 0.911 G x WR = 4.075			Genotypes = 0.952 Water regime = 0.301 G x WR = 1.346			Genotypes = 0.355 Water regime = 0.112 G x WR = 0.247			Genotypes = 1.572 Water regime = 0.497 G x WR = 2.223			Genotypes = 3.069 Water regime = 0.979 G x WR = 4.379	

Table 1.
Mean performance under different water regimes and percent reductions for yield and yield parameters under water stress in common bean (Phaseolus vulgaris L.).

Parameter	Common bean			
	Susceptible lines		Tolerant lines	
	WB-285575	WB-335	WB-216	WB-451
Canopy temperature depression (°C)	−1.07	−1.84	−0.68	−2.58
Relative water content (%)	18.82	9.70	10.69	11.54
Membrane stability index	0.456	0.592	0.459	0.656
Chlorophyll stability index	0.675	0.336	0.295	0.174
Proline content (μmol/g)	20.65	22.63	36.72	37.33

Table 2.
Comparative performance of tolerant susceptible cultivars for various physio- biochemical parameters under stress in common bean.

obtained using membrane stability and other biochemical parameters in tolerant and sensitive wheat cultivars, according to Chakraborty and Pradhan (2012). However, only a qualitative assessment of DAB was performed in this work. Ghahfarokhi et al. (2016) performed an experiment to examine the effects of drought stress caused by withholding irrigation at the vegetative stage (4-5 leaves) and reproductive stage on crop production, physiological, and biochemical features in hybrids of maize (*Zea mays* L.) (anthesis). Results indicated that these traits were significantly impacted by drought stress (**Table 2**). Under water stress, both the yield and its constituent parts significantly reduced. The main causes of the yield reduction were a decrease in the quantity of grain ear^{-1} and the weight of 1000 grains. In comparison to other hybrids, short maturity hybrids had a larger yield reduction. These results suggested that water stress lead to the production of reactive oxygen species (ROS), which caused an increased membrane permeability and oxidative stress in the maize plants. The reliability of the DAB test can be further validated by conducting a quantitative assessment of hydrogen peroxide evolution in common bean under water stress.

5. Conclusion

ROS are produced by electron transport activities of mitochondria, chloroplast, plasma membrane or as a byproduct of various metabolic pathways localized in different cellular compartments. Study of formation and fate of ROS using advanced and analytical techniques help in developing broader view of the role of ROS in plants. DAB assay is employed to delineate genotypic response in terms of qualitative differentiation of oxidative damage as indicated by differential staining under DAB treatment. All the genotypes revealed almost similar staining in irrigated conditions. While as, under drought conditions, genotypes which showed better resilience to water stress in terms of higher yield and drought had significantly lesser staining as compared to susceptible ones. Therefore, DAB staining can be used as complementary method for differentiating genotypes to water stress.

Author details

Asmat Ara[1*], Mahroofa Jan[2], Parvaze A. Sofi[1], Munezeh Rashid[1], Ajaz Ahmad Lone[1], Zahoor Ahmad Dar[1], Mohd. Ashraf Rather[1] and Musharib Gull[1]

1 Division of Genetics and Plant Breeding, Wadura, Sopore, India

2 Department of Botany, University of Kashmir, Srinagar, India

*Address all correspondence to: mirasmat35@gmail.com

References

[1] Upadhyaya H, Khan MH, Panda SK. Hydrogen peroxide induces oxidative stress in detached leaves of *Oryza sativa* L. Genetics and Plant Physiology. 2007;**1**(2):83-95

[2] Omae H, Kumar A, Egawa Y, Kashiwaba K, Shono M. Genotypic differences in plant water status and relationship with reproductive responses in snap bean (*Phaseolus vulgaris* L.) during water stress. Japan Journal of Tropical Agriculture. 2005;**49**:1-7

[3] Mittler R. Oxidative stress, antioxidants and stress tolerance. Trends in Plant Science. 2002;**7**(9):405-410

[4] Shah K, Kumar RG, Verma S, Dubey RS. Effect of cadmium on lipid peroxidation, superoxide anion generation and activities of antioxidant enzymes in growing rice seedlings. Plant Science. 2001;**161**:1135-1144

[5] Sharma P, Dubey RS. Drought induces oxidative stress and enhances the activities of antioxidant enzymes in growing rice seedlings. Plant Growth Regulation. 2005;**46**(3):209-221

[6] Maheshwari R, Dubey RS. Nickel-induced oxidative stress and the role of antioxidant defence in rice seedlings. Plant Growth Regulation. 2009;**59**(1):37-49

[7] Mishra S, Jha AB, Dubey RS. Arsenite treatment induces oxidative stress, upregulates antioxidant system, and causes phytochelatin synthesis in rice seedlings. Protoplasma. 2011;**248**(3):565-577

[8] Srivastava S, Dubey RS. Manganese-excess induces oxidative stress, lowers the pool of antioxidants and elevates activities of key antioxidative enzymes in rice seedlings. Plant Growth Regulation. 2011;**42**:1-16

[9] Šnyrychová I, Ayaydin F, Hideg É. Detecting hydrogen peroxide in leaves in vivo–a comparison of methods. Physiologia Plantarum. 2009;**135**(1):1-18

[10] Asada K. The water-water cycle in chloroplasts: Scavenging of active oxygen and dissipation of excess photons. Annual Review of Plant Biology. 1999;**50**:601-639

[11] Foyer CH, Harbinson J. Oxygen Metabolism and the Regulation of Photosynthetic Electron Transport in Causes of Photooxidative Stresses and Amelioration of Defense Systems in Plants. Boca Raton, Fla, USA: CRC Press; 1994. pp. 1-42

[12] Daudi A, O'Brien JA. Detection of hydrogen peroxide by DAB staining in Arabidopsis leaves. Bio-Protocol. 2012;**2**(18):263

[13] Foyer CH, Noctor G. Managing the cellular redox hub in photosynthetic organisms. Plant Cell Environment. 2012;**35**(2):199-201

[14] Apel K, Hirt H. Reactive oxygen species: Metabolism, oxidative stress, and signal transduction. Annual Review of Plant Biology. 2004;**55**:373-399

Section 3

Mechanisms of Plant Responses to Biotic Stresses

Chapter 11

Physiological Mechanisms of Tolerance to Drought and Heat in Major Pulses for Improving Yield under Stress Environments

Partha S. Basu, Sushil Kumar Chaturvedi, Pooran Mall Gaur, Biswajit Mondal, Surendra Kumar Meena, Krishnashis Das, Vaibhav Kumar, Kalpana Tewari and Kusum Sharma

Abstract

Reduction in biomass and pollen fertility are the two major constraints resulting in poor grain yield in major pulses grown under rainfed agrosystem. Generally, pulses are encountered into both heat and drought stresses during terminal reproductive stages. Though pulses have many adaptive features to counter the adverse effects of various abiotic stresses but yield is substantially reduced when the magnitude of these stresses is very high. The factors have been identified to enhance grain yield under stress environments which include promotion of biomass in the above ground part enabling crops to reserve a maximum amount of photosynthesis and water in the plant system itself before the onset of drought and heat stresses during reproductive stages. Various physiological mechanisms and fertility enhancement components including genetic diversity in key traits have been discussed here to improve yield of pulses under stressed conditions.

Keywords: pollen fertility, root, antioxidant, osmotic adjustment, photosynthesis, remobilization, chlorophyll fluorescence, pod

1. Introduction

The earth's climate system is experiencing a warmer phase. Increase in temperature and atmospheric CO_2 concentration are the major effects of climate, besides increase or decrease in the local rainfall. The climate modeling systems envisage that as the twenty-first century progresses, there will be rise in temperature from 2 to 6°C. Rainfed agriculture accounts for more than 40% of total food grain production, and nearly 90% of it is occupied by cultivation of pulses. Thus, country like India which is a major producer and consumer of pulses, there are about 100 million hectare cultivated areas dependent on rainfall. In general, crop productivity is adversely affected by high temperature and drought [1]. The rise in global mean temperature and drought has affected agricultural productivity worldwide [2]. According to IPCC (2013), decreasing water availability and increasing temperature are posing a great threat to food security. Therefore, it is an urgent need to identify tolerant plant species for these stresses [3]. The pulses or food legumes are rainfed crops grown under diverse soil types and agro-ecosystems in low rainfall areas of semi-arid regions including India, Pakistan, Bangladesh, Myanmar, and Nepal. The pulses find an extremely important place in the agricultural system as they require less chemical fertilizers and limited moisture. The legumes or pulse crops include chick pea (*Cicer arietinum* L.), pigeon pea (*Cajanus cajan* L.), green gram (*Vigna radiata* L.), black gram (*Vigna mun*go L.), lentil (*Lens culinaris* L.), as major pulse crops (**Figure 1**) while field pea (*Pisum sativum* L.), kidney bean or rajmash (*Phaseolus vulgaris* L.), lathyrus, cowpea, horsegram, etc. are considered as minor pulses because of their lesser consumption by people. These pulses are broadly categorized into cool-season legumes constituting chick pea, lentil, field pea, rajmash, and lathyrus while warm-season legumes include pigeon pea, green gram, black gram, cowpea, horsegram, etc. The pulse seeds have high nutritive values with a low glycemic index comprising of approximately 18–30% protein which varies among legume species. The pulses also supplement essential amino acids, resistant starches, complex oligosaccharides, fibers, minerals, vitamins, phenolics, tannins, phytic acids, antioxidants, and folic acids which have enormous health benefits [4]. Additionally, pulses improve soil health by contributing soil nitrogen through fixing atmospheric nitrogen by a symbiotic association of N_2 fixing *Rhizobium* in their root nodules and host a number

Figure 1.
Major pulses of cool season (A–F). Chick pea leaf with pods (A), chickpea seeds (B); lentil leaf with pods (C), lentil seeds (D); field pea leaf bearing pods (E), field pea seeds (F): major pulses of summer season (G–L). Green gram leaf with podding cluster (G), green gram seeds (H); black gram leaf with podding clusters (I), black gram seeds (J); pigeon pea leaf with pods (K), pigeon pea seeds (L).

of beneficial microbes in their rhizosphere such as phosphate solubilizing bacteria (PSB) that solubilize insoluble phosphates in the soil. They are hardy crops thriving well with their vigorous root system and have strong ability to proliferate root system into the deep soil layers. Keeping in view of the growing demand, additional lands are required for expanding the areas of cultivation of pulses. The productivity of pulses had been always low as they are threatened by great challenges of climatic aberrations such as drought, heat, salinity, frost, and cold. Among several abiotic stresses, drought and heat are considered to be the major yield limiting factor followed by unprecedented high temperature during the reproductive phase [5, 6]. High temperature stress affects various physiological processes and alters the plant-water relationship [7]. The rise in temperature beyond 35°C may cause irreversible damage to growth and development of plants [8]. It is therefore imperative to analyze critically the implications of climate change on pulses. There are several knowledge gaps in our understanding toward low productivity of pulses particularly when drought and heat are combinedly imposed on crops [9]. Efforts are needed to make pulses more climate-resilient and productive [10]. Our current understanding is not sufficient to counter the challenges imposed by climate adversities [11]. Therefore, systematic approaches are to be taken to hit the primary targets of achieving higher productivity with limited resources and sustain our ecosystem, soil health, protecting environment keeping in view of the fact that pulses play a central role in sustaining our agro-ecosystem under the climate change scenario.

2. Effects of major abiotic stresses on pulses

2.1 Drought

Yield reduction in pulses has been directly correlated with the intensity of drought [5]. Among rainfed pulses, *Vigna* species green gram and black gram being warm season crops require four irrigations till crop maturity. Sources of drought tolerance in *Vigna* crops are very rare, though green gram is relatively better adapted to warm climates. The highest yield reduction was observed during the reproductive stage of many pulses. However, yield reduction in pigeon pea under drought was relatively lower than that of lentils, groundnut, cowpea, and green gram. In extra-short-duration pigeon pea, the impact of drought stress is evident at the vegetative, flowering, and pod-filling stages [12]. The germination and seedling growth, seedling vigor, hypocotyl length is adversely affected by drought stress in pigeon pea. Tolerance to drought in short-duration pigeon pea has been ascribed to the crop's ability to maintain total dry matter, a small pod size, few seeds in the pod, high seed mass, and low flowering synchronization [13]. The important mechanisms of drought tolerance in pigeon pea were included with high root resistance to water flow; slow shoot development; limited initial root development at depth; partitioning of assimilates into vegetative parts; leaflet movement during water stress; dehydration tolerance; and osmotic adjustment [14]. The lethal leaf water potential, i.e. the lowest water potential experienced by the last viable leaf, was a key measure of dehydration tolerance. The pigeon pea has more dehydration tolerance than others. The ability of cells to continue metabolism at low leaf water status is termed dehydration tolerance [15]. Membrane disorder is often measured as a leakage of solutes from the cell [16]. Water status parameters like relative water content may be a good indicator of drought tolerance in pigeon pea under semi-arid conditions [17]. Accumulation of proline in cell in response to water deficit is another mechanism protecting protein structures as cell dehydrate, and as an organic nitrogen source. The relative drought tolerance in a range of pulses is based upon the physiological traits such as osmotic adjustment, root system, and lethal leaf water potential. Thus, the order of drought tolerance in different pulses could be Pigeon pea > chick pea > lentil > black gram > green gram.

2.2 High temperature

High temperature results in an overall reduction in plant growth including roots, leaf area, and dry weight [18]. It has been predicted that an average 1°C increase will reduce yield by at least 3–4% in many crops [19]. The average surface temperatures are expected to rise by 3–5°C by the end of this century, posing a major threat to the production of many food crops including legumes worldwide, especially in the semi-arid tropics [20]. Moreover, increase in temperature will have more adverse effects on cool-season pulse crops (e.g. chick pea, lentil, and field pea) than the summer and rainy-season crops like green gram, black gram, and pigeon pea [21]. A temperature increase of 1–2°C above the threshold level is sufficient to reduce yield in many leguminous crops such as cowpea [22], groundnut [23], common bean [24], lentil [25], and chick pea [26, 27]. Heat stress causes considerable reduction in biomass production and grain yield in several crops [28]. A rise in temperature may limit the development of various yield components [29]. The majority of the food legumes including chick pea, lentil, pigeon pea, green gram, and black gram are grown by developing countries of south Asia under rainfed situation. They are often heavily influenced by climate variables. Crops grown at lower latitudes are often exposed to high temperatures above 40°C. Grain yield reduction in heat stress has been reported to be associated with a decrease in photosynthetic capacity because of altered membrane stability [30, 31] and enhanced maintenance respiration [32] along with a reduction in radiation-use efficiency. Photosynthesis is the most sensitive physiological process impaired by heat stress [33] which could be due to structural and functional disruptions of chloroplasts, reduction of chlorophyll, inactivation of chloroplast enzymes [34], or both stomatal and nonstomatal limitations [8]. High temperatures adversely affect starch and sucrose synthesis through a reduction in the activity of sucrose phosphate synthase and ADP-glucose pyrophosphorylase [35]. Crops exposed to high temperature are often subjected to oxidative stress-producing reactive oxygen species (ROS), which are highly toxic to cellular functions in plants because they damage nucleic acids and cause protein oxidation and lipid peroxidation; this oxidative damage eventually causes cell death [36, 37]. ROS toxicity during various stresses is considered to be one of the major causes of low crop productivity worldwide [38]. An increase in the activity of antioxidant enzymes, such as guaiacol peroxidase (GPX) and catalase (CAT), plays a significant role in minimizing the toxic effects of stress-induced ROS production [39]. High temperatures adversely affect nitrogen fixation in chick pea [40]. Increased day temperatures ranging from 32 to 35°C have shown reduction in nodule formation and nitrogen fixation in chick pea (**Figure 2**). It has been reported that high temperature delays nodulation, retards nitrogen fixation, and impairs nodule function and structure in chick pea [41–43].

2.2.1 Effect of high temperature on reproductive and seed development in pulses

The reproductive phase of major pulses is highly sensitive to temperature extremities [44]. High temperature inhibits flower set and flower retention, impair the

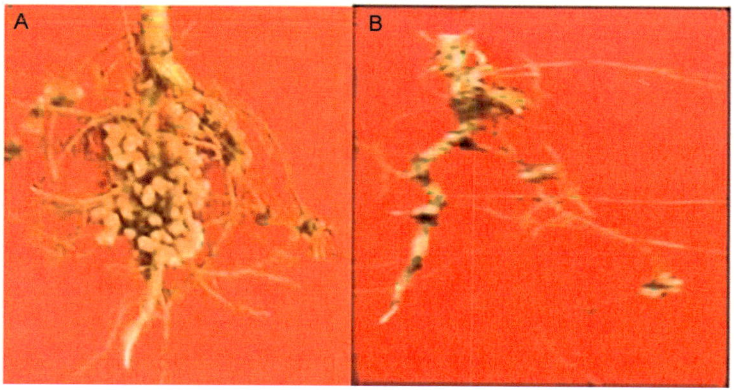

Figure 2.
Effect of high temperature on chick pea cv K 850 nodulation. (A) Normal sown chickpea nodule (25/15°C max/min), (B) late sown chick pea nodule (40/27°C max/min).

normal development of male and female gametophytes leading to ovule abortion, reduce pod settings, and impair grain filling which eventually results in significant yield loss. Several reports have indicated that heat stress caused reproductive failure due to impaired sucrose metabolism in the leaves, developing grains, and inhibition of sucrose transporters those results in decreased carbon-source to the anthers and developing pollen grains [45]. The heat stress results in drastic yield losses due to the decline of relative tissue water content (RWC) and leaf water potential (LWP) [46] pollen or ovule inactivity, flower abortion, and postfertilization impaired the growth and development of embryos or seeds in many pulses [47]. Ultrastructural studies revealed that pollen sterility could be due to degenerated tapetum owing to heat stress [48]. Temperature extremities have specifically detrimental effects on male gametophyte causing disrupted meiosis, tapetal hypertrophy, stunted development of pollen grains, anther protein degradation, pollen sterility, and pollen tube deformation [49] while heat stress adversely affects female gametophyte causing reduced size of style and ovary, disrupted meiosis, reduced stigma receptivity, callose deposits in style, damaged embryo sac components, and fertilization arrest. Flowering is either early or delayed, flowers abscised and become distorted and shedding occurs. Grain-filling process is impaired due to altered source-sink relations which lead to seed abortion and yield loss. However, the relative heat sensitivity varies for different crops [50]. High temperatures reduce yield and yield attributes such as dry matter accumulation and partitioning [51], pod set, pod weight, and harvest index in snap beans [52]. Temperatures above 40°C resulted in reduced pod set, seed production, and yield in soybean [53]. The water scarcity in floral parts and leaves due to high temperature driven increased transpiration causes heavy yield losses in snap bean (*P. vulgaris*) [54]. High temperature also adversely affects pollen germination and pollen tube growth. In cool-season legume lentils, pollen germination and pollen tube growth have been found to be retarded above 35°C [55]. Stressful temperatures often lead to impaired microsporogenesis and megasporogenesis at the pre-fertilization stage in various legumes such as chick pea (*C. arietinum*) [56]; and *P. vulgaris* [48, 57]. Loss of pollen viability and pollen germination have been reported at high temperatures in *C. arietinum* [58]; and *P. vulgaris* [57]; *Arachis hypogea* [59], loss of stigma receptivity, loss of ovule viability and flower abscission in *C. arietinum* [58]; *P. vulgaris*: [48]. Most of the pulse crops such as chick pea and lentil are sensitive to heat stress when day temperature exceeds above 35°C resulting in reduced pollen germination, and decreased pod number and seed size. Arrest of fertilization and reduced embryogenesis have been reported at extreme temperatures in chick pea [60] and *Glycine max* [61]. The pigeon pea often experiences high temperature during the reproductive phase which results in pollen sterility, retards germination and pollen load on stigma, reducing the pollen receptivity on stigma and length of pollen tubes [58]. Consequently, heat stress has detrimental effects on both microsporogenesis and megasporogenesis causing incomplete cell division and eventually loss of viability of pollens and ovules [45]. Failure of anthesis would be a likely cause of flower abscission and also discontinuity of sustained photoassimilate mobilization to the developing grains results in pod abortion [62]. Identifying mechanisms of reproductive temperature tolerance in various legumes could be achieved through screening diversified germplasm available in gene banks. Warm-season legumes such as green gram, pigeon pea, black gram, and cowpea, though having higher temperature tolerance limits, are also affected negatively leading to a reduction in pod set in response to moderately-high night temperatures [63]. Thus, various legumes are sensitive to temperature extremities to different degrees from the vegetative to reproductive stage, resulting in metabolic and reproductive dysfunction and finally result in low yields.

2.3 Effect of combined stresses of drought and heat

There is a need to identify the tolerance mechanism of pulses in response to simultaneously occurring heat and drought stresses [64]. It is considered that the demand of water will proportionately increase with rise in temperature due to increase in soil-plant evapotranspiration water loss. To improve pulses for dual tolerance, traits that confer the tolerance to both heat and drought must be studied. The photosynthetic electron transport rate has been drastically declined when drought and heat superimposed simultaneously in chick pea. A high irradiance level had been found to be more deleterious for photosynthesis in chick pea under combined drought and heat stresses; however, genotypic variation was evident in light response of photosynthesis when different chick pea genotypes were subjected to combined stress (**Figure 3**). Similarly, it is urgently needed to inherit combined tolerance in pulses also. When both high temperature and drought stresses are superimposed, their combined effects are more detrimental and negative. The combined effects could be deleterious additive effects on phenology, growth, chlorophyll content, photosynthesis, grain number, fertility, grain-filling duration, and grain yield in several crops [65]. For example, the combined drought and heat stresses decrease leaf chlorophyll content by 49% while drought or heat alone reduce it by 9% or 27%, respectively [66]. Reproductive stages are more sensitive to combined heat and drought stresses than individual ones [1]. However, in some ways, drought and heat stresses are interlinked. Both the stresses significantly reduce photosynthetic efficiency, stomatal conductance, leaf area, and water-use efficiency in many crops, e.g. wheat, rice, chick pea [67]. Heat stress increases evapotranspiration from soil and plant canopy that leads to a disruption of water relations and plants often are succumbed to drought stress driven by high temperatures [68]. Crops may respond to drought and heat stresses individually or combinedly in a different manner [69] which eventually lead to yield reduction. Pigeon pea, chick pea, lentil, and field pea are often exposed to drought and high temperatures simultaneously particularly during reproductive stages [6]. The basic physiological mechanisms of tolerance to drought and heat may differ or be common with respect to some traits. For example, reduction of photosynthesis is a common

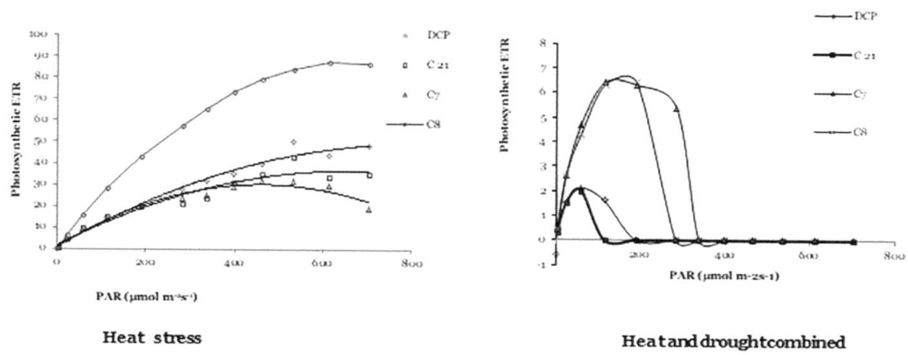

Figure 3.
Genotypic variation in the light response of different chick pea genotypes under combined stress of drought and heat.

phenomenon under drought and heat; however, photosynthetic inhibition due to drought is largely attributed by stomatal closure, reduced biomass production, and decreased leaf size. Drought limits gaseous exchange by triggering stomatal closure and in this way modifies plant-water relations whereas heat stress inhibits photosynthetic electron transport and PSII activity, conformational changes in thylakoid proteins, and distortion of membranes, and alters Rubisco enzyme and change in the solubility of CO_2 and O_2 inside mesophyll cells. Heat stress has detrimental effects on reproductive parts such as microsporogenesis and megasporogenesis of male and female gametophytes, loss of pollen viability and retards pollen germination and pollen tube growth, decreased stigma receptivity and pollen load on stigma, cell division, and accelerated rates of respiration and inhibition of carbohydrate metabolism in developing grains. However, there are common tolerance or avoidance mechanisms against both heat and drought stresses such as generation of harmful reactive oxygen species (ROS) and to counteract the deleterious effects of ROS, both the stresses are induced to produce antioxidative enzyme complex that enables scavenging of superoxide radicals being generated when plants are exposed to high temperatures in combination with drought and heat. Several heat shock proteins (HSPs) are expressed in response to high temperatures that helps to protect the membrane and vital molecules from denaturation and allows cellular machinery to operate normally. Both heat and drought stresses adversely affect water relation characteristics of plant which include decline in leaf RWC, water potential, osmotic, and turgor potential.

3. Response of major food legumes to abiotic stresses

3.1 Cool-season legumes

3.1.1 Chick pea

Chick pea (*C. arietinum* L.) is the second largest grown food legume of the world after beans. India, Australia, Pakistan, Turkey, Myanmar, Ethiopia, Iran, Mexico, Canada, and USA are the major chick pea producing countries. Chick pea is a good source of protein (20–22%) and rich in carbohydrates (~60%) [70], dietary fiber, and minerals [71]. The demand of chick pea is increasing keeping in view of its enormous health benefits which include prevention of cardiovascular diseases, type 2 diabetes, digestive diseases, and even cancer [72]. Chick pea fixes atmospheric nitrogen through symbiotic nitrogen fixer rhizobium resulting in lesser dependence on chemical fertilizers and residual nitrogen left in the soil after harvest benefits the subsequent crops. Drought and heat are the major abiotic stresses and constraints limiting chick pea production globally which together account for about 50% of the yield losses [73]. Chick pea is largely grown under rainfed on residual soil moisture after withdrawal of monsoon. The crop often experiences terminal drought if winter rain fails and the soil moisture starts receding and reaching below to a critical level at grain-filling stages causing heavy yield losses primarily due to water limitation. Exposure to heat stress (≥35°C) at flowering and podding in chick pea results in drastic reductions in seed yields [74]. High temperature adversely affects seed germination, photosynthesis, respiration, membrane stability, fertilization, fruit maturation, quality of seeds, nutrient absorption, protoplasmic movement, transport of materials and also modulated levels of hormones, and primary and secondary metabolites [8, 75]. It has been observed that lower grain yields with greater exposure to hot days (30–35°C), during the reproductive period [7]. Heat stress at reproductive stages is thus increasingly becoming a serious constraint to chick pea production due to climate change. The optimal temperatures for chick pea growth range between 15°C and 30°C [76].

3.1.2 Lentil

Lentil (*Lens culinaris* Medik) is another cool-season food legume grown widely and consumed for edible purposes and intensifying the cereal-based cropping systems. Lentil often experiences high temperatures (>35°C) during flowering and pod filling stages, which leads to forced maturity and consequently affects seed yield and quality [77]. The delayed sowing of lentil coincides with terminal heat stress. Consequently, a large portion of cultivated areas (~11.7 million ha) in India remains fallow after the late harvest of rice [78]. In Australia, ~70% yield losses in lentil were observed due to a 6-day heat wave with a maximum temperature of 35°C or above [77]. In recent years, heat stress as a result of global warming has become a major challenge to crop production and productivity in general [79]. Development of heat-tolerant lentil cultivars is required to sustain production and productivity of lentil for semi-arid regions. These issues could be addressed by distinguishing the heat-tolerant and sensitive lentil genotypes at critical temperature [80]. Thus, identification of key physiological traits that impart heat tolerance can help to facilitate a breeding program for developing heat-tolerant lentil cultivars, leading to a reduction the yield losses under a changing climate scenario [81]. High and low temperatures cause photo damage to PSII [82] which could be due to damage of proteins that are involved in a photo damage-repair cycle [83]. However, cyclic electron transport around PSII constitutes an effective protective mechanism against photo-inhibitory damage [83], and some phenolic compounds have been identified in this protection [84]. In lentil, pollen and leaf traits could also be helpful in identifying heat-tolerant genotypes [47].

3.2 Warm-season legumes

3.2.1 Pigeon pea

Pigeon pea (*C. cajan* L.) Millsp. is a major grain legume of the arid and semi-arid regions of the world [85]. Drought and high temperature during the reproductive stage are becoming a recurrent phenomenon in these regions resulting in significant yield loss in pigeon pea. Among pulses, pigeon pea is the hardiest crop which is cultivated in a wide range of climatic conditions from tropics to subtropics between 30°N and 30°S latitude. It is well adapted to semi-arid, arid, and marginalized regions receiving even less than 600 mm rainfall. This crop even tolerates as high as a temperature of 35°C if soil moisture is optimally available. Among pulses, the pigeon pea is an inherently drought and heat-tolerant crop to some extent [86]. High temperature leads to excessive water loss from crop canopy and soil through increased evapotranspiration. The decrease in soil moisture below a certain threshold level and the rise of temperature exceeding 35°C or more during the grain-filling stage often lead to poor yield in pigeon pea [87]. The high temperature causes oxidative damage, affects cell division, and may cause severe damage to the membranes and proteins and their synthesis, along with inactivation of major enzymes [88]. Even exposure to the high temperature for a shorter period during the seed filling can result in accelerated filling and eventually it results in incomplete grain development, poor quality, and reduction in the yield. High temperature beyond 35°C often leads to flower shedding, pod abortion, and incomplete grain development, inhibits photosynthesis [89] with increased respiration which cumulatively imbalance the source-sink relation [6]. Drought tolerance in pigeon pea is due to its deep-rooting tap root reaching up to 6 feet (2 m) in depth that helps to improve water infiltration into the deep soil [90]. Inherent drought tolerance in pigeon pea could be associated with the prolific root system with a higher number of thin lateral roots, higher hydraulic resistance to restrict flow of water or tends to conserve available water more efficiently, smaller but high stomatal density that regulates transpiration water loss to minimum level but maintain photosynthesis with lower order but not completely inhibited. Among all legumes, drought tolerance characters of pigeon pea also involve high osmotic adjustment and very low (more negative) lethal leaf water potential indicating a higher degree of dehydration postponement and dehydration tolerance. Genes expressing heat shock proteins (HSPs), dehydration responsive element-binding DREB, and cyclophilin have been found to be responsive to combined stresses of drought and heat in pigeon pea [91, 92]. Signaling effects of abscisic acid and reactive oxygen species (ROS), calcium, calcium-regulated proteins have been well characterized for their role in signal transmission under stress [93].

3.2.2 Black gram or urdbean

Black gram (*V. mungo* L. Hepper) is a popular food legume grown in many Asian countries including India, Pakistan, Myanmar, Bangladesh, Thailand, and China. India is the largest producer and consumer of black gram. It is a warm-season food legume, which requires 25–35°C temperature along with a high humidity for its normal growth and development. However, prevailing high temperature (>40°C) during flowering results in deformation of flower parts or flower drop leading to negative impact on yield. Nutritionally, urdbean is dense with protein (21–28%), dietary fiber (161–187 g/kg), iron (16–255 mg/kg), zinc (5–134 mg/kg), and other micronutrients like other pulses [94]. Urdbean is grown in different ecological conditions and seasons across the growing regions. In India, it is grown mainly in the rainy season (July-October) and in the southern part it is also cultivated as a winter season crop (November-February). However, its cultivation is not wide in the summer season due to excessive heat stress and a lack of humidity in the atmosphere. Thus, availability of heat-tolerant cultivars can bring more areas under urdbean cultivation. Urdbean is a close relative of mungbean, which is extensively cultivated in identical ecological conditions. In this crop as well as in another *Vigna* pulse crop, e.g. cowpea, sources of heat tolerance have already been identified [49]. Knowledge of genetics underlying key traits imparting heat tolerance helps the breeder to make genetic improvements more precisely. In recent years, molecular markers helped to decipher the genetics of complex key morpho-physiological traits imparting heat tolerance in several crops [95].

3.2.3 Green gram or mungbean

Among all food legumes, green gram (*V. radiata* L. Wilczek), also known as mungbean, contains high amount of easily digestible seed proteins ranging between 24 and 28% which is higher than chick pea, lentil, and pigeon pea. The crop is grown during summer or rainy season and thrives most effectively at temperatures between 30°C and 40°C; however, significant flower shedding occurs at temperatures beyond 40°C [47]. It has been reported that the abscission of reproductive organs is the primary determinant of yield under heat stress in mungbean [24]. This is a short duration crop with a yield potential of about 1200–1500 kg/ha. During the reproductive stage, high temperatures cause flower drop, induce male sterility, impair anthesis, and shorten the grain-filling period. The productivity and adaptability of mungbean are seriously affected by a range of abiotic stresses including heat and drought. In mungbean, high temperature increases flower shedding [96] pollen sterility and dehiscence of anthers [22]. Being a summer crop, it is often exposed to temperature exceeding 40°C resulting in high turnover of sterile pollens, infertility, and flower abortion, and grain filling is adversely affected. High night temperature also affects grain filling and carbon partitioning. Further increase in the day maximum temperature to 44°C or above causes production of smaller and hard seeds. The pollen viability and germination were extremely sensitive to high temperature (>40°C) in mungbean, though a wide genotypic variation in the pollen germinability was observed. Seed size reduces; however, a majority of genotypes had reduced, shriveled, or deformed grains at high temperatures exceeding 40°C. The critical temperature range for damage of reproductive organs was found somewhere in between 40 and 45°C; however, sensitivity varied among genotypes. Earlier reports suggest that brief exposure of plants to high temperatures during seed filling accelerates senescence, diminishes seed set and seed weight, and reduces yield [97]. In mungbean, remobilization of pre-anthesis reserve carbohydrates and nitrogen in leaves, podwall, and stems contribute significantly towards grain filling. In the changing scenario of climate, sudden rise in the temperature beyond 35°C causes increase in the respiration rates and unusually high degradation of stored starch as major chloroplasts carbon source was observed. As a result, failure to set pods, reduced or incomplete grain development at high temperature could be partly due to inadequate supply of carbon and nitrogen from leaves or by decrease in the activity of sucrose synthase, the key enzyme playing a crucial role in grain development. Poor partitioning of carbon and nitrogen at high temperature leads to low harvest index and low productivity in mungbean. The productivity and adaptability of green gram are adversely affected by several abiotic stresses including heat, drought, salinity, and water-logging, which affect crop growth and development by altering physiological processes and the plant-water relationship [98]. Several studies have reported a reduction in the growth and development of legumes because of high-temperature stress [99].

4. Mechanism of drought and heat tolerance in pulses

Knowledge of key traits imparting heat and drought tolerance can help to improve the grain yield of food legumes [81]. Therefore, understanding of physio-biochemical mechanisms associated with these key traits imparting tolerance is essential for large-scale phenotyping of pulse germplasm under both field and controlled conditions [100]. In several crops, various physiological and biochemical traits such as accumulation of phenolic compounds, organic acids, photosynthetic activity, water-use efficiency, canopy temperature, rooting length, osmotic adjustment, membrane stability, and pollen viability [47, 101] have been used to identify heat and drought tolerant genotypes and a significant genetic variability has been reported for key physiological traits under stress conditions [102].

Under rainfed or water-limiting regions, pulses experience drought, heat stress, or both at the terminal end that is at the reproductive stage. There is enough scope to improve grain yield in pulses when drought or heat appears particularly during reproductive or grain-filling stages. This type of terminal stresses is better defined as there was no moisture limitation at initial growth stages or temperature remains conducive to support normal physiological activities. The crop productivity is largely determined by initial crop vigor, biomass, number of podding nodes, water retention capacity in the tissue through regulated ways of gaseous exchange making a balance between photosynthetic carbon gain and minimum water loss through transpiration. However, if drought or heat during germination or intermittent drought accompanied by unusual temperature prevails at initial growth stages, it could be more detrimental for the crop to cope up the situation even most tolerant species introduced under such conditions. Therefore, drought or heat tolerance in pulses is practically limited to the conditions when they appear at the later or terminal end of the crop. There are two types of mechanisms that are operating in pulses toward improving yield in pulses under terminal drought or heat. One of the important requirements of the plant is to avoid terminal stresses or to develop a mechanism to postpone dehydration with depletion of soil moisture. Another strategy is to develop cellular resistances to counter the stresses such as stable membrane system, higher water-use efficiency, efficient photosynthetic system, expression of heat shock proteins, etc. Different models have been proposed for drought resistance of pulses which includes traits associated with productivity such as dry matter, harvest index, and water-use efficiency and other drought resistance mechanisms comprising morpho-physiological traits conferring resistance to drought [103, 104].

4.1 Matching phenology

Matching the phenology to the water supply is the primary way in which chick pea crop yield can be improved in water-limited environments [105]. This involves the identification of plant with an early flower and pod initiation, rapid but short growth period.

4.2 Early biomass accumulation

Early extensive biomass accumulation and setting of reproductive organs before the onset of terminal drought are largely considered as drought escape mechanisms. In environments in which terminal drought is likely, selection for shorter time to

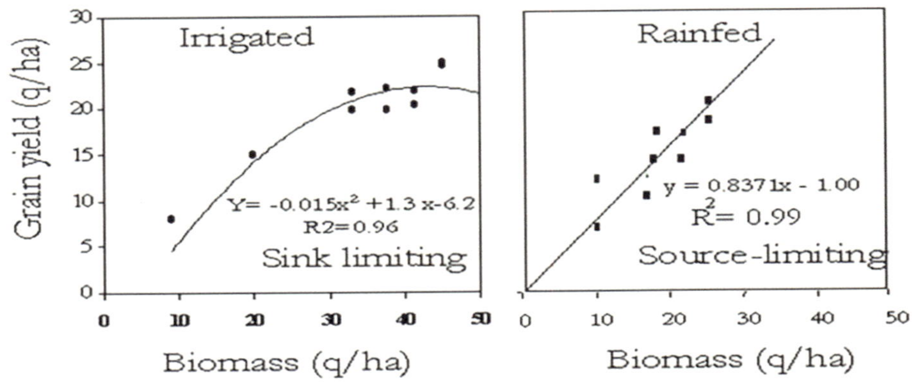

Figure 4.
Biomass-yield relationship of chick pea under rainfed (stress) and irrigated (non-stress) conditions.

flowering has been highly successful [106] . Vegetative biomass in most of the pulses has been found to be the most sensitive to water stress. The leaf expansion is affected even under mild water stress with a leaf water potential declined to ≤1.2 MPa. The biomass showed a linear relationship with yield under rainfed indicating sources (leaves & stems) are the major limiting factor for low yield. Contrary to this, increasing biomass beyond a certain limit did not further increase in the grain yield in irrigated fields that suggests the sink limiting condition of irrigated crops (no stress) **Figure 4**.

4.3 Importance of plant types to improve yield under stress

Higher biomass does not always lead to contribute higher yield in pulses. Two distinct plant types in chickpea had shown significant yield difference in chick pea. The plant with multiple shoots having uniform height emerged at the seedling stage leads to profuse primary and secondary branches with a higher number of podding nodes (**Figure 5**). On the contrary a different plant type with a single erect branch,

Figure 5.
Two distinct plant types in chick pea. High yielding plant type (left) with multiple primary branches and low yielding plant type with single stem (right).

with suppressed lateral branches at the base, has been observed to be a poor yielder due to lack of podding branches although these types of plant had shown vigorous root system and are better adapted to drought conditions (**Figure 5**). Therefore, it is essential to tag both attributes together that plant types with profuse branching and extensive root system.

4.4 High biomass with profuse branching leads to higher yield

Terminal drought markedly reduces leaf photosynthesis during grain filing in chick pea. Prestored assimilate reserve in the stems and leaves during the vegetative stage plays a major role in contributing carbon (C) and nitrogen (N) to the developing grains. The storing capacity of photosynthates (C and N reserve) is thus directly proportional to the biomass which includes stem and leaf reserves. The carbon and nitrogen fixed during vegetative growth is a potential source of assimilates for the developing seeds as photosynthetic gain during reproductive stages is substantially reduced due to stomatal closure or other nonstomatal factors as a result of progressive increase in the water stress. Thus, grain yield in chick pea and pigeon pea under terminal drought is largely influenced by remobilization of prestored carbon and nitrogen in the stems and leaves. Depending upon the cultivars, about 10–20% carbon and more than 60% nitrogen (N) in the seeds of water-stressed plants have been reported to be remobilized from stems and leaves of chick pea [107]. The soluble sugars including transportable sucrose and amino acids in leaves of water-stressed plants markedly increased as a result of starch and protein breakdown. The significant increase in the activation state of sucrose-phosphate synthase in leaves of water-stressed plants had been shown to faster conversion of sucrose in the leaves [108]. The increase in the low-molecular weight compounds like sugars, amino acids, organic acids, and sugar alcohols as a result of degradation of macromolecules, e g. starch and proteins in water-stressed plants, also plays a significant role in enhancing low-molecular weight osmotic solutes which could likely to help in maintaining transient water balance during stress.

4.5 Role of pod wall to contribute photoassimilate under water stress

Increasing photosynthetic ability as a whole is essential for acquiring higher crop yields. Non-leaf green organs (NLGOs) make important contributions to photosynthate formation, especially under stress conditions [109]. The pod wall of pea (*P. sativum* L.) was shown to contain two distinct photosynthetic layers. The outer, comprising chlorenchyma of the mesocarp, captured CO_2 from the outside atmosphere; the inner, a chloroplast-containing epidermis lining the pod gas cavity, was involved in photoassimilation of the CO_2 released from respiring seeds [110]. The respiratory CO_2 released from the embryo of developing grains of legume pods is refixed by a layer of cells on the inner pod wall [111]. This refixation of respiratory CO_2 plays a significant role to gain carbon by developing seeds under drought when leaf photosynthesis is drastically declined. It has been reported that the excised embryo, seed coat, and pod wall in chick pea are all photosynthetically competent, but the pod wall alone is capable of net O_2 evolution over and above respiration. More than 80% of CO_2 is fixed by this tissue when provided to the pod interior. Moreover, seed coat cell turgor in chick pea is independent of changes in plant and pod water potential [112] which indicates that seed coat water status is less influenced under drought though photosynthesis of pod wall and leaves may be affected markedly but seed coat photosynthesis maintains normal photosynthesis and significantly contributes photoassimilates to the growing seeds.

4.6 Remobilization of carbon and nitrogen from leaves and stems

The onset of pod filling of chick pea occurs when net photosynthesis and nitrogen fixation are low as a result of the onset of terminal drought. Remobilization of carbon (C) and nitrogen (N) from vegetative parts to developing seeds may be an important alternative source of C and N for seed filling. Genotypic variation in remobilization ability has been observed in desi and kabuli cultivars. The better drought tolerance in desi genotypes could be a consequence of better remobilization and higher pod number. Thus, C and N assimilated prior to podding can supplement the supply of current assimilates to the filling seed in both well-watered and water-stressed chick pea. Remobilization of pre-podding N is an essential source of N for seed filling irrespective of environmental stress. Terminal drought decreases the rate of net photosynthesis of leaves and N fixation during seed filling [41, 113]. A high demand for assimilate from filling seeds when the supply of current assimilate is decreasing often results in an assimilate shortfall [114]. Consequently, alternative sources of assimilate are required to maintain seed filling and seed size. In a number of species, the remobilization of pre-anthesis stored reserves of C and N can be an important assimilate source for seed filling, particularly when plants are subjected to water deficit [115–117]. In cereals, remobilization has been reported to contribute between 3% and 64% of seed dry matter (DM). The contribution of remobilized DM in pulses ranges from 2 to 42% [118, 119]. In chick pea, a decrease of DM in the stems, leaves, and pod walls of field-grown plants suggests that DM remobilization is important as an alternative assimilate source to current photosynthate [120]. Using this method, it was estimated that one-third of the pod DM was derived from the remobilization of DM from vegetative tissue in field-grown chick pea [120]. Chick pea has an indeterminate growth habit with flowers, pods, and seeds developing sequentially along branches [120]. Under terminal drought, pods and seeds that form later in the upper parts of the canopy are likely to be subject to a more severe water deficit during seed filling than pods formed early in the lower parts of the canopy. Consequently, the contribution of remobilized C and N is expected to be higher in seeds filling later when competition for assimilates is high and current C assimilation is very low. There have been no studies examining how remobilized assimilates are distributed among seeds of different ages and this requires investigation. Similar to other indeterminate species, chick pea responds to water deficit by reducing stem elongation, the rate of leaf emergence and expansion, and the establishment of new reproductive nodes [121, 122]. In a previous study by Davies et al. [107], terminal drought led to decreased rates of flower and pod production with a significant decrease in pod and seed number and seed yield in both desi and kabuli chick pea. However, desi chick pea tolerated drought better than kabuli chick pea and was better able to maintain yield relative to the irrigated plants. As there were no substantial genotypic differences in the decrease in photosynthetic rate with water deficit, this difference in drought tolerance may be accounted for by variation in remobilization of assimilates between desi and kabuli chick pea types.

4.7 Root characteristics

A wide genetic variability among root characters such as density and depth, total root biomass has been reported in chick pea genotypes (**Figure 6**). Chick pea lines

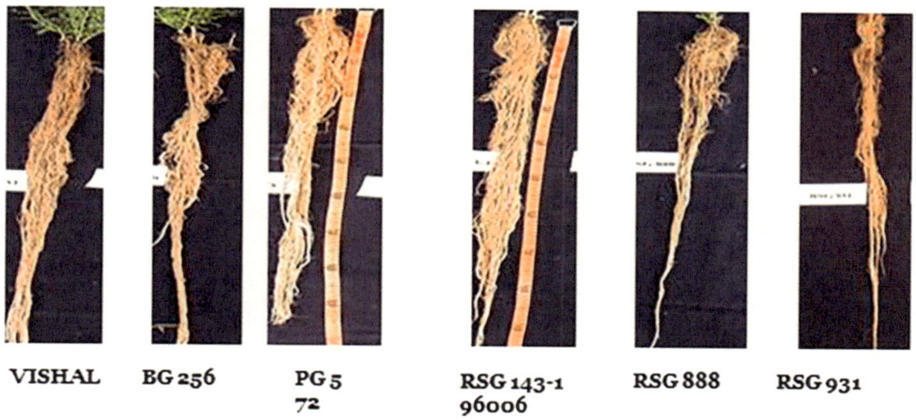

Figure 6.
Genotypic variability in root biomass in chick pea.

with increased drought tolerance through increased root biomass have identified by several workers [123]. A drought resistant genotype ICC 4958 had 30% higher root dry weight than the standard control cultivar "Annigeri" which is relatively more sensitive to drought stress [124].

Although vigorous root system plays a major role toward drought adaptation, however root architecture differs across the locations depending upon the drought situations. In a case study, majority of the chick pea grown under cool environment, root system tends to proliferate just 30 cm below the surface of ground allowing to trap surface moisture received by dew precipitation during winter as well as glandular leaf trichomes use to trap dew water quite heavily. As a result, root length density (RLD) at different depths of the soil profile varies in chick pea depending upon the environmental conditions. In chick pea cultivars adapted to warmer environment roots usually go into the deep soil for mining water efficiently due to lack of surface soil moisture as a result of high temperature-dependent faster evaporative water loss coupled with a very negligible amount of dew precipitation. Therefore, root architecture is very important to consider to make a breeding strategy in developing drought tolerant cultivars. Near infrared reflectance (NIR) pattern in an RIL (recombinant inbred lines) population derived from ICC 4958 (high root) × Annigeri (low root) cross revealed a significant NIR reflectance depicting different amount of water content in leaves (**Figure 7**).

4.8 Water-use efficiency, canopy temperature, and transpiration under stress

The specific leaf area is negatively associated with SPAD Chlorophyll meter reading (SCMR) and water-use efficiency measured by Carbon isotope discrimination ($^{13}C/^{12}C$ ratio). Therefore, SLA represents an indirect way of measuring water-use efficiency of plants, the low SLA indicated higher WUE and SCMR which could be useful to identify efficient genotypes with high water-use or transpiration-use efficiency (**Figure 8**) which is defined as more photosynthetic gain per unit loss of water through transpiration, or transpiration efficiency is a ratio between biomass and transpiration. The transpiration-use efficiency could be one of best strategies toward improved adaptation under stress environment. Improving water-use efficiency is associated with stomatal density, efficiency of RuBisco, and other physiological

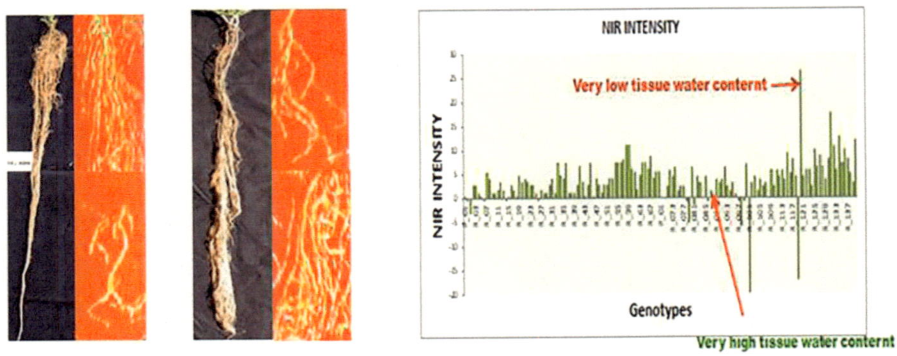

Figure 7.
(A) Chick pea genotype with high root length density (RLD) at the top soil surface layer adapted to cooler environment. (B) Chick pea genotype with high root length density (RLD) at the bottom or deep soil layer, adapted to warmer environment. (C) Different NIR (near infrared reflectance) values shown by RIL population having different levels of tissue water content which is likely to be associated with different amounts and depths of root biomass.

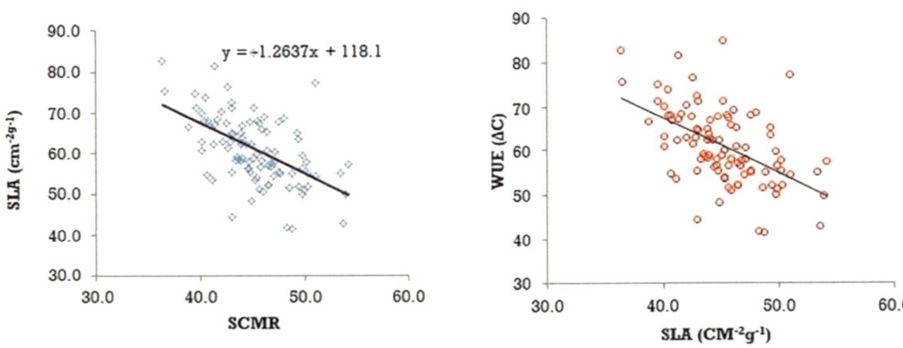

Figure 8.
Associations of specific leaf area (SLA) and water-use efficiency, $^{13}C/^{12}C$ ratio (WUE).

parameters of mesophyll cells of leaves. Significant genotypic variation has been observed in chick pea and pigeon pea in water-use efficiency [125]. A decrease in transpiration rate due to drought may result in an increase of leaf internal temperature due to decreased evaporative cooling of leaf and therefore drought stressed plants show higher leaf or canopy temperatures than well-watered plants [126]. Cool canopies are associated with better yield output as a result of higher transpiration and photosynthesis [127]. To maintain canopy cool under drought and heat, plants should have attributed with deeper root system to access water from the deep soil layers. Key physiological traits like osmotic adjustments and root architecture need to be modified to make canopy cool under heat and drought stresses. Osmolyte accumulation also alters photoassimilate allocation between roots and shoots. Drought induced osmolyte accumulation tends to allocate more photosynthates toward roots; as a result, plants are able to adapt under drought by extracting more water from soil.

Figure 9.
Plant microclimate under drought: A. Pigeon pea dwarf/closed canopy; B. Thermal image of A; C. Pigeon pea open canopy and D. Thermal image of C; E. Semi-erect dense chick pea canopy; F. Thermal image of E; G. Spreading open chick pea canopy; H. Thermal image of G. Canopy temperature can be assessed by different colors as shown by the thermal scale given at the right side.

4.9 Plant architecture for conserving moisture

Conserving moisture is one of the strategies of plants to withstand drought. The plant architecture plays an important role in configure plant microclimate for conserving moisture and to make the canopy cool. Evaluation of contrasting pigeon pea genotypes by a thermal imaging technique showed that dwarf and closed canopy of pigeon pea (**Figure 9A** and **B**) remained cooler as compared to tall and open canopy (**Figure 9C** and **D**). Similarly dense and semi-spreading chick pea (**Figure 9E** and **F**) may have the advantage to keep the canopy cooler than spreading and open chick pea plant type (**Figure 9G** and **H**). However, plant types with deeper root system maintain the canopy cool by means of extracting soil moisture efficiently from deeper soil layers which in turn allowing plant to transpire under drought conditions. The transpiration cooling under limiting soil moisture prevents stomatal closure to some extent and simultaneously restores photosynthesis to sustain dry matter production.

5. Dehydration postponement

The second mechanism is the postponement of drought by reducing water loss primarily by stomatal control of transpiration, by increasing water use through deep and vigorous root system, or by the accumulation of solutes to maintain turgor as water shortages develop (osmotic adjustment).

5.1 Stomatal mechanism

The open stomata result in transpirational cooling of leaf hence canopy of a particular genotype having higher stomatal conductance should show lower temperature than those having close stomata. Techniques are now available to measure canopy temperature by infrared thermometry; hence, screening of pulse germplasm with varying stomatal responses to drought is possible. The reduction in the stomatal density and an increase in the leaf reflectance through production of glandular hairs in chick pea are considered to be adaptive traits through which leaves reduce water loss and intercepted non-photosynthetic radiation.

5.2 Osmotic adjustment

Osmotic adjustment (OA) is the process of active accumulation of solutes in leaves in response to increasing water deficits in the soil thereby maintaining turgor. Osmotic adjustment has been shown to maintain stomatal conductance and photosynthesis at low leaf water potential in many crops [128], delay leaf senescence, reduce flower abortion, and improve root growth and water extraction from the soil [129]. Intra and interspecies difference in osmotic adjustment and its range in grain legumes are given in **Table 1**. From the table it is clear that among pulses chick pea, pigeon pea, and peanut are tolerant to drought as compared to others. The genotypic variation in OA ranging from 0.45 to 1.25 MPa was significant at LWP (leaf water potential) below −2.0 MPa in chick pea and maximum was noticed in cultivar Tyson. The significant genetic variation in OA from 0 to 1.3 MPa has been reported earlier in chick pea at very low LWP. Comparative evaluation of two cultivars such as Tyson with high OA (1.25 MPa) and Kaniva with low OA (0.45) revealed that relative water content (RWC) varied periodically with progressive increase in water stress as crop growth advances toward maturity while RWC declined linearly in Kaniva with less OA (**Figure 10**). The expression of high OA in chick pea cultivars was usually observed in selective chick pea having inherent ability to osmotically adjust when crop was subjected to severe stress with decline in leaf water potential to −3.0 MPa and above, and this characteristic enables the cultivar

Species (pulses)	Range osmotic adjustment (M Pa) in leaves	Degree of dehydration postponement	Species (cereals/ vegetables)	Range in osmotic adjustment (M Pa)	Degree of dehydration postponement
Groundnut	0.2–1.6	Very high	Sorghum	0.8–1.7	Very high
Pigeon pea	0.1–1.3	High	Wheat	0.2–1.5	High
Soybean	0.3–1.0	High	Barley	0.2–0.5	Moderate
Chickpea	0.0–1.3	High	Maize	0.1–0.4	Moderate
Lentil	0.0–0.6	Moderate	Potato	0.0–0.25	Low/sensitive
Green gram	0.3–0.4	Moderate			
Black gram	0.1–0.5	Moderate			
Cowpea	0.0–0.4	Moderate			
Lupin	0.1–0.5	Moderate			
Field pea	0.0–0.4	Moderate			
Faba bean	0.0–0.2	Low/sensitive			
Lathyrus	0.0–0.1	Low/sensitive			

Table 1.
Range of osmotic adjustment in grain legumes.

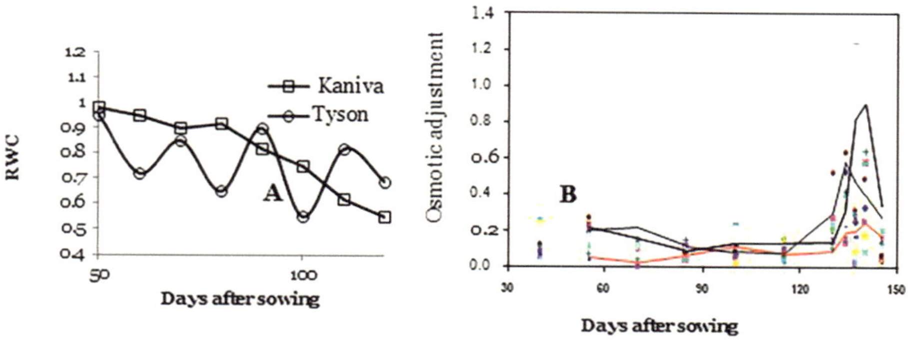

Figure 10.
Linear decrease of RWC in chick pea cv Kaniva having low OA while periodical changes of RWC in chick pea cv Tyson having high OA were observed with progressive increase in drought as crop growth advances. (B) High expression of OA in different chick pea cultivars occurred when the drought level was intensified at the terminal end of crop growth and OA declined after reaching a threshold level.

to survive even under severe moisture limiting conditions. The mild water stress helps in conversion of starch into reducing sugars followed by increase in sucrose phosphate synthase (SPS) activity in leaves which led to transient accumulation of sucrose contributing osmotic solutes. The non-transportable carbohydrate, starch, is converted to transportable form of sugar such as sucrose in the process of osmotic adjustment, thereby facilitates remobilization of prestored photosynthates from leaf to various sinks. When sucrose accumulated a threshold level, it increases the water flux into the leaf cells to regain the turgidity and improve LWP or RWC. The leaf carbohydrates or sucrose and others as osmotic solutes started remobilizing as a consequence of regaining the leaf turgidity. Subsequently partial recovery of RWC (stress relief) led to decrease in the SPS activation state, partial restoration of photosynthesis, and reactivation of starch synthesis. Thus, it is evident that osmotic adjustment prevents lowering of RWC below a critical level by efficient water uptake, restores photosynthesis, and maintains positive carbon balance for longer stress periods. The increase in cellular osmolarity allows influx of water into cells and maintains the necessary turgor for cell expansion, maintains membrane integrity and prevents protein denaturation under adverse environmental conditions such as drought, high or low temperatures [130], and confers protection against oxidative damage [131]. The accumulation of mannitol in chloroplasts leads to increase resistance to oxidative stress in tobacco [132]. The accumulation of osmolyte in response to drought has a critical role in the mitigation or avoiding/delaying the adverse effects of stress. The accumulation of osmoregulatory solutes has been considered as a unique biochemical trait which could be manipulated by breeding or transformation technologies.

5.3 Dehydration tolerance

5.3.1 Lethal leaf water potential

Sinclair and Ludlow [133] considered that the lethal leaf water potential, i.e. the lowest water potential experienced by the last viable leaf, was a key measure of dehydration tolerance. The lethal leaf water potentials in a range of grain legumes have been worked out, which shows that soybean has more dehydration tolerance than others.

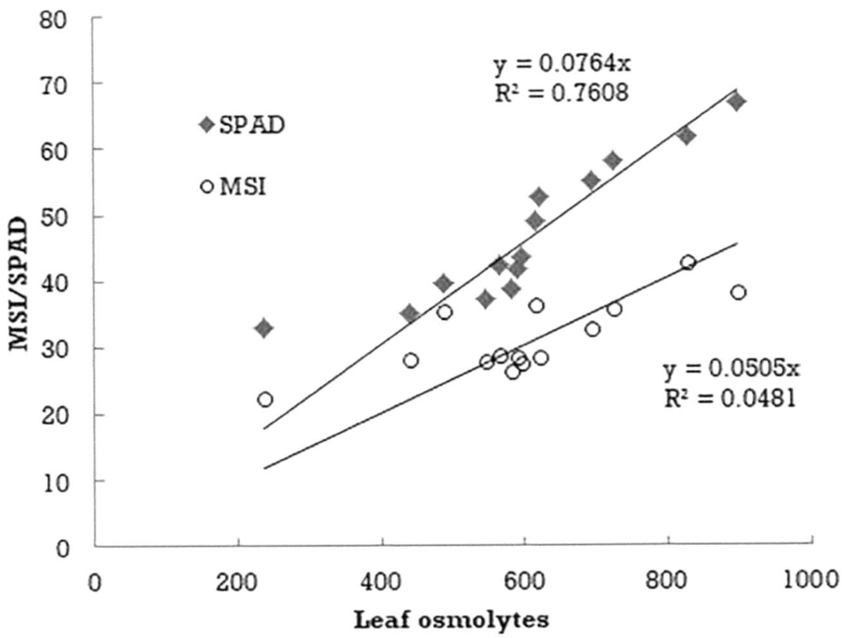

Figure 11.
Significance of osmotic solutes in imparting drought tolerance by improving membrane stability (MSI) and chlorophyll content (SPAD).

5.3.2 Membrane stability

The drought tolerant mechanisms involve stability of the membrane, which could be assessed through electrolyte leakage from desiccated tissue (**Figure 11**). The ability of cells to continue metabolism at low leaf water status is termed dehydration tolerance. Membrane disorder is often measured as leakage of solutes from the cell [134]. Electrolyte leakage measured by a conductivity meter has been used as a screening technique for heat and dehydration tolerance.

5.3.3 Proline accumulation

Accumulation of proline in cell in response to water deficit is another mechanism protecting protein structures as cell dehydrate, and as an organic nitrogen source. Neither membrane injury index nor proline accumulation is being used as a screening technique against dehydration tolerance.

6. Pollen and ovule fertility at high temperature

Cool-season pulses such as chick pea, lentil, field pea, French beans, etc. are often exposed to very high day temperature during reproductive stages exceeding 40°C. The upper limit of cool-season legumes has been reported to be within the range of 30–35°C. The heat stress in cool-season legumes has been considered as major abiotic stress leading to substantial yield loss, reduction in seed size, flower abortion, leaf senescence, forced maturity, failure of pod setting, retarded pollen germination,

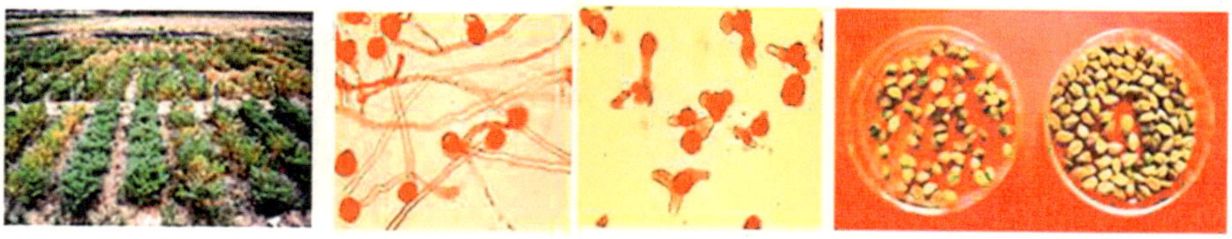

Figure 12.
(A) Forced maturity of chick pea at high temperature with distinct genotypic variation; (B) normal pollen tube growth in chick pea at 25°; (C) reduced and abnormal pollen tube emergence in chick pea at high temperature 41°C; and (D) reduction in the seed size in chick pea when day temperature exceeding >40°C.

and ovule deformation (**Figure 12**) [135]. Though significant genotypic variation has been reported in all these reproductive traits. While warm-season pulses such as green gram, black gram, cowpea, and pigeon pea are less affected as their upper temperature limit of tolerance is relatively higher ranging between 35 and 45°C. Many legumes exhibit a high sensitivity to heat stress during flowering. One of the major yield determinants in food legumes is pollen fertility and flower shedding at high temperature. The pollen sap may be altered and becomes more viscous by progressive increase in the temperature beyond 37°C. The transformation of pollen sap into a dense and viscous fluid probably hinders the smooth movement of male gametes. A reduction in the pollen tube length was observed in the heat-tolerant chick pea ICCV 92944 and green gram genotype EC 398889 at 40°C and beyond. In contrast, multiple pollen abnormalities including emergence of multiple tubes, single or coiled forms, bursted pollen tubes were observed in heat-sensitive green gram LGG 460 at high temperatures. Earlier reports on rice have also indicated that an increase in temperature could limit yield by affecting pollen germination and grain formation [136]. The male gametophyte is particularly sensitive to high temperatures at all stages of development, while the pistil and the female gametophyte are considered to be more tolerant [137]. The pollens are most sensitive to high temperature; the crop yield is affected when temperature rises during pollen development [138]. High temperature decreases pollen viability and leads to sterile pollens and decrease of pod set and yield [139]. In legumes, heat stress during post-anthesis results in poor pollen germination on the stigma and reduced pollen tube growth in the style [140]. Under high temperatures (>30°C), flower sterility has been correlated with diminished anther dehiscence, poor shedding of pollens, and germination of pollens on stigma [6]. The reduction in photosynthesis under high temperature may also restrict supply of photoassimilates such as sucrose, hexoses, and starch in the developing pollens resulting in decrease in pollen fertility [49]. The role of sugars and invertase/sucrose synthase activity in anther development and pollen germination has been reported in several crops [141]. Some of the thermotolerant food legumes flowered and set pods at high temperature. Therefore, assessment of pollen viability and pollen germination at high temperature proved to be potential screening tools for heat tolerance.

6.1 Acquired thermotolerance

Acquired thermotolerance is a mechanism naturally occurring in plants and has been extensively used in thermotolerant line identification [142]. The cell viability after heat shock was tested using 2%Triphenyl tetrazolium chloride (TTC). Cells were considered viable if tissue turns purple color when treated with TTC; on the

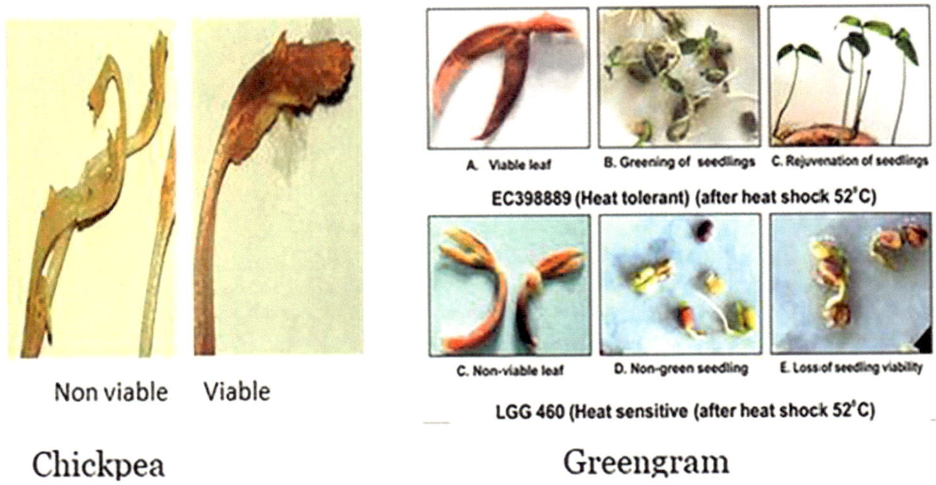

Figure 13.
Non-viable (TTC negative) and viable seedlings (TTC positive) of two contrasting chick pea genotypes after heat shock at 52°C. Acquired thermotolerance (ATT) in green gram in HT line EC 398889 (A–C) top and lack of ATT in HS green gram LGG 460 (C–E).

other hand faint purple color or if tissue does not take stain, it is considered as dead. Thus, TTC (triphenyl tetrazolium chloride) test for tissue viability and chlorophyll accumulation after heat shock appeared to have some promise to identify thermotolerant genotypes having ability to acquire thermotolerance. Heat-sensitive green gram genotype LGG 460 lost cell viability after heat shock and thus identified as TTC negative. Thus, higher membrane thermostability and cell viability after heat stress could be monitored by the TTC test and the technique has been widely used for assessment of heat tolerance [143]. The TTC reduction assay measures the level of mitochondrial respiration activity, which serves as an indicator of cell viability [144]. Variability was detected among the 56 green gram genotypes for acquired thermotolerance ranging from 14.1% to 61.3% while it ranges from 20 to 40% in chick pea **Figure 13**.

6.2 Membrane stability (MSI)

Under stress conditions, a sustained function of cellular membranes is considered crucial for maintaining cellular processes such as photosynthesis and respiration [145]. The integrity and function of cell membranes are sensitive to high temperatures, as heat stress alters structures of membrane proteins leading to increased permeability of membranes resulting in increased loss of ions or solutes. The increased solute leakage is closely associated with cell membrane thermostability [146], and various attempts have been made to use this method as an indirect measure of heat tolerance in diverse plant species such as food legumes [147], soybean [81], potato, cotton, and tomato [148], and wheat [149]. Study conducted using green gram germplasm for assessing membrane stability index (MSI) and chlorophyll content or greenness index showed that both MSI and chlorophyll remained higher in heat-tolerant green gram line EC 398889 as compared to sensitive line LGG 460 when plants of these two contrasting genotypes grown under high thermal regimes 42/28°C maximum/minimum temperature [49].

6.3 Photosynthesis and chlorophyll fluorescence

Photosynthesis is the most thermosensitive plant function [150] which can occur optimally at wide temperature ranges between 15°C and 35°C, although adversely affected at temperatures ranging 35–40°C and above. The cultivars can be distinguished based on the photosynthetic performance under high temperatures. Photosynthesis at high temperatures can be assessed through gaseous exchange or the chlorophyll fluorescence imaging technique. The fluorescence imaging technique visualizes the activity or effects of stress on PSII, photosynthetic membrane system, and electron transport rates (ETR). The ETR in pretreated leaves (40°C) of heat-tolerant green gram EC 398889 was less affected at high irradiances whereas heat-sensitive genotype LGG 460 with similar treatment showed complete reduction of photosynthetic ETR (**Figure 14**). Reduced electron transport and damaged photosystems caused by high temperature have been reported in poplar [89]. Chloroplast stroma and thylakoid membranes are damaged by high temperatures [151]. Photosystem (PS) II in the light reaction [152] and Rubisco (ribulose1, 5-bisphosphate carboxylase/oxygenase) activase in the Calvin cycle [153] are both thermolabile. Heat stress thus impairs the electron transport chain and affects the activation and activity of the enzyme Rubisco [154]. Although PSI and PSII are both adversely affected by high temperatures, PSII is more sensitive to heat stress than is PSI [155]. Chlorophyll fluorescence is a rapid and non-invasive, high-resolution technique to determine changes in photochemistry through monitoring the fluorescence emission of photosystem II (PSII) *in situ* [156]. High temperature affects membrane stability, cell viability, and the quantum efficiency of photosystem II, as measured by chlorophyll fluorescence [157]. Chlorophyll fluorescence as affected by heat stress causes a decrease of *Fv*/*F*m ratio in susceptible wheat compared with tolerant lines [158]. The quantum yield is one of the powerful non-destructive parameters to differentiate

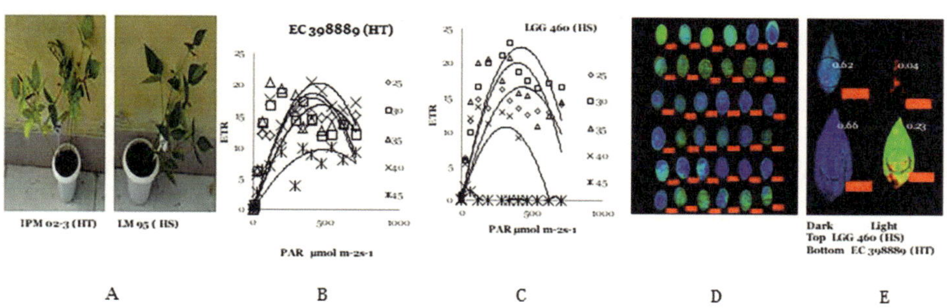

Figure 14.
A. Heat-tolerant green gram genotype IPM 02-3 abled to set pods at high temperature 43/28°C whereas heat-sensitive genotype LM 95 failed to set pods or flower at the same temperature regime. B. Light response of photosynthetic electron transport rate (ETR) at different temperatures in heat-tolerant (HT) green gram exotic accession EC 398889. High photosynthesis (ETR) detected even at 40°C coupled with high irradiance levels; however, ETR was inhibited beyond 40°C. C. Light response of photosynthetic electron transport rate (ETR) at different temperatures in heat-sensitive (HS) green gram exotic accession LGG 460. Photosynthesis (ETR) inhibited even at 40°C coupled with high irradiance levels; however, ETR was inhibited completely at 45°C at all irradiances. D. Large-scale phenotype of heat tolerance in green gram genotypes based on heat treatment of leaves at 40°C for 1 h followed by capturing quantum yield images. Deep blue color represents high values of quantum yield (Fv/Fm) and indicated tolerant lines. Whereas color deviated to green or pale green represented reduction in the quantum yield or heat-sensitive lines. E. Fluorescence imaging (quantum yield; Fv/Fm) of high temperature treated (40°C for 1 h) and light-adapted leaves distinguished HT (EC 398889) and HS (LGG 460) green gram accessions.

tolerance levels of a large number of genotypes with respect to particular abiotic stress. QTLs have been reported for chlorophyll fluorescence in drought- or heat-stressed wheat. The fluorescence parameters such as minimal Fo, maximal fluorescence, Fm, and variable fluorescence Fv change when plants are subjected to various abiotic stresses and eventually quantum yield of PSII (ratio of variable to maximum fluorescence, Fv/Fm) is affected. Using a large number of different fluorescence data, modifications of photosynthetic process under stresses can be assessed precisely, which includes thylakoid membrane organization, electron transport, and carbon assimilation [159]. The images captured for effective PSII quantum yield (YII) in high temperature treated leaves under high irradiances would be able to distinguish heat-tolerant and susceptible genotypes. The numerical values of different fluorescence parameters such as Fo, Fm, and Fv/Fm were converted to image format to visualize the adverse effects of heat on photosynthetic machinery at the chloroplast level. The image transformation with specific color code was shown in **Figure 7**. The intensity of the color depicts increase or decrease in the numerical values of fluorescence parameters. Similarly, the light response of electron transport rate (ETR) based on calculation using quantum yield (Fv/Fm) and PAR was also able to distinguish the genotypes based on their sensitivity to heat stress. Overall, this technique involving chlorophyll fluorescence imaging has proved to be effective and a precise phenotyping method for screening germplasm of pulses in a big way for stress tolerance. One of the mungbean line EC 398889 has been identified as highly heat tolerant based on chlorophyll fluorescence imaging and pollen fertility, and this line was used as one of the donor parents for developing short duration mungbean variety 'Virat'. The first distinct change in both the structure and function of photosystem II (PSII) reported to be occurred at 40–50°C in barley [160]. The first temperature induced transient changes had been shown at 42–48°C with a disruption of the PSII donor side and corresponding loss of oxygen evolution [161] followed by changes in thylakoid membranes at about 60°C and loss of electron transport through PSII [162] representing a denaturation of the PSII reaction centers. At about 75°C, a denaturation of light-harvesting complex of PSII (LHCII) has been observed [162]. The modification of chlorophyll florescence in response to heat stress has been reported in numerous crops, and heat tolerance of plant species can be quantified by measuring chlorophyll florescence [163]. When leaves are exposed to high temperatures and high light intensity simultaneously, the combined effects are more detrimental to photosynthesis. The relative assessment of fluorescence images for quantum yield (Fv/Fm) in high temperature treated in several green gram genotypes revealed that light-adapted leaves of the heat-tolerant green gram genotype EC 398889 exhibited higher quantum yield than the heat-sensitive genotype, LGG 460. The photosynthetic system partially or completely collapsed in light-adapted leaves of LGG 460. The fluorescence images combined with the light curve of ETR strongly could precisely differentiate varying sensitivity of photosynthesis to heat stress in the two contrasting genotypes, e.g. heat-tolerant green gram EC 398889 and sensitive ones LGG 460 [49]. The differential degree of membrane thermostability may distinguish the genotypes toward different sensitivity to heat stress. Chloroplast-targeted AtFtsH11 protease plays critical roles for maintaining the thermostability and structural integrity of photosystems under high temperatures [164]. Therefore, the photosynthetic efficiency may be modified under heat stress by improving FtsH11 protease in photosystems, hence, to improve plant productivity. Molecular characterization of heat tolerance in green gram using specific marker CEDG 147 could also be able to distinguish heat-tolerant green gram genotypes from sensitive ones (**Figure 15**).

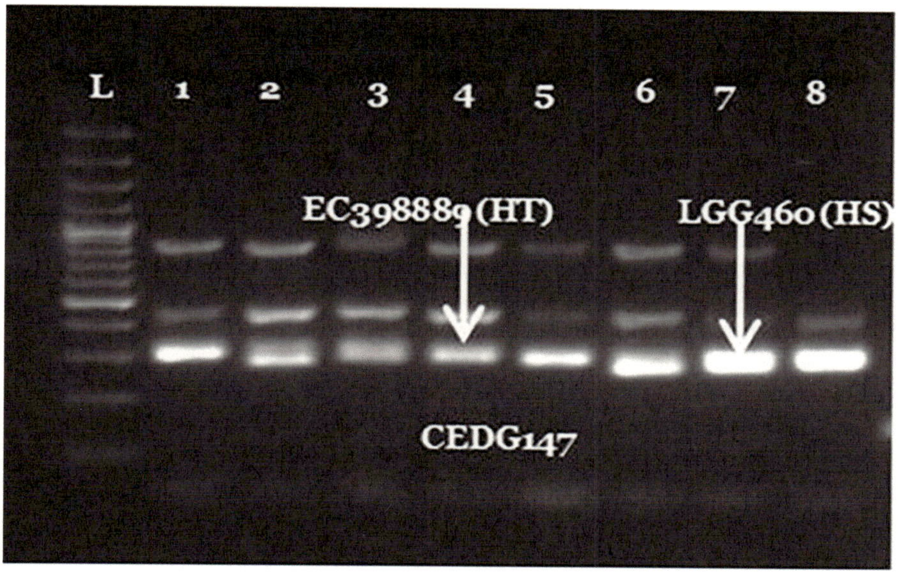

Figure 15.
Molecular characterization of heat tolerance in green gram. Marker CEDG147 (C) differentiated heat-tolerant (HT) and heat-sensitive (HS) green gram genotypes, L- 100 bp ladder, 1 (HUM12), 2 (Ganga-8), 3 (EC398889), 4 (IPM-02-3), 5 (IPM-02-14), 6 (LGG460), 7 (Kopergaon), 8 (NSB 007).

6.4 Oxidative stress

The plants produce antioxidant enzyme systems as a defensive mechanism which involve superoxide dismutase (SOD), catalase (CAT), and peroxidase (POX) for scavenging ROS under stressed conditions [165]. Heat-sensitive chick pea and black gram genotypes were induced to express more SOD and POX under heat treatment as a defensive mechanism of protection from ROS; however, heat-tolerant genotype is inherently tolerant to stress and therefore producing less ROS (**Figure 16**). During prolonged stress exposure, photosynthetic activity is further inhibited by excessive accumulation of reactive oxygen species (ROS), causing damage to the membranes, proteins, and chlorophyll molecules of the photosynthetic apparatus [2, 166]. Plants use a complex antioxidant system to regulate ROS levels and avoid toxicity, but

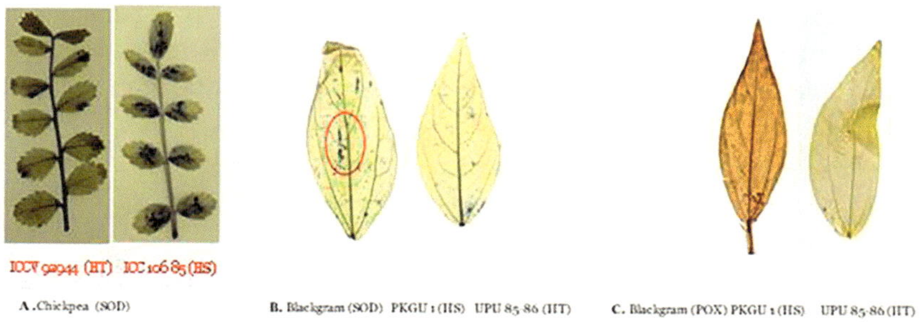

Figure 16.
Superoxide radical SOD in heat-tolerant (HT) and sensitive (HS) chick pea (A) and black gram (B) and peroxidase (POX) in black gram (C) when both HT and HS grown under high temperature regimes 43/25°C max/min day/night.

changes in redox status are also perceived by plants as a signature of a specific stress that will result in a corresponding acclimation response [167]. ROS scavenging is commonly induced under drought and heat stresses through enhancing antioxidant activities, and this is correlated with tolerance to stress [168]. In some wheat genotypes, tolerance to drought or heat stress was associated with increased antioxidant capacity and reduced oxidative damage in some wheat genotypes [169]. Under prolonged stress exposure, photosynthetic activity is further inhibited by excessive accumulation of ROS, causing damage to the membranes, proteins, and chlorophyll molecules of the photosynthetic apparatus [166].

6.5 Trait introgression for combined tolerance: Use of wild accessions

The pulses have narrow genetic diversity to accept the challenges of global warming and associated changes in the water availability [170]. Therefore, it is necessary to exploit wild species and land races for extensive gene mining for useful adaptive traits/genes to incorporate into our present cultivars. The wild species are the rich sources of many useful genes [171] as they have evolved under natural selection to survive climatic extremes and can potentially provide further genetic gains [172]. Therefore, wild species need to be exploited in genetic improvement programs to alleviate the challenges of global warming and its related effects in pulses. Two wild accessions of *Vigna* were identified as photo-thermo insensitive. These include one accession each of *V. glabrescens* (IC 251372) and *Vigna umbellata* (IC 251442) based upon viable pollen and normal pollen tube formation, podding, and seed set at high temperature up to 44°C and low temperature up to 4.4°C. Distant hybridization programme for climate resilience in mungbean was initiated using these wild *Vigna* species. Wild relatives of pigeon pea are known to have many useful traits for tolerance to various abiotic stresses. The wild species of pigeon pea like *Cajanus sericeus, C. scarabaeoides,* and *Cajanus acutifolius* showed a high degree of osmotic adjustment, OA (−2.5 to −5.0 MPa) imparting drought and heat tolerance based on the chlorophyll fluorescence imaging and pollen fertility test. These wild species showed fertile pollens and pod setting under severe drought. The photo-thermo insensitive accession of wild *C. scarabaeoides* ICP 15671 demonstrated high OA and normal pollen germination at temperature as high as 47°C. The heat-tolerant ICP 15671 showed normal pollen fertility and ability to set pods even at 47°C (**Figure 17**). The accumulation of osmolytes with higher order plays a major role as thermo and desiccation protectant preventing damage of membrane and vital macromolecules. This wild species is photo-thermo insensitive as well as heat and drought tolerant, and therefore, this wild *Cajanus sps* (ICP 15671) appeared to be a potential reservoir of genes for multiple abiotic stress tolerance. Many wild derivatives developed from crosses *C. cajan cajan* × *C. scarabaeoides* (ICP 15671) showed significant genetic variations in the combined

Figure 17.
Combined tolerance (heat and drought) in pigeon pea wild accession Cajanus scarabaeoides (L.). One of the accessions of C. scarabaeoides ICP 15671 flowered and set pods even temperature as high as 47°C with normal pollen germination.

tolerance to both drought and heat stresses. The pre-breeding line ICP 711 developed from a cross between cultivated vs. wild chick pea *C. arietinum* × *C. judaicum* showed an increased number of primary branches, pods per plant, and green seeds for further use in chick pea improvement program [173].

7. Conclusion

Drought and heat stresses or combined effects of both are the major challenges for pulses which adversely affect grain yield to the extent of about 60%. Underlying mechanisms of drought tolerance in major pulses involve extensive and deep root systems, osmotic adjustment, high water-use efficiency, remobilization of prestored photosynthates, stomatal control for regulated water loss, and internal moisture conservation, etc. While stable photosynthetic system, membrane stability, and high pollen fertility have been considered as potential physiological considerations for heat tolerance. However, the fact is that very little success has been achieved so far to improve grain yield through introgression of these traits. The major yield limiting factors for low yield in pulses under these stresses could be due to poor biomass accumulation which is the most sensitive component as affected by stresses. No potential traits have so far been identified by which biomass enhancement can be achieved, if biomass accumulation is adequate at initial stages of growth, the adverse effects of these stresses can be minimized. Therefore, plant type concept has been emerged as a potential strategy to improve yield under abiotic stresses. At the initial growth stage, plants must be configured to achieve more branches and pod bearing nodes to store more moisture in the vegetative parts as it is an inherent character of pulses to retain water for longer time. Pods are semi-autonomous organs to photosynthesize and gaining carbon and capable of refixing respiratory release of CO_2 internally, and moreover the water status of pod wall/seed coat is partially independent of water status of plants. High biomass also contributes proportionately higher carbon and nitrogen mobilization to developing sinks. The early, phenology, quick biomass accumulation with profuse branching and podding nodes are considered to be important in determining yield under drought and heat. The physiological trait introgression needs to be considered important on the background of high yielding plant types. Extensive efforts have been made to identify genes/QTLs in chick pea, pigeon pea, and other pulse crops [174]. Recently, genomics tools are becoming an integral part of the current conventional breeding which could be applied for genetic improvement for climate-smart pulses [175]. Genome sequences of major pulses including pigeon pea and chick pea are now available [176].

Acknowledgements

Authors are grateful to the Indian Council of Agricultural Research, New Delhi, India for funding this work under the project National Innovations on Climate resilient Agriculture (NICRA). The facilities provided by Director, ICAR-Indian Institute of Pulses Research, Kanpur, India to carry out this investigation are thankfully acknowledged.

Author details

Partha S. Basu[1*], Sushil Kumar Chaturvedi[2], Pooran Mall Gaur[3], Biswajit Mondal[1], Surendra Kumar Meena[4], Krishnashis Das[1], Vaibhav Kumar[1], Kalpana Tewari[1] and Kusum Sharma[1]

1 ICAR-Indian Institute of Pulses Research, Kanpur, Uttar Pradesh, India

2 Rani Lakshmi Bai Central Agriculture University, Jhansi, Uttar Pradesh, India

3 International Crops Research Institute for the Semi-Arid Tropics (ICRISAT), Hyderabad, India

4 ICAR-IIPR, Arid Pulses Research Centre, Bikaner, Rajasthan, India

*Address all correspondence to: psbsu59@gmail.com

References

[1] Barnabás B, Jager K, Feher A. The effect of drought and heat stress on reproductive processes in cereals. Plant Cell and Environment. 2008;**31**:11-38

[2] Awasthi R, Kaushal N, Vadez V, Turner NC, Berger J, Siddique KHM, et al. Individual and combined effects of transient drought and heat stress on carbon assimilation and seed filling in chick pea. Functional Plant Biology. 2014;**41**:1148-1167

[3] Zandalinas SI, Mittler R, Balfagon D, Arbona V, Gomez-Cadenaz A. Plant adaptations to the combination of drought and high temperatures. Physiolgia Plantarum. 2018;**162**:2-12

[4] Singh J, Kanaujia R, Singh NP. Pulse phytonutrients: Nutritional and medicinal importance. Journal of Pharmacy and Nutrition Sciences. 2016;**6**:160-160. DOI: 10.6000/1927-5951.2016.06.04.5

[5] Daryanto S, Wang L, Jacinthe PA. Global synthesis of drought effects on food legume production. PLoS ONE. 2015;**10**:e0127401. DOI: 10.1371/journal.pone.0127401

[6] Fahad S, Bajwa AA, Nazir U, Anjum SA, Farooq A, Zohaib A, et al. Crop production under drought and heat stress: Plant responses and management options. Frontiers in Plant Science. 2017;**8**:1147. DOI: 10.3389/fpls.2017.01147

[7] Summerfield RJ, Hadley R, Minchin EH, Awsthorne FRS. Sensitivity of chick pea (*Cicer arietinum*) to hot temperatures during the reproductive period. Experimental Agriculture. 1984;**20**:77-93

[8] Wahid A, Gelani S, Ashraf M, Foolad MR. Heat tolerance in plants: An overview. Environment and Experimental Botany. 2007;**61**:199-223. DOI: 10.1016/j.envexp bot. 2007.05.011

[9] Morison JI, Baker NR, Mullineaux PM, Davies WJ. Improving water use in crop production. Philosophical transactions of the Royal Society of London. Series B, Biological Sciences. 2008;**363**:639-658. DOI: 10.1098/rstb.2007.2175

[10] Acevedo M, Pixley K, Zinyengere N, Meng S, Tufan H, KarenCichy K, et al. A scoping review of adoption of climate-resilient crops by small-scale producers in low-and middle-income countries. Nature Plants. 2020;**6**:1231-1241. DOI: 10.1038/s41477-020-00783-z

[11] FAO. The Future of Food and Agriculture—Trends and Challenges. Rome; 2017

[12] Nam NH, Chauhan YS, Johansen C. Effect of timing of drought stress on growth and grain yield of extra-short duration pigeon pea lines. Journal of Agricultural Science. 2001;**136**:179-189. DOI: 10.1017/S0021859601008607

[13] Lopez FB, Johansen C, Chauhan YS. Effects of timing of drought stress on phenology, yield and yield components of short duration pigeon pea. Journal of Agronomy and Crop Science. 1996;**177**:311-320

[14] Sayan S. Adaptive mechanisms of Black gram (*Vigna mungo* (L.) Hepper) and pigeon pea (*Cajanus cajan* (L.) Millsp.) to water stress at different growth states [PhD thesis]. The University of Queensland: School of Land, Crop and Food Sciences; 1991. DOI: 10.14264/uql.2017.579

[15] Turner NC, Wright G, Siddique KHM. Adaptation of grain

legumes (pulses) to water-limited environments. Advances in Agronomy. 2001;**71**:193-231. DOI: 10.1016/S0065-2113(01)71015-2

[16] Leopold AC, Musgrave ME, Williams KM. Solute leakage resulting from leaf desiccation. Plant Physiology. 1981;**68**:1222-1225

[17] Kimani PM, Benzioni A, Ventura M. Genetic variation in pigeon pea (*Cajanus cajan* (L.) Mill sp.) in response to successive cycles of water stress. Plant and Soil. 1994;**158**:193-201. DOI: 10.1007/BF00009494

[18] Incrocci L, Pardossi A, Vernieri P, Tognoni F, Serra. Effects of heat stress and hypoxia on growth, water relations and ABA levels in bean (*Phaseolus vulgaris* L.) seedlings. Acta Horticulturae. 2000;**516**:31-39. DOI: 10.17660/ActaHortic.2000.516.3

[19] Mishra B. Challenges and preparedness for increasing wheat production in India. Journal of Wheat Research. 2007;**I**(1 and 2):1-12

[20] Kumar S. Climate change and crop breeding objectives in the twenty first century. Current Science. 2006;**90**:1053-1154

[21] Van Der Maesen LJG. Cicer L.: A monograph of the genus, with special reference to the chickpea (*Cicer arietinum* L.) its ecology and cultivation. Mededelingen Landbouwhogeschool Wageningen; 1972

[22] Hall AE. Breeding for heat tolerance. Plant Breeding Review. 1992;**10**:129-168

[23] Prasad PVV, Craufurd PQ, Summerfield RJ. Fruit number in relation to pollen production and viability in groundnut exposed to short episodes of heat stress. Annals of Botany. 1999;**84**:381-386. DOI: 10.1006/anbo.1999.0926

[24] Rainey K, Griffiths PD. Evaluation of *Phaseolus acutifolius* A. Gray plant introductions under high temperatures in a controlled environment. Genetic Resource and Crop Evolution. 2005;**52**:117-120. DOI: 10.1007/s10722-004-1811-2

[25] Barghi SS, Mostafaii H, Peighami F, Zakaria RA. Path analysis of yield and its components in lentil under end season heat condition. International Journal of Agriculture Research Review. 2012;**2**:969-974

[26] Gaur PM, Kumar J, Gowda CLL, Pande S, Siddique KHM, Khan TN, et al. Breeding chick pea for early phenology: Perspectives, progress and prospects. In: Kharkwal MC, editor. Food Legumes for Nutritional Security and Sustainable Agriculture. Vol. 2. New Delhi, India: Indian Society of genetics and Plant Breeding; 2008. pp. 39-48

[27] Devasirvatham V, Gaur PM, Mllikarjuna N, Tokachichu RN, Trethowan RM, Tan DKY. Effect of high temperature on reproductive development of chick pea genotypes under controlled environments. Functional Plant Biology. 2012;**39**:1009-1018. DOI: 10.1071/FP12033

[28] Giaveno C, Ferrero J. Introduction of tropical maize genotypes to increase silage production in the central area of Santa Fe, Argentina. Crop Breeding and Applied Biotechnology. 2003;**3**:89-94

[29] Boote KJ, Allen LH, Prasad PVV, Baker JT, Gesch RW, Snyder AM, et al. Elevated temperature and CO_2 impacts on pollination, reproductive growth, and yield of several globally important crops. Journal of Agricultural Meteorology. 2005;**60**:469-474

[30] Horváth I, Glatz A, Nakamoto H, Mishkind ML, Munnik T, Saidi Y. Heat shock response in photosynthetic organisms: Membrane and lipid connections. Progress in Lipid Research. 2012;**51**:208-220. DOI: 10.1016/j.plipres.2012.02.002

[31] Rakavi B, Sritharan N. Physiological response of green gram under heat stress. Journal of Pharmacognosy and Phytochemistry. 2019;**SP1**:181-185

[32] Reynolds MP, Pierre CS, Saad ASI, Vargas M, Condon AG. Evaluating potential genetic gains in wheat associated with stress-adaptive trait expression in elite genetic resources under drought and heat stress. Crop Science. 2007;**47**:S-172-S-189. DOI: 10.2135/cropsci.2007.10.0022 IPBS

[33] Wang Q-L, Chen J-H, He N-Y, Guo F-Q. Metabolic reprogramming in chloroplasts under heat stress in plants. International Journal of Molecular Sciences. 2018;**19**:849. DOI: 10.3390/ijms19030849

[34] Langjun C, Jianlong L, Yamin F, Sheng X, Zhen Z. High temperature effects on photosynthesis, PSII functionality and antioxidant activity of two Festuca arundinacea cultivars with different heat susceptibility. Botanical Studies. 2006;**47**:61-69

[35] Zhao FC. Effects of heat stress during grain filling on sugar accumulation and enzyme activity associated with sucrose metabolism in sweet corn. Acta Agronomica Sinica. 2013;**39**:1644-1651. DOI: 10.3724/SP.J.1006.2013.01644

[36] Suzuki N, Koussevitzky S, Mittler R, Miller G. ROS and redox signalling in the response of plants to abiotic stress. Plant Cell Environment. 2012;**35**:259-270. DOI: 10.1111/j.1365-3040.2011.02336.x

[37] Meriga B, Krishna Reddy B, Rajender Rao K, Ananda Reddy L, Kavi Kishor PB. Aluminium-induced production of oxygen radicals, lipid peroxidation and DNA damage in seedlings of rice (*Oryza sativa*). Journal of Plant Physiology. 2004;**161**:63-68. DOI: 10.1078/0176-1617-01156

[38] Vadez V, Soltani A, Krishnamurthy L, Sinclair TR. Modelling possible benefit of root related traits to enhance terminal drought adaption of chick pea. Field Crops Research. 2012;**137**:108-115. DOI: 10.1016/j.fcr.2012.07.022

[39] Hassan M, Mansoor S. Oxidative stress and antioxidant defense mechanism in mung bean seedlings after lead and cadmium treatments.© TÜBİTAK. Turkish Journal of Agriculture and Forestry. 2014;**38**:55-61. DOI: 10.3906/tar-1212-4

[40] Rodrigues CS, Laranjo M, Oliveira S. Effect of heat and pH stress in the growth of chick pea Mesorhizobia. Current Microbiology. 2006;**2006**(53):1-7

[41] Kurdali F. Nitrogen and phosphorus assimilation, mobilization and partitioning in rainfed chick pea (*Cicer arietinum* L.). Field Crops Research. 1996;**47**:81-92

[42] Minchin FR, Summerfield RJ, Hadley P, Roberts EH, Rawsthorne S. Carbon and nitrogen nutrition of nodulated roots of grain legumes. Plant, Cell & Environment. 1981;**4**:5-26

[43] Rawsthorne S, Hadley P, Roberts EH, Summerfield RJ. Effects of supplemental nitrate and thermal regime on the nitrogen nutrition of chick pea (*Cicer arietinum* L.) П: Symbiotic development and nitrogen assimilation. Plant and Soil. 1985;**83**:279-293

[44] Hedhly A, Hormaza JI, Herrero M. Global warming and plant sexual

reproduction. Trends Plant Science. 2008;**14**:30-36

[45] Kaushal N, Bhandari K, Siddique KHM, Nayyar H. Food crops face rising temperatures: An overview of responses, adaptive mechanisms, and approaches to improve heat tolerance. Cogent Food and Agriculture. 2016;**2**:1134380. DOI: 10.1080/23311932.2015.1134380

[46] Omae H, Kumar A, Egawa Y, Kashiwaba K, Shono M. Genotypic differences in plant water status and relationship with reproductive responses in snap bean (*Phaseolus vulgaris* L.) during water stress. Japanese Journal of Tropical Agriculture. 2005;**49**:1-7

[47] Sita K, Sehgal A, Bindumadhava HR, Nair RM, Prasad PVV, Kumar S, et al. Food legumes and rising temperatures: Effects, adaptive functional mechanisms specific to reproductive growth stage and strategies to improve heat tolerance. Frontiers in Plant Science. 2017;**8**:1658. DOI: 10.3389/fpls.2017.01658

[48] Suzuki K, Takeda H, Tsukaguchi T, Egawa Y. Ultrastructural study on degeneration of tapetum in anther of snap bean (*Phaseolus vulgaris* L.) under heat stress. Sexual Plant Reproduction. 2001;**13**:293-299. DOI: 10.1007/s004970100071

[49] Basu PS, Pratap A, Gupta S, Sharma K, Tomar R, Singh NP. Physiological traits for shortening crop duration and improving productivity of green gram (*Vigna radiata* L. Wilczek) under high temperature. Frontiers in Plant Science. 2019;**10**:1508. DOI: 10.3389/fpls.2019.01508

[50] Sung D-Y, Kaplan F, Lee K-J, Guy CL. Acquired tolerance to temperature extremes. Trends in Plant Science. 2003;**8**:179-187

[51] Omae H, Kumar A, Kashiwaba K, Shono M. Influence of temperature shift after flowering on dry matter partitioning in two cultivars of snap bean (*Phaseolus vulgaris*) that differ in heat tolerance. Plant Production Science. 2007;**10**:14-19

[52] Omae H, Kumar A, Kashiwaba K, Shono M. Influence of high temperature on morphological characters, biomass allocation, and yield components in snap bean (*Phaseolus vulgaris* L.). Plant Production Science. 2006;**9**:200-205. DOI: 10.1626/pps.9.200

[53] Board JE, Kahlon CS. Soybean yield formation: What controls it and how it can be improved? In: ElShemy HA, editor. Soybean Physiology and Biochemistry. Rijeka: InTech Open Access; 2011. pp. 1-36

[54] Tsukaguchi T, Kawamitsu Y, Takeda H, Suzuki K, Egawa Y. Water status of flower buds and leaves as affected by high temperature in heat-tolerant and heat-sensitive cultivars of snap bean (*Phaseolus vulgaris* L.). Plant Production Science. 2003;**6**:4-27

[55] Barghi SS, Mostafaii H, Peighami F, Zakaria RA, Nejhad RF. Response of in vitro pollen germination and cell membrane thermostability of lentil genotypes to high temperature. International Journal of Agriculture Research Review. 2013;**3**:13-20

[56] Kumar U, Joshi AK, Kumari M, Paliwal R, Kumar S, Roder MS. Identification of QTLs for stay green trait in wheat (*T. aestivum* L.) in the Chirya 3′9 Sonalika' population. Euphytica. 2010;**174**:437-445. DOI: 10.1007/s10681-010-0155-6

[57] Porch TG, Jahn M. Effects of high-temperature stress on microsporogenesis in heat-sensitive and heat-tolerant genotypes of *Phaseolus vulgaris*. Plant Cell Environment. 2001;**24**:723-731. DOI: 10.1046/j.1365-3040.2001.00716.x

[58] Kumar S, Thakur P, Kaushal N, Malik JA, Gaur PM, Nayyar H. Effect of varying high temperatures during reproductive growth on reproductive function, oxidative stress and seed yield in chick pea genotypes differing in heat sensitivity. Archives of Agronomy and Soil Science. 2013;**59**:823-843. DOI: 10.1080/03650340.2012.683424

[59] Kakani VG, Reddy KR, Koti S, Wallace TP, Prasad PVV, Reddy VR, et al. Differences in *in vitro* pollen germination and pollen tube growth of cotton cultivars in response to high temperature. Annals of Botany. 2005;**96**:59-67. DOI: 10.1093/aob/mci149

[60] Clarke HJ, Siddique KHM. Response of chick pea genotypes to low temperature stress during reproductive development. Field Crops Research. 2004;**90**:323-334. DOI: 10.1016/j.fcr.2004.04.001

[61] Ohnishi S, Miyoshi T, Shirai S. Low temperature stress at different flower developmental stages affects pollen development, pollination, and pod set in soybean. Environment and Experimental Botany. 2010;**69**:56-62

[62] Pang J, Turner NC, Khan T, Du YL, Xiong JL, Colmer TD, et al. Response of chick pea (*Cicer arietinum* L.) to terminal drought: Leaf stomatal conductance, pod abscisic acid concentration, and seed set. Journal of Experimental Botany. 2017;**68**:1973-1985. DOI: 10.1093/jxb/erw153

[63] Thiaw S, Hall AE. Comparison of selection for either leaf-electrolyte-leakage or pod set in enhancing heat tolerance and grain yield of cowpea. Field Crops Research. 2004;**86**:239-253. DOI: 10.1016/j.fcr.2003.08.011

[64] Priya M, Dhanker OP, Siddique KHM, Bindumadhava H, Nair R, Pandey S, et al. Drought and heat stress-related proteins: An update about their functional relevance in imparting stress tolerance in agricultural crops. Theoretical and Applied Genetics. 2019;**132**. DOI: 10.1007/s00122-019-03331-2

[65] Perdomo JA, Capó-Bauçà S, Carmo-Silva E, Galmés J. Rubisco and Rubisco activase play an important role in the biochemical limitations of photosynthesis in rice, wheat, and maize under high temperature and water deficit. Frontiers in Plant Science. 2017;**8**:490

[66] Pradhan GP, Prasad PVV, Fritz AK, Kirkham MB, Gill BS. Effects of drought and high temperature stress on synthetic hexaploid wheat. Functional Plant Biology. 2012;**39**:190-198

[67] Farooq M, Hussain M, Ul-Allah S, Siddique KHM. Physiological and agronomic approaches for improving water-use efficiency in crop plants. Agricultural Water Management. 2019;**219**:95-108

[68] Lamaoui M, Jemo M, Datla R, Bekkaoui F. Heat and drought stresses in crops and approaches for their mitigation. Frontiers in Chemestry. 2018;**6**:26-30

[69] Rivero RM, Mestre TC, Mittler R, Rubio F, Garcia-Sanchez F, Martinez V. The combined effect of salinity and heat reveals a specific physiological, biochemical and molecular response in tomato plants. Plant Cell and Environment. 2014;**37**:1059-1073

[70] FAO. Statistical Database. Rome, Italy: Food and Agriculture Organization of the United Nations; 2016. Available from: http//www.apps.fao.org

[71] Jukanti A, Gaur PM, Gowda CLL, Chibbar R. Nutritional quality and health benefits of chick pea (*Cicer arietinum* L.): A review. The British Journal of Nutrition. 2012;**108**(Suppl. 1):S11-S26. DOI: 10.1017/S0007114512000797

[72] Ryan JG. A global perspective on pigeon pea and chick pea sustainable production systems: Present status and future potential. In: Asthana AN, Ali M, editors. Recent Advances in Pulses Research. Kanpur: Indian Institute of Pulses Research; 1977. pp. 1-31

[73] Gaur PM, Gowda CLL, Knights EJ, Warkentin TD, Acikgoj N, Yadav SS, et al. Breeding achievements. In: Yadav SS, Redden B, Chen W, Sharma B, editors. Chick Pea Breeding and Management. Wallingford: CABI; 2007. pp. 391-416

[74] Wang J, Gan YT, Clarke F, McDonald CL. Response of chick pea yield to high temperature stress during reproductive development. Crop Science. 2006;**46**:2171-2178. DOI: 10.2135/cropsci2006.02.0092

[75] Fowden I, Manstield T, Stoddart J, editors. Plant adaptation to environmental stress. Landan. Edition I. Netherlands: Springer, Chapman and Hall; 1993:109-132

[76] Kalra N, Chakraborty D, Sharma A, Rai HK, Jolly M, Cher S, et al. Effect of increasing temperature on yield of some winter crops in Northwest India. Current Science. 2008;**94**:82-88

[77] Delahunty A, Nuttall J, Nicolas M, Brand J. Genotypic heat tolerance in lentil. In: Proceedings of the 17th ASA Conference, Hobart, Australia. 2015. pp. 20-24

[78] Subbarao GV, KumarRao JVDK, Kumar J, Johansen C, Deb UK, Ahmed I, et al. Spatial Distribution and Quantification of Rice-Fallows in South Asia-Potential for Legumes. Patancheru, India: ICRISAT; 2001

[79] Kaur R, Bains TS, Bindumadhava H, Nayyar H. Responses of mungbean (*Vigna radiata* L.) genotypes to heat stress. Effects on reproductive biology, leaf function and yield traits. Scientia Horticulturae. 2015;**197**:527-541. DOI: 10.1016/j.scienta.2015.10.015

[80] Gaur P, Saminen S, Krishnamurthy L, Kumar S, Ghane M, Beebe S, et al. High temperature tolerance in grain legumes. Legume Perspectives. 2015;7:23-24

[81] Scafaro AP, Haynes PA, Atwell BJ. Physiological and molecular changes in *Oryza meridionalis* Ng., a heat-tolerant species of wild rice. Journal of Experimental Botany. 2010;**61**:191-202. DOI: 10.1093/jxb/erp294

[82] Murata N, Takahashi S, Nishiyama Y, Allakhverdiev SI. Photoinhibition of photosystem II under environmental stress. Biochimica et Biophysica Acta (BBA)–Bioenergetics. 2007;**1767**:414-421. DOI: 10.1016/j.bbabio.2006.11.019

[83] Allakhverdiev SI, Murata N. Environmental stress inhibits the synthesis de novo of proteins involved in the photodamage–repair cycle of photosystem II in Synechocystis sp. PCC 6803. Biochimica et Biophysica Acta (BBA)—Bioenergetics. 2004;**1657**:23-32. DOI: 10.1016/j.bbabio.2004.03.003

[84] Allakhverdiev SI, Klimov VV, Carpentier R. Evidence for 55 the involvement of cyclic electron transport in the protection of photosystem II against photoinhibition: Influence of a new phenolic compound. Biochemistry. 1997;**36**:4149-4154. DOI: 10.1021/bi962170

[85] Nene YL, Sheila VK. c. In: Nene YL, Hall SD, Sheila VK, editors. Pigeon Pea. Wallingford, UK: CAB Intl; 1990. pp. 1-14

[86] Upadhyaya HD, Kashiwagi J, Varshney R, Gaur PM, Saxena K, Krishnamurthy L, et al. Phenotyping chick peas and pigeonpeas for adaptation to drought. Frontiers in Physiology. 2012;**3**:179. DOI: 10.3389/fphys.2012.00179

[87] Basu PS, Singh U, Kumar A, Praharaj CS, Shivran RK. Climate change and its mitigation strategies in pulses production. Indian Journal of Agronomy. 2016;**61**:S71-S82

[88] Smertenko A, Draber P, Viklicky V, Opatrny Z. Heat stress affects the organization of microtubules and cell division in *Nicotiana tabacum cells*. Plant Cell Environment. 1997;**20**:1534-1542. DOI: 10.1046/j.1365-3040.1997.d01-44.x

[89] Song Y, Chen Q, Ci D, Shao X, Zhang D. Effects of high temperature on photosynthesis and related gene expression in poplar. BMC Plant Biology. 2014;**14**:111. DOI: 10.1186/1471-2229-14-111

[90] Valenzuela H, Smith J. Pigeon pea: Sustainable Agriculture Green Manure Crops Aug. 2002, SA-GM-8 Cooperative State Research, Education, and Extension Service, U.S. Department of Agriculture, and the Agricultural Experiment Station, Utah State University, under Cooperative Agreement 98-ESAG-1-0340. Portions of this text were adapted from the USDA. Natural Resources Conservation Service Hawaii Field. 2002

[91] Zhang L, Li FG, Liu CL, Zhang CJ, Zhang XY. Construction and analysis of cotton (*Gossypium arboreum* L.) water-related cDNA library. BMC Research Note. 2009;**2**:120

[92] Sekhar K, Priyanka B, Reddy VD, Rao KV. Isolation and characterization of a pigeon pea cyclophilin (*CcCYP*) gene, and its over-expression in *Arabidopsis* confers multiple abiotic stress tolerance. Plant, Cell and Environment. 2010;**33**:1324-1338. DOI: 10.1111/j.1365-3040.2010.02151.x

[93] Mazars C, Thuleau P, Lamotte O, Bourque S. Cross-talk between ROS and calcium in regulation of nuclear activities. Molecular Plant. 2010;**3**:706-718

[94] Sen Gupta D, Singh U, Kumar J, Shivay YS, Dutta A, Sharanagat VS, et al. Estimation and multi-variate analysis of iron and zinc concentration in a diverse panel of urdbean (*Vigna mungo* L. Hepper) genotypes grown under differing soil conditions. Journal of Food Composition and Analysis. 2020;**93**:103605. DOI: 10.1016/j.jfca.2020.103605

[95] Roy SJ, Tucker EJ, Tester M. Genetic analysis of abiotic stress tolerance in crops. Current Opinion in Plant Biology. 2011;**14**:232-239. DOI: 10.1016/j.pbi.2011.03.002

[96] Sinha SK. Food legumes: Distribution adaptability and biology of yield. In: Food and Agriculture Organisation of the United Nations, Plant Production and Protection Paper No. 3. Rome: FAO; 1977. p. 124

[97] Siddique KHM, Loss SP, Regan KL, Jettner RL. Adaptation and seed yield of cool season grain legumes in Mediterranean environments of 5 South-Western Australia. Australian Journal of Agricultural Research. 1999;**50**:375-387. DOI: 10.1071/A98096

[98] Landi S, Hausman JF, Guerriero G, Esposito S. Poaceae vs. abiotic stress: Focus on drought and salt stress, recent insights and perspectives. Frontiers in Plant Science. 2017;**8**:1214. DOI: 10.3389/fpls.2017.01214

[99] Bindumadhava HR, Nair RM, Nayyar H. Salinity and high temperature tolerance in mungbean [*Vigna radiata* (L.) Wilczek] from a physiological perspective. Frontiers in Plant Science. 2016;7:957. DOI: 10.3389/fpls.2016.00957

[100] Gaur PM, Samineni S, Thudi M, Tripathi S, Sajja SB, Jayalakshmi V, et al. Integrated breeding approaches for improving drought and heat adaptation in chick pea (*Cicer arietinum* L.). Plant Breeding. 2019;**138**:389-400. DOI: 10.1111/pbr.12641

[101] Asseng S, Ewert F, Martre P, Rötter RP, Lobell DB, Cammarano D, et al. Rising temperatures reduce global wheat production. Nature Climate Change. 2015;**5**:143-147. DOI: 10.1038/nclimate2470

[102] Challinor A, Wheeler T, Garforth C, Craufurd P, Kassam A. Assessing the vulnerability of food crop systems in Africa to climate change. Climate Change. 2007;**83**:381-399. DOI: 10.1007/s10584-007-9249-0

[103] Passioura JB. Grain yield, harvest index, and water use of wheat. Journal of the Australian Institute of Agricultural Science. 1977;**43**:117-121

[104] Turner NC. Further progress in crop water relations. Advances in Agronomy. 1997;**528**:293-338

[105] Yadav S, Sharma KD. Molecular and morphophysiological analysis of drought stress in plants. In: Rigobelo EC, editor. Plant Growth. London: IntechOpen; 2016. DOI: 10.5772/65246

[106] Subbarao G, Johansen C, Slinkard AE, Rao N, Saxena NP, Chauhan Y. Strategies for improving drought resistance in grain legumes. Critical Reviews in Plant Sciences. 1995;**14**:469. DOI: 10.1080/713608125

[107] Davies S, Turner NC, Palta J, Siddique K, Plummer JA. Remobilisation of carbon and nitrogen supports seed filling in chick pea subjected to water deficit. Australian Journal of Agricultural Research. 2000;**51**:855-866. DOI: 10.1071/Ar00018

[108] Basu PS, Ali M, Chaturvedi SK. Osmotic adjustment increases water uptake, remobilization of assimilates and maintains photosynthesis in chick pea under drought. Indian Journal of Experimental Biology. 2007;**45**:261-267

[109] Wang H, Zhou Q, Mao P. Ultrastructural and photosynthetic responses of pod walls in alfalfa to drought stress. International Journal of Molecular Sciences. 2020;**21**:4457. DOI: 10.3390/ijms21124457

[110] Atkins CA, Kuo J, Pate JS. Photosynthetic pod wall of pea (*Pisum sativum* L.): Distribution of carbon dioxide-fixing enzymes in relation to pod structure. Plant Physiology. 1977;**60**:779-786. DOI: 10.1104/pp.60.5.779

[111] Furbank R, White R, Palta J, Turner NC. Internal recycling of respiratory CO2 in pods of chick pea (*Cicer arietinum* L.): The role of pod wall, seed coat, and embryo. Journal of Experimental Botany. 2004;**55**:1687-1696. DOI: 10.1093/jxb/erh190

[112] Kenneth A, Shackel TNC. Seed coat cell turgor in chick pea is independent of changes in plant and pod water potential. Journal of Experimental Botany. 2000;**51**:895-900. DOI: 10.1093/jexbot/51.346.895

[113] Hooda RS, Rao AS, Luthra YP, Sheoran IS, Singh R. Partitioning and utilization of carbon and nitrogen for dry matter and protein production in chick pea (*Cicer arietinum* L.). Journal of Experimental Botany. 1986;**37**:1492-1502

[114] Egli DB, Crafts-Brandner SJ. Soybean. In: Zamski E, Schaffer AA, editors. Photoassimilate Distribution in Plants and Crops. Source-sink Relationships. New York: Marcel Dekker, Inc.; 1996. pp. 595-623

[115] Bidinger F, Musgrave RB, Fischer RA. Contribution of stored pre-anthesis assimilates to grain yield in wheat and barley. Nature. 1977;**270**:431-433. DOI: 10.1038/270431a0

[116] Palta JA, Kobata T, Turner NC, Fillery IR. Remobilization of carbon and nitrogen in wheat as influenced by postanthesis water deficits. Crop Science. 1994;**34**:118-124. DOI: 10.2135/cropsci1994.0011183X003400010021x

[117] Foster EF, Pajarito A, Acosta-Gallegos J. Moisture stress impact on N partitioning, N remobilization and N-use efficiency in beans (*Phaseolus vulgaris*). Journal of Agricultural Science (Cambridge). 1995;**124**:27-37

[118] Rawson HM, Evans LT. The contribution of stem reserves to grain development in a range of wheat cultivars of different height. Australian Journal of Agricultural Research. 1971;**22**:851-863. DOI: 10.1071/AR9710851

[119] Bushby HVA, Lawn RJ. Accumulation and partitioning of nitrogen and dry matter by contrasting genotypes of mungbean (*Vigna radiata* (L.) Wilczek). Australian Journal of Agricultural Research. 1992;**43**:1609-1628

[120] Khanna-Chopra R, Sinha SK. The chickpea book chapter: Physiological aspects of growth and yield. In: Saxena MC, Singh KB, editors. The Chick Pea. Wallingford, UK: CAB International; 1987:163-187. ISBN: 9780851985718

[121] Muchow R, Sinclair T, Bennett J, mHammond L. Response of leaf growth, leaf nitrogen, and stomatal conductance to water deficits during vegetative growth of field-grown Soybean1. Crop Science. 1986;**26**:1190-1195. DOI: 10.2135/cropsci1986.0011183X002600060024x

[122] Mwanamwenge J, Loss SP, Siddique KHM, Cocks PS. Effect of water stress during floral initiation, flowering and podding on the growth and yield of faba bean (Vicia faba L.). European Journal of Agronomy. 1999;**11**:1-11

[123] Krishnamurthy L, Johansen C, Ito O. Genotypic variation in root system development and its implications for drought resistance in chick pea. In: Conference Paper presented in Dynamics of roots and nitrogen in cropping systems of semi-arid tropics held at ICRISAT. Patancheru, India. January 1996. pp. 236-250

[124] Saxena NP, Krishnamurthy L, Johansen C. Registration of a drought resistant chick pea germplasm. Crop Science. 1994;**33**:1424

[125] Kashiwagi J, Krishnamurthy L, Crouch JH, Serraj R. Variability of root length density and its contributions to seed yield in chick jpea (*Cicer arietinum* L.) under terminal drought stress. Field Crops Research. 2006;**95**:171-181

[126] Reynolds MP, Manes Y, Izanloo A, Langridge P. Phenotyping for physiological breeding and gene discovery in wheat. Annals of Applied Biology. 2009;**155**:309-320

[127] Pinto RS, Reynolds MP. Common genetic basis for canopy temperature depression under heat and drought stress associated with optimized root distribution in bread wheat. Theoretical and Applied Genetics. 2015;**128**:575-585. DOI: 10.1007/s00122-015-2453-9

[128] Ludlow MM, Santamaria JM, Fukai S. Contribution of osmotic

adjustment to grain yield in *Sorghum bicolor* (L.) Moench under water-limited conditions. II. Water stress after anthesis. Australian Journal of Agricultural Research. 1990;**41**:67-78

[129] Morgan JM, Condon AG. Water use, grain yield and osmoregulation in wheat. Australian Journal of Plant Physiology. 1986;**13**:523-532

[130] Crowe JH, Hoekstra FA, Crowe LM. Anhydrobiosis. Annual Review of Physiology. 1992;**54**:579-599

[131] Rhodes D, Hanson AD. Quaternary ammonium and tertiary sulfonium compounds in higher plants. Annual Review of Plant Physiology and Plant Molecular Biology. 1993;**44**:357-384

[132] Shen B, Jensen RG, Bohnert HJ. Increased resistance to oxidative stress in transgenic plants by targeting mannitol biosynthesis to chloroplasts. Plant Physiology. 1997;**113**:1177-1183

[133] Sinclair TR, Ludlow MM. Influence of soil water supply on the plant water balance of four tropical grain legumes. Australian Journal of Plant Physiology. 1986;**13**:329-341. DOI: 10.1071/PP9860329

[134] Blum A. Plant Breeding for Stress Environments. Boca Raton, FL: CRC Press; 1988

[135] Prasad PVV, Craufurd PQ, Kakani VG, Wheeler TR, Boote KJ. Influence of high temperature during pre- and post-anthesis stages of floral development on fruit-set and pollen germination in peanut. Australian Journal of Plant Physiology. 2001;**28**:233-240

[136] Wassmann R, Jagadish SVK, Heuer S, Ismail A, Redoña E, Serraj R, et al. Climate change affecting rice production: The physiological and agronomic basis for possible adaptation strategies. Advances in Agronomy. 2009;**101**:59-122

[137] Hedhly A. Sensitivity of flowering plant gametophytes to temperature fluctuations. Environmental and Experimental Botany. 2011;**74**:9-16. DOI: 10.1016/j.envexpbot. 2011.03.016

[138] Ploeg Van der A, Heuvelink E. Influence of sub-optimal temperature on tomato growth and yield: A review. The Journal of Horticultural Science and Biotechnology. 2005;**80**:652-659. DOI: 10.1080/14620316.2005.11511994

[139] Hasanuzzaman M, Nahar K, Alam MM, Roychowdhury R, Fujita M. Physiological, biochemical, and molecular mechanisms of heat stress tolerance in plants. International Journal of Molecular Science. 2013;**14**:9643-9684. DOI: 10.3390/ijms14059643

[140] Talwar HS, Yanagahira S. Physiological basis of heat tolerance during flowering and pod setting stages in groundnut Arachis hypogaea *L*. In: JIRCAS Working Report, No. 14. Tsukuba, Japan: JJRCAS; 1999. pp. 47-65

[141] García CC, Guarnieri M, Pacini E. Inter-conversion of carbohydrate reserves from pollen maturation to rehydration in a chili pepper. American Journal of Plant Science. 2013;**4**:1181-1186. DOI: 10.4236/ajps.2013.46146

[142] Song L, Jiang Y, Zhao H, Hou M. Acquired thermotolerance in plants. Plant Cell, Tissue and Organ Culture (PCTOC). 2012;**111**:265-276. DOI: 10.1007/s11240-012-0198-6

[143] Gupta S, Kaur S, Sehgal S, Sharma A, Chhuneja P, Bains NS. Genotypic variation for cellular thermotolerance in Aegilops tauschii Coss., the D genome progenitor of

wheat. Euphytica. 2010;**175**:373-381. DOI: 10.1007/s10681-010-0185-0

[144] Berridge M, Herst P, Tan A. Tetrazolium dyes as tools in cell biology: New insights into their cellular reduction. Biotechnology Annual Review. 2005;**11**:127-152. DOI: 10.1016/S1387-2656(05)11004-7

[145] Blum A. Drought resistance, water-use efficiency, and yield potential—Are they compatible, dissonant, or mutually exclusive? Australian Journal of Agricultural Research. 1998;**56**:1159-1168. DOI: 10.1071/AR05069

[146] Ilík P, Špundová M, Šicner M, Melkovičová H, Kučerová Z, Krchňák P, et al. Estimating heat tolerance of plants by ion leakage: A new method based on gradual heating. The New Phytologist. 2018;**218**(3):1278-1287. DOI: 10.1111/nph.15097. Epub 2018 Mar 24

[147] Srinivasan A, Takeda H, Senboku T. Heat tolerance in food legumes as evaluated by cell membrane thermostability and chlorophyll fluorescence techniques. Euphytica. 1996;**88**:35-45. DOI: 10.1007/BF00029263

[148] Hu X, Li Y, Li C, Yang H, Wang W, Lu M. Characterization of small heat shock proteins associated with maize tolerance to combined drought and heat stress. Journal of Plant Growth Regulation. 2010;**29**:455-464. DOI: 10.1007/s00344-010-9157-9

[149] Blum A, Klueva N, Nguyen HT. Wheat cellular thermotolerance is related to yield under heat stress. Euphytica. 2001;**117**:117-123. DOI: 10.1023/A:100408330 5905

[150] Kim K, Portis AR. Temperature dependence of photosynthesis in Arabidopsis plants with modifications in Rubisco activase and membrane fluidity. Plant & Cell Physiology. 2005;**46**:522-530. DOI: 10.1093/pcp/pci052

[151] Wang GP, Zhang XY, Li F, Luo Y, Wang W. Overaccumulation of glycine betaine enhances tolerance to drought and heat stress in wheat leaves in the protection of photosynthesis. Photosynthetica. 2010;**48**:117-126. DOI: 10.1007/s11099-010-0016-5

[152] Heckathorn SA, Ryan SA, Baylis JA, Wang D, Hamilton EW, Cundiff L. In vivo evidence from an Agrostis stolonifera selection genotype that chloroplast small heat-shock proteins can protect photosystem II during heat stress. Functional Plant Biology. 2002;**29**:935-946. DOI: 10.1071/PP01191

[153] Crafts-Brandner SJ, Salvucci ME. Rubisco activase constrains the photosynthetic potential of leaves at high temperature and CO_2. Proceedings of the National Academy of Sciences. 2000;**97**:13430-13435. DOI: 10.1073/pnas.230451497

[154] Ahmad A, Diwan H, Abrol YP. Global climate change, stress and plant productivity. In: Pareek A, Sopory SK, Bohnert HJ, Govindjee, editors. Abiotic Stress Adaptation in Plants: Physiological, Molecular and Genome Foundation. Dordrecht: Springer Science Business Media BV; 2010. pp. 503-521. DOI: 10.1007/978-90-481-3112-9_23

[155] Moustaka J, Ouzounidou G, Sperdouli I, Moustakas M. Photosystem II is more sensitive than photosystem I to Al3+induced Phytotoxicity. Materials. 2018;**11**:1772. DOI: 10.3390/ma11091772

[156] Murchie EH, Lawson T. Chlorophyll fluorescence analysis: A guide to good practice and understanding some new applications. Journal of Experimental

Botany. 2013;**64**:3983-3998. DOI: 10.1093/jxb/ert208

[157] Mohammed A-R, Tarpley L. Impact of high night time temperature on respiration, membrane stability, antioxidant capacity, and yield of rice plants. Crop Science. 2009;**49**:313-322

[158] Izanloo A, Condon AG, Langridge P, Tester M, Schnurbusch T. Different mechanisms of adaptation to cyclic water stress in two south Australian bread wheat cultivars. Journal of Experimental Botany. 2008;**59**:3327-3346

[159] Kalaji HM, Rastogi A, Živčák M, Brestic M, Daszkowska-Golec A, Sitko K, et al. Prompt chlorophyll fluorescence as a tool for crop phenotyping: An example of barley landraces exposed to various abiotic stress factors. Photosynthetica. 2018;**56**:953-961. DOI: 10.1007/s 11099-018-0766-z

[160] Lípová L, Krchňák P, Josef KJ, Ilík P. Heat-induced disassembly and degradation of chlorophyll-containing protein complexes in vivo. Biochimica et Biophysica Acta. 2010;**1797**:63-70. DOI: 10.1016/j.bbabio.2009.08.001

[161] Cramer WA, Whitmarsh J, Low PS. Differential scanning calorimetry of chloroplast membranes: Identification of endothermic transition associated with the water-splitting complex of photosystem II. Biochemistry. 1981;**20**:157-162. DOI: 10.1021/bi00504a026

[162] Smith KA, Ardelt BK, Huner NPA, Krol M, Myscich E, Low PS. Identification and partial characterization of the denaturation transition of the light harvesting complex II of spinach chloroplast membranes. Plant Physiology. 1989;**90**:492-499

[163] Willits DH, Peet MM. Measurement of chlorophyll fluorescence as a heat stress indicator in tomato: Laboratory and greenhouse comparisons. Journal of American Society of Horticultural Science. 2001;**126**:188-194. DOI: 10.21273/JASHS.126.2.188

[164] Chen J, Burke JJ, Xin Z. Chlorophyll fluorescence analysis revealed essential roles of FtsH11 protease in regulation of the adaptive responses of photosynthetic systems to high temperature. BMC Plant Biology. 2018;**18**:11. DOI: 10.1186/ s12870-018-1228-2

[165] Chen J, Xu W, Velten J, Xin Z, Stout J. Characterization of maize inbred lines for drought and heat tolerance. Journal of Soil and Water Conservation. 2012;**2012**(67):354-364

[166] Redondo-Gómez S. Abiotic and biotic stress tolerance in plants. In: Rout GR, Das AB, editors. Molecular Stress Physiology of Plants. New Delhi: Springer India; 2013. pp. 1-20

[167] Choudhury FK, Rivero RM, Blumwald E, Mittler R. Reactive oxygen species, abiotic stress and stress combination. The Plant Journal. 2017;**90**:856-867

[168] Suzuki N, Rivero RM, Shulaev V, Blumwald E, Mittler R. Abiotic and biotic stress combinations. New Phytology. 2014;**203**:32-43. DOI: 10.1111/nph.12797

[169] Kiani R, Arzani A, Mirmohammady Maibody SA, Rahimmalek M, Razavi K. Morpho-physiological and gene expression responses of wheat by Aegilops cylindrica amphidiploids to salt stress. Plant Cell, Tissue and Organ Culture (PCTOC). 2021;**144**:619-639

[170] Rana JC, Gautam NK, Gayacharan MS, Yadav R, Tripathi K, Yadav SK, et al. Genetic resources of pulse crops in India: An overview. Indian Journal of Genetics and Plant Breeding. 2016;**76**:420-436

[171] Choudhary AK, Sultana R, Vales MI, Saxena KB, Kumar RR, Ratnakumar P. Integrated physiological and molecular approaches to improvement of abiotic stress tolerance in two pulse crops of the semi-arid tropics. Crop Journal;**6**:99-114

[172] Maxted N, Kell SP. Establishment of a Global Network for the *in Situ* Conservation of Crop Wild Relatives: Status and Needs. FAO Commission on Genetic Resources for Food and Agriculture: Roman, Italy; 2009

[173] Brumlop S, Reichenbecher W, Tappeser B, Finckh M. What is the SMARTest way to breed plants and increase agrobiodiversity? Euphytica. 2013:194. DOI: 10.1007/s10681-013-0960-9

[174] Kumar J, Choudhary A, Gupta DS, Kumar S. Towards exploitation of adaptive traits for climate-resilient smart pulses. International Journal of Molecular Sciences. 2019;**20**:10.3390/ijms20122971

[175] Kole C, Muthamilarasan M, Henry R, Edwards D, Sharma R, Abberton M, et al. Application of genomics-assisted breeding for generation of climate resilient crops: Progress and prospects. Frontiers in Plant Science. 2015;**6**. Available from: https://www.frontiersin.org/article/10.3389/fpls.00563DOI=10.3389/fpls.2015.00563

[176] Varshney RK, Pandey MK, Bohra A, Singh VK, Thudi M, Saxena RK. Toward the sequence -based breeding in legumes in the post-genome sequencing era. Theoretical and Applied Genetics. 2019;**132**:797-816

Chapter 12

Role of Microorganisms in Alleviating the Abiotic Stress Conditions Affecting Plant Growth

Talaat El Sebai and Maha Abdallah

Abstract

Agriculture is one of the main sectors that participate in building up world economy, and offers the main source of food, income, and employment to their rural populations. Despite the necessity of doubling agricultural production, quantitatively and qualitatively, to cope with the worsening increase in the global population and to meet the increasing humanitarian needs, the agricultural sector faces many abiotic stress conditions. Additionally, the great climate changes lead to an increase in the negative impact of these stressors. There are many conventional and nonconventional ways that could directly or indirectly mitigate the adverse effects of these stressors, each of them has its advantages and disadvantages. The biological tool is one of the promising methods; it depends on the effective use of beneficial microorganisms to alleviate stress conditions that affect plant growth, development, and therefore productivity. This method is economically inexpensive and eco-friendly toward the environment. Beneficial soil microorganisms such as PGPRs and AMF colonize the root zone of many plant species and help to enhance plant growth and development. Thus, this chapter is aiming to highlight the role of microorganisms in alleviating the abiotic stress conditions affecting in plant growth.

Keywords: environmental stress, mitigating, plant productivity, PGPR, sustainable agriculture, climatic changes

1. Introduction

Agriculture is the backbone of developed and particularly developing countries, with more than 60% of the population of the developing countries depending on it for their livelihood. Increasing food production to fulfill the needs of increasing world population becomes a major concern. By the year 2050, it is expected that the human population will rise up to 10 billion. Hence, it is necessary to produce 70% more food for meeting the need of additional population. Furthermore, fighting poverty and hunger, consuming limited natural resources with more efficiencies, and acclimatizing to global warming must be taken into account to attain sustainable development [1]. Therefore, to make sure nourishment security, crop production will have to be doubled, and produced in more environmentally sustainable means [2]. However, improvements in the agriculture production process, land and water use are essential to realizing food security, poverty reduction, and total sustainable development. This can be realized by increasing cultivable land area and/or by increasing efficiently the productivity of land and water units. Really, several other factors cause a further reduction in crop productivity resulting in a lack of food security, particularly in developing countries. Of them, the availability of agricultural land, freshwater resources, ever-increasing abiotic and biotic stresses, and low economic activity in agricultural sector are the main factors. Moreover, Agriculture sector is categorized as one of the most exposed sectors to climate change. Plant productivity, principally in arid and semi-arid zones is fronting growing stresses triggered by natural and human's activities issues. Augmented occurrence of both abiotic and biotic stresses has become the principal cause for declining productivity in main crops. There is evidence of yield drops in several crops in many parts of the world due to increasing drought, salinity stress, reduction in precipitation rate and elevated air temperature. Abiotic stresses can directly or indirectly disturb the physiological status of an organism by changing its metabolism, growth, and development. It is generally thought that abiotic stresses are considered to be the main source of yield reduction [3].

Abiotic stresses affect plants in various ways and are causes of reducing crop productivity (**Figure 1**). To enhance plant production, it necessities to apply

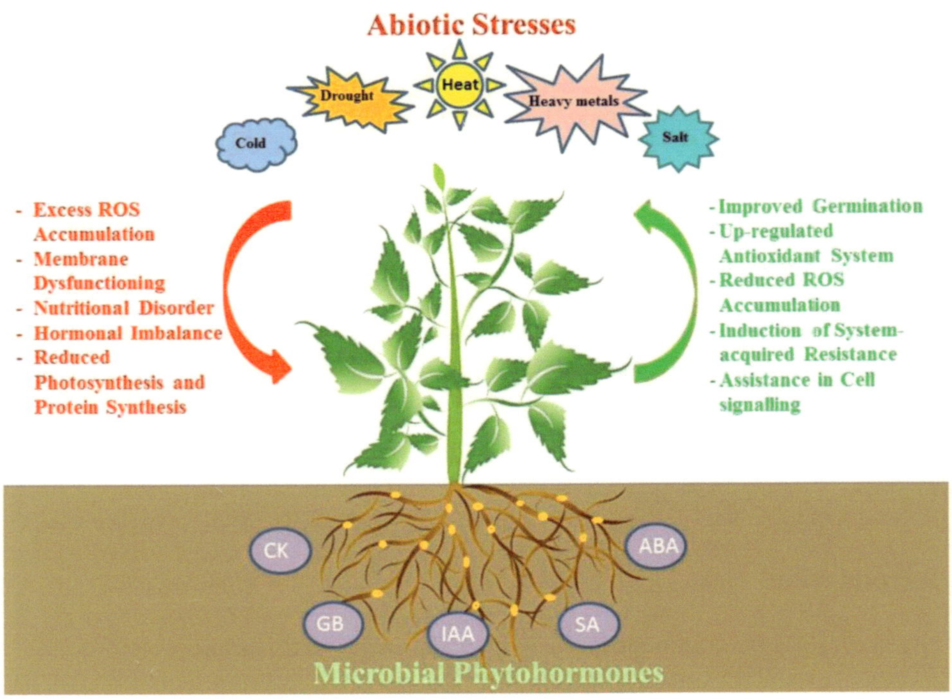

Figure 1.
Adverse effects of abiotic stress on plants and the role of PGPRs in alleviation of these stresses. This figure illustrates an overview of mechanisms in microbial phytohormone-mediated plant stress tolerance. Several root associated microbes produce cytokinin (CK), gibberellin (GB), indole-3-acetic acid (IAA), salicylic acid (SA), and abscisic acid (ABA), which help plants to cope with stress by improving its antioxidant potential, by up-regulation of the antioxidant system and by accumulation of compatible osmolytes therefore reducing oxidative stress-induced damage; improving photosynthetic capacity and membrane stability; promoting cell division and stomatal regulation; stimulating growth of root system, and acquisition of water and nutrients. (Adapted from [4]).

cost-effective technologies to control stress conditions. Soil microorganisms, living in the soil under normal and harsh conditions, have shown great properties, which, if exploited can help agriculture for improving and sustaining crop productivity. Whereas it is well recognized that beneficial microbes can stimulate growth and increase productivity through mechanisms like increasing nutrient availability, hormone production and disease controlling, it is also becoming increasingly clear that their effects may be more far-reaching.

Soil microorganisms (SMs) are very important in naturally occurring populations that play a significant role in soil fertility, plant growth, and maintaining healthier environment. This microbial population may comprise number of microorganisms like bacteria, actinomycetes, cyanobacteria, and fungi. Some of these are considered efficient owing to their growth enhancing abilities. Among these naturally occurring populations, plant growth promoting rhizobacteria (PGPR) have been investigated widely due to their positive effect on plant growth and protecting the environment from various hazards. PGPR are free living bacteria that enhance plant growth by root colonization [5]. These are also noted as plant health promoting bacteria (PHPB) or nodule promoting bacteria (NPB) [6] and can be characterized as intracellular PGPR (iPGPR) and extracellular PGPR (ePGPR) on the basis of their proximity in related to the host plant [7]. **Figure 2** shows the degree of nearness and influence of the plant-microbe interactions.

In the present chapter, we attempt an overview of current knowledge on how plant-PGPMs (Rhizobacteria, fungi, Arbuscular Mycorrhizal Fungi (AMF), Blue Green Algae or CyanoBacteria (BGA, CB), Actinomycetes or Actinobacteria, etc.) interactions help in alleviating abiotic stress conditions in different crop systems, which can be used for sustainable agriculture.

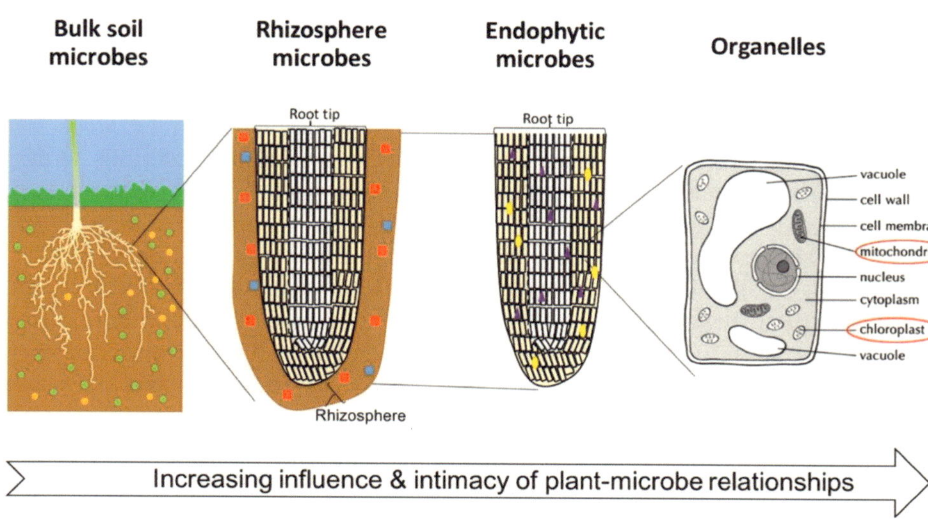

Figure 2.
The extent of proximity and influence of the plant-microbes interactions, small colored shapes (blue, green, red, purple and yellow) represent soil microbes. Diversity and density of microbes are variable according to soil organic contents and types, distance from plant roots, plant species, and plant tissue. (Adapted from [8]).

2. Stress definition and types

Stress conditions are a set of either abiotic or biotic factors that are unsuitable for plant growth of which the plant may be exposed during its various growth stages (one or more) from germination to fruiting, which may not only negatively affect its growth and productivity but may lead to entirety stopping its growth and thus its productivity. To which the plant may respond by making physiological and/or molecular and/or morphological changes or all of the previously. The plant stresses are defined as responses describing a suite of molecular and cellular processes prompted by the detection by a plant of some form of stress. These processes may be accompanied by the plant's induces for a reduction or an increase in some plant metabolites leading to an increase in plant resistance or tolerance. These stresses can be abiotic stress such as nutrient deficit, drought (water deficit or salinity), water-logging or flooding, extreme cold, frost, heat, sodicity, and metal and metalloid toxicity or biotic stress which are responsible for the damage done to an organism by other living organisms like herbivores or pathogens, bacteria, viruses, fungi, parasites, beneficial and harmful insects, weeds, and cultivated or native plants.

3. Adverse effects of abiotic stress conditions on plant growth and productivity

Various abiotic stress conditions such as salinity, drought, flooding, temperature (heat, cold), nutrient elements deficiency, alkalinity, organic and inorganic pollutants and heavy metals adversely affect crop plants growth, development and productivity [9] as shown in **Figure 3**.

3.1 Adverse effects of salinity stress conditions

Excessive salinity is one of the most important abiotic factors influencing the world's agricultural lands [11]. Also, it is one of the principle reasons that limit agricultural productivity [12]. It delays plant development by shifting numerous physiological, biochemical, and metabolic processes. Excessive accumulation of sodium chloride (NaCl) and other salts persuades water-deficient conditions owing to uncontainable stomata closure causing osmotic stress to plant roots. It results in ionic inequity which causes reduction in shoot and leaf growth, untimely leaf death, and necrosis [13, 14]. Reduced water absorption and augmented salts accumulation like Na^+, K^+, Mg^+_2, Ca^+_2, and Cl^- inside the cell and as a result increased ion toxicity. The reduced growth of the plants under salinity is due to nutrient disturbances, affecting the availability, mobilization, and distribution of nutrients. This may be attributed to the competition of sodium (Na^+) and chloride (Cl^-) with nutrients such as potassium (K^+), calcium (Ca^+_2) and nitrate (NO^-_3) [15]. Under higher accumulation of salts, the activity of nitrogenase enzyme encompassed in biological nitrogen fixation (BNF) is reduced then the nodulation process highly diminished [16, 17]. Currently, 50% of all irrigation patterns are impacted by salinity.

3.2 Adverse effects of drought stress conditions

Drought stress is one the greatest stressors for plants which can occur when the availability of water to the roots is insufficient or when the transpiration rate is too

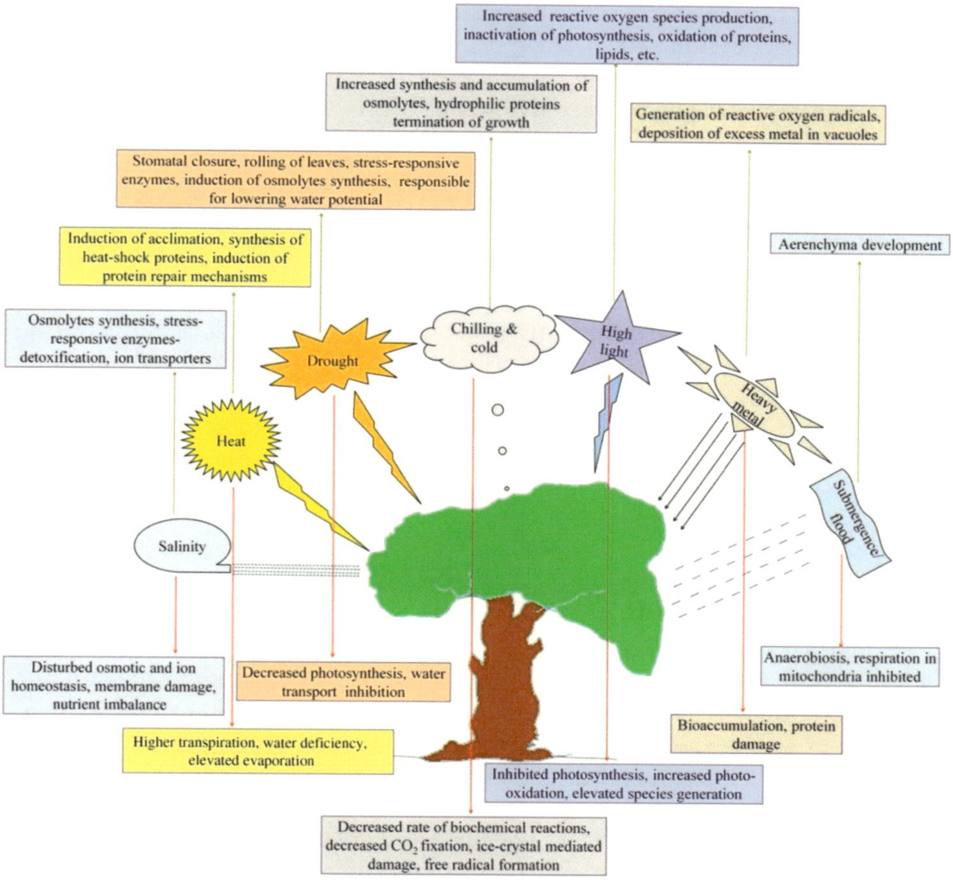

Figure 3.
Diverse abiotic stresses and the strategic defense mechanisms adopted by the plants. This figure shows diverse abiotic stresses and the strategic defense mechanisms adopted by the plants. although the consequences of salinity, heat, drought, and chilling are different, the biochemical responses seem more or less similar. High light intensity and heavy metal toxicity also generate similar impact, but submergence/flood situation leads to degenerative responses in plants where aerenchyma are developed to cope with anaerobiosis. It is, therefore, clear that adaptive strategies of plants against variety of abiotic stresses are analogous in nature. It may provide an important key for mounting strategic tolerance to combined abiotic stresses in crop plants. (Adapted from [10]).

high. These two conditions regularly coincide with tropical (arid) and sub-tropical (semi–arid) climates. Water deficit restricted photosynthesis activity due to imbalance between light capture and its utilization as a consequence oxidative stress occurred [18]. Drought stress prompted a remarkable decreasing in photosynthesis, which is reliant on photosynthesizing tissue and photosynthetic pigments [19, 20]. Through stresses, active solute buildup (*i.e.*, TSS, proteins, and FAAs) is claimed to be an effective stress tolerance mechanism [21]. Drought stress conditions lead to a decrease in the metabolic and physiological performance of plants and consequently the plant growth and productivity negatively affects. Additionally, drought stressor limits biological nitrogen fixation, and pigment content [13] as well as it reduces nutrients accessibility and their passage. Likewise, it greatly increases reactive oxygen species (ROS) concentration leading to an increase in oxidative stress, which take place because of an inequity created between the rate of electron transport and reducing power activity for metabolic consumption [22, 23]. Reactive oxygen species further prompt modifications in tissue construction and performance, enzyme stability, and lipid peroxidation [24].

3.3 Adverse effects of temperature stress conditions

Climatic changing conditions result in an increase of the intensity of heat and cold stress. The temperature stress causes alterations in membrane, water potential, and photosynthetic activity in plants. The optimum temperature for third carbon plants' (C3 plants') growth is stated 15–25°C by a number of scientists [25–27]. Up and down the optimum temperature, the plant performance was limited. Heat stress restricts cool-season plant development in summer in many positions of the world. Throughout the warm season, heat stress limited photosynthesis and carbohydrate buildup, augmented cell membrane damages triggered protein folding and even cell death in C3 plants [27]. The same damages have been recorded in warm-season plants, fourth carbon plants' (C4 plant species), in the winter. Also, the C4 species uptake less water and needed to alter themselves to be able to absorb mineral elements with low solubility [28].

3.4 Adverse effects of nutrient element deficiency stress conditions

Nutrient elements are considered fundamental for plant growth, development, and survival. 17 essential elements are necessary to maintain plant growth and development. Three of them (C, H and O) are derived from the air and water whereas the rest (N, P, K, Ca, Mg, S, Fe, Mn, Cu, Zn, Cl, B, Mo, and Co) are supplied either from soil or by adding fertilizers. Each of them plays a special role in plant life cycle and their necessity varies with the plant species and growth phases. Both the shortage and surplus of these nutrients lead to negative impacts on plant growth and development (**Figure 4**). Further, to make sure the efficient utilization of the nutrients, the environmental factors should be satisfactory. The plants absorb these elements in ionic form and its ability to absorb them is related to their quantities and distribution in the soil.

3.5 Adverse effects of alkalinity stress conditions

Alkalinity achieves its specific negative effect characteristics on crop plants in alkaline soils and disturbs plants at biological and physiological level. In addition to sodium chloride (NaCl) stress, there are other salts like sodium carbonate ($NaCO_3$) and sodium hydrogen carbonate ($NaHCO_3$) which are harmful to crops at excessive accumulations. High pH (more than eight) in alkaline soils diminishes the nutrient availability of crucial macro- and micro-nutrients, such as phosphorus (P), manganese (Mn), zinc (Zn), copper (Cu), and iron (Fe) causing nutrient deficiency and osmotic stress [29].

3.6 Adverse effects of contaminants stress conditions

Organic and inorganic pollutants are repeatedly being used in our environment via human interfering comprising industrial effluent discharge and agricultural practices, e.g., unreasonable and undue application of mineral elements and plant protective materials (pesticides) to soil. These chemical pollutants are causing major dangers to human health and their environment and may be directly or indirectly affecting on crop growth, development and productivity

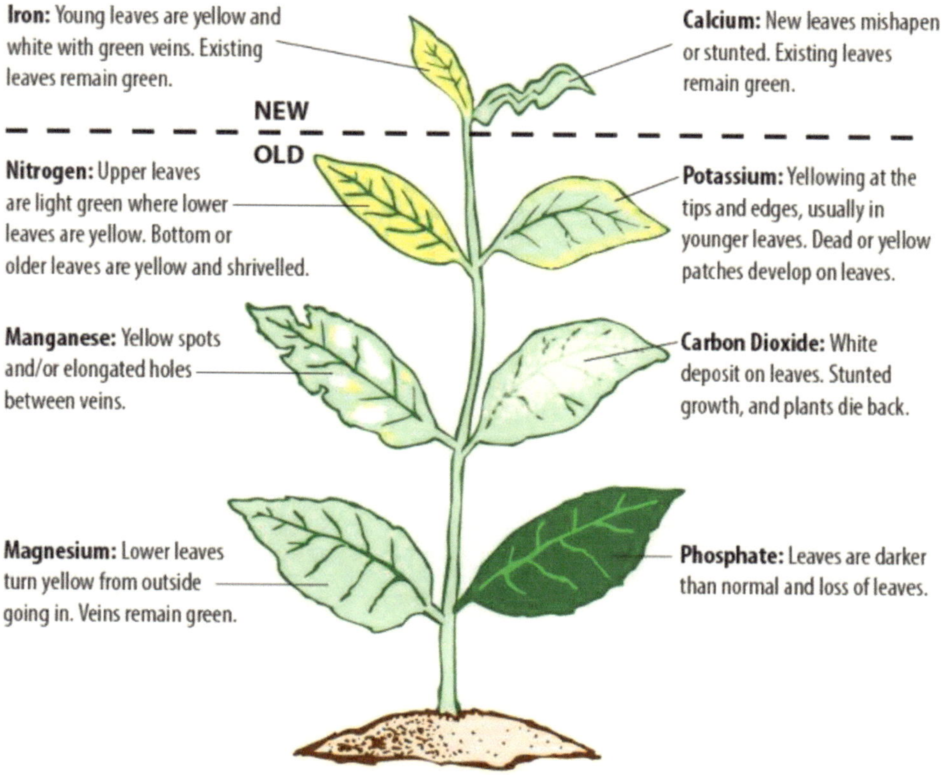

Figure 4.
The signs of essential nutrient elements deficiency in plants.

4. Plant behavior under stress conditions

A deficit of one or more of the vital nutrient elements caused several alterations that may be occurred at morphological, physiological, and also molecular levels of crop plants. The data presented in **Table 1** summaries these changes and in addition the symptoms that result from the deficiency of these essential nutrient elements on plants.

5. Role of microorganisms in mitigating abiotic stress conditions

The rhizosphere contains the tiny parts of soil inherent to roots of plants. The average count of microorganisms at the plant root region is very high as compared with the rest of the soil. So, it is clear that plant roots have an assortment of mineral, nutrient, and metabolite components, which are considered the principle factor for captivating microorganisms to assemble and link together. Root exudate of plants is a critical factor for microbial settlement in the rhizosphere. Shifting of microorganisms regarding the root exudates has an important role in pulling force of the microbial population to colonize the plant roots.

The interactions between microbial community and crop plants are vital to the modification and endurance of both in any abiotic environment. Induced Systemic Tolerance (IST) is the expression exploited for microbe-negotiated triggers of abiotic

ND	Plant responses at		Symptoms	References
	Physiological level	**Morphological level**		
N	A decrease in the activity of nitrate reductase and RO scavenger enzymes like SOD and POD, Decrease in chlorophyll content, and photosynthesis rate induces the chloroplast disintegration and loss of chlorophyll. An increase in production of phenolic compounds as secondary metabolites.	An elevated root shoot ratio with shortened lateral branches leaf area, High decreased in biomass production	Early, the older leaves show chlorosis comparing to the newer, Necrosis occurs. At later phases stunted growth and plant death if nitrogen deficiency continues.	[30, 31]
P	An increase in production of phenolic compounds as secondary metabolites and more organic acid was released.	A reduction in plant growth rate and remobilization of phosphorus happen	Optical disorders like red or purple color leaves occur because of anthocyanin accumulation.	[30, 32]
K	An increase in production of phenolic compounds as secondary metabolites.	A reduction in plant growth rate	Shortens inter nodes followed by bushy appearance, chlorosis and necrosis.	[30]
Ca	A significant decrease in chlorophyll content up to 50%, and protein, photosynthesis, Limited in translocation of photosynthesis compounds from source to sink significant increase in soluble nitrogen content of the plant.	Great reduction in growth rate, Less protein–N, RNA and DNA, A significant increase in soluble nitrogen content of the plant	A reduced amount of root and shoot branches.	[33–35]
Mg	Accumulation of sugar in source leaves was detected before reduction of photosynthesis and chlorophyll biosynthesis.		Leaf vines chlorosis.	[36, 37]
S			Habitually no visual symptoms, Reduction in total crop yield. Leaves begin to develop chlorosis. The chlorosis started from the leaf's edge and spread over intercostal area, but the zones beside the veins permanently remain green. Chlorosis happens, but it never turns into necrosis.	[38, 39]
Fe	Reduction in plant growth and even may stop some plant function An efficient decrease in photosynthesis.		Chlorosis of young leaves Reduction in crop production.	
Zn		A decrease in biomass production	Initial early senescence of the old leaves or slight yellowing of the newer leaves to the formation of the yellow chlorotic or even necrotic areas on the leaves.	

ND	Plant responses at		Symptoms	References
	Physiological level	**Morphological level**		
Mn	Necrotic spots or marginal necrosis may also develop. In dicotyledons the chlorosis develops first on the distal portions of the affected leaf blades, whereas in cereals, the leaf bases are first affected.	Diffuse interveinal chlorosis on the young, expanded leaf blades	Wheat leaves became mottled, Decrease in chlorophyll content, plant appearance turned yellow.	[40–42]
Cu	A decrease in activity of the cytochrome oxidase. This enzyme has a role in plant root nodule cells recovery under low oxygen stress for nitrogen fixation.		New leaves margin necrosis, lateral shoot death, unformed leaf margin, bleeding in main node stem and low lignification value in vessels.	[43]
Mo	A decrease in nitrogen fixation, Shoot and nodule dry weights.			
B	Accumulation of the phenolic compound A reduction in lignin biosynthesis.	Limitation of lateral bud growth in some plants	The inhibition or cessation of the roots and shoots elongations.	[44–48]
Cl		Less cluster formation and fewer yields	Reduction in leaf area and plant biomass. Leaves wilting, chlorotic mottling, bronzing, and tissue necrosis.	[49, 50]

Table 1.
Summaries the plant response at physiological and morphological levels and some symptoms under nutrient deficiency stress.

stress reactions. The duty of microorganisms in altering abiotic stresses in plants attracted the attention of several researchers [51–53]. The intrinsic metabolic of microbes and genetic aptitudes, participate to reduce abiotic environmental stresses in the plants [54]. The function of numerous rhizospheric microbes inhabitants with the genera *Azospirillum* [55], *Azotobacter* [56, 57], *Bacillus* [58–60], *Bradyrhizobium* [61], *Burkholderia* [62], *Enterobacter* [60], *Methylobacterium* [63], *Rhizobium* [60, 64], *Pantoea* [60, 65], *Pseudomonas* [60, 66], *Trichoderma* [67], and cyanobacteria [68] in elevation and control of growth in plant grown under different kinds of abiotic stresses has been reported.

In this regard, [69] reported that *Streptomyces* sp. strain PGPA39 alleviates salinity stress and stimulates the growth of "Micro-Tom" tomato plants and *Arabidopsis* [70]. *Burkholderia phytofirmans* strain PsJN overcome drought stress in maize [71] and wheat [72]. The data presented in **Tables 2** and **3** outline some examples of beneficial microorganisms that play a pivotal role in alleviation the adverse effects of abiotic stresses.

5.1 Role of microorganisms in mitigating salinity stress conditions

Endophytes and rhizobacteria as PGPB have potent in mitigating salinity stress. Their direct actions involve stimulation of phytohormones production, improvement of nutrient uptake, promotion of siderophore production, and nitrogen fixation. Some other indirect roles have resembled to actions in water-deficit stress as osmotic stability, which is pivotal in both conditions, such as accumulation of osmolytes

Stress Type	PGPRs	Plant	References
Salt	*Azospirillum brasilense*	*Hordeum vulgare, Lactuca sativa,Pisum sativum. Cicer arietinum*	[55, 73–75]
Salt	*Bacillus amylolequifaciens*, *B. insolitus Microbacterium* sp. *Pseudomonas syringae*	*Triticum aestivum*	[76]
Salt	*P. fluorescens*	*Arachis hypogea*	[77]
Salt	*B. subtilis*	*Arabidopsis thaliana*	[78]
Salt, drought	*Achromobacter piechaudii*	*Lycopersicon esculentum*	[79, 80]
Drought	*Azospirillum brasilense Pseudomonas* spp.	*Zea mays*	[81, 82]
Drought	*Rhizobium* sp., *P. putida* P5	*Helianthus annus*	[83, 84]
Drought	*Bacillus, P. mendocina*	*Lactuca sativa*	[85, 86]
Drought	*B. megaterium*	*Trifolium*	[87]
Drought	*Pseudomonas* sp., *Variovorax paradoxus*	*Pisum sativum*	[88, 89]
Drought	*Paenibacillus polymyxa, Rhizobium tropici*	*Vigna radiata*	[90]
Drought	*Pseudomonas* spp.	Asparagus	[91]
Drought	*Azospirillum* sp., *B. safensis, Ochrobactrum pseudogregnonense*	*Triticum aestivum*	[92, 93]
Flooding	*Enterobacter cloacae, P. putida*	*Lycopersicon esculentum*	[94]
Heat	*Pseudomonas* sp. AMK-P6	*Sorghum bicolor*	[66]
Cold	*P. putida*	*Brassica napus*	[95]
Cold and heat	*Burkholderia phytofirmans*	*Vitis vinifera*	[96]
Heavy metals	*Sanguibacter* sp., *Pseudomonas* sp.	*Nicotiana tabacum*	[97]
Heavy metals	*B. subtilis, Pantoea agglomerans*	*Avena sativa*	[98]
Heavy metals	*P. fluorescens, Microbacterium* sp.	*Brassica napus*	[99]
Ni and Cd	*Methylobacterium oryzae, Burkholderia* sp.	*Lycopersicon esculentum*	[100]
Iron toxicity	*B. subtilis*, *Bacillus* sp., *B. megaterium*	*Oryza sativa*)	[101]

Table 2.
Soil microorganisms (endophyte or rhizobacteria acting as PGPR and conferring the plants' abiotic stress tolerance.

(glycine betaine, proline, trehalose, EPS, and volatile organic compounds accumulation). These compounds elevate plant growth via perpetuate ion homeostasis. PGPR improves plant tolerance to salinity stress via induced systemic tolerance (IST) [16, 122]. In this connection, [123] proved that the application of plant growth-promoting bacteria, PGPB, producing ACC deaminase enzyme or transgenic plants revealed the corresponding acdS gene, growth evolution, seeds productivity, and enhancement of *Camelina sativa* quality on plants grown in marginal land which not suitable for cultivation due to high salinity.

5.2 Role of microorganisms in mitigating drought stress conditions

Plant Growth Promoting Bacteria (PGPB) supports the antioxidant apparatus of plants via managing antioxidant enzyme level, consequently, increasing the plant

Stress type	Arbuscular mycorrhizal fungi (AMF)	Plant	References
Drought	*G. mosseae*	*Poncirus trifoliata*	[102]
Drought	*G. deserticola*	Pepper	[103]
Drought	*G. intraradices*	*Rosa hybrida* L., *Lactuca sativa, Cicer arietinum*	[104–106]
Drought	*G. etunicatum, G. versiform*	*Cicer arietinum*	[106]
Salt	*Glomus etunicatum*	*Carthamus tinctorius*	[107]
Salt	*G. intraradices*	*Zea mays,Trigonella foenum-graecum, Fragaria ananassa*	[108–110]
Salt	*G. viscosum*	*Medicago sativa* L.	[111]
Salt	*G. etunicatum*	*Brachiaria humidicola*	[112]
Nutrient deficiency, Heat	*G. mosseae*	*Dalbergia sissoo, Acacia nilotica, Poncirus trifoliata*	[113–115]
Heat and cold	*G. mosseae, G.* sp. R10 *G. aggregatum, G. fasciculatum Gigaspora margarita*	*Fragaria ananassa*	[116]
Heavy metals	*G. mosseae*	*Piper nigrum*	[117]
Heavy metals	*G. mosseae, Aculaospora laevis*	*Zea mays*	[118]
Heavy metals	*G. intraradices*	*Thlaspi* sp.	[119]
Heavy metals	*G. etunicatum, G. intraradices*	–	[120]
Heavy metals	*G. macrocarpum*	*Zea mays*	[121]

Table 3.
Arbuscular Mycorrhizal Fungi (AMF) that act as PGP and conferring the plants abiotic stress tolerance.

resistance to abiotic stresses [124]. Plant growth-promoting rhizobacteria mitigate the water deficit condition by altering several physiological and biochemical processes in plants via a rhizobacterial-induced drought endurance and resilience (RIDER). This procedure includes secretion of exo-polysaccharides (EPS), management of endogenous phytohormones and antioxidants, and coordinated organic solutes, e.g., sugars, amino acids, and polyamines, and/or fabricating of volatile organic constituents, dehydrins, and heat shock protein [125]. These techniques help plants to sustain water deficit by preserving plant growth, membrane stability, and enzyme constancy and effectively controlling the water and mineral uptake by increasing the surface area of root [16, 126].

5.3 Role of microorganisms in mitigating temperature stress conditions

Adapted microbes to high or low temperatures could alleviate their harmful effects. Microbes have explicit enzymatic structures that manage their metabolism to overcome the changing temperature and preserve their membrane and enzyme stability. Under these conditions, heat and cold shock proteins are established. These molecular chaperones contribute resistance to adjacent high-temperature stress [16, 127]. These severe conditions caused protein denaturation, which is handled with trehalose through formation of a gel-like web to save plants from dehydration [128]. Cold-adapted microbes found at high-altitude agro-ecosystem, have a vast prospect to assist plants in alleviating unfavorable climatic conditions. In cold desert of the Himalayas, India psychrophilic and psychro-tolerant bacteria exhibited plant growth-stimulating characteristics, including *Arthrobacter, Aeromicrobium, Aeromonas, Bacillus, Bosea, Burkholderia, Brevundimonas, Citricoccus, Exiguobacterium, Janibacter, Janthinobacterium, Jeotgalicoccus, Kocuria, Methylobacterium, Pseudomonas, Providencia, Psychrobacter, Pantoea, Plantibacter, Rhodococcus, Sanguibacter, Sporosarcina, Staphylococcus, Sphingobacterium,* and *Variovorax* [129]. Correspondingly, the isolation of bacteria associated with heat-tolerant plants from wheat exhibited improvement in traits of plant growth and development under heat stress. They encompassed bacterial genera like *Alcaligenes, Arthrobacter, Bacillus, Delftia, Methylobacterium,* and a number of pseudomonads [130].

5.4 Role of microorganisms in mitigating alkalinity stress conditions

Application of encouraging phytoremediation technology depends on the integrated effect of plants and associated microbes. It has a valuable strategy to clean up the biodegradation of organic pollutants and heavy metal-polluted soils.

5.5 Role of microorganisms in mitigating contaminants stress conditions

Application of encouraging phytoremediation technology depends on the integrated effect of plants and associated microbes. It has a valuable strategy to clean up biodegradable organic pollutants and heavy metal-polluted soils [131]. PGPB responds to heavy metal stress via different mechanisms involving bioaccumulation, enzymatic detoxification, metal mobilization, immobilization, volatilization, and EPS complexation as well as accumulation of phytohormone, solubilization of phosphate, siderophore, ACC-deaminase, and NF [132, 133]. Metal solubility and accessibility in the soil were influenced by microbes. Any metal pollutants cannot be easily degraded, so they must be either stabilized or extracted from the soil. Metal-chelating siderophores and enzyme mechanisms involved in phosphate solubilization expedite heavy metal uptake under stress conditions [134]. Growth-promoting microbes build up chelating compounds such as siderophores which may decrease soil pH and promote metal solubility via complex formation. Also, the production of organic acids, such as citric, gluconic, and oxalic, may promote metal mobilization, and uptake consequently, accumulation in plant shoots, by phytoextraction. Redox processes promote bioavailability of metals as reduction of Mn (IV) to Mn (III) and Fe (III) to Fe (II) so, become less toxic. Moreover, the bioavailability could increase using bio-surfactants and phyto-chelatins via formation of the complex with heavy metals [134–138]. Phyto-stablization through growth-enhancing bacteria and plant development may reduce metal availability in highly metal-polluted soils. This may occur via the formation of new specific metals, altered metal adsorption on plant cell walls, or ejection through downfall. Phyto-management is a combination of several phyto-technologies, a sustainable application and cost valid can contribute enormous assistance in repair of metal-polluted soils [139].

6. Mechanisms of microorganisms for alleviating abiotic stress conditions

The bio-fertilizers, bio-stimulators, and bio-control effects of PGPRs (**Table 4**) are contingent on their natural ability, as well as the interaction manner and militant endurance circumstances. GPB promotes plant proliferation with direct and/or

PGPR forms	Definition	Mechanism of action	References
Bio-fertilizer	An ingredient that has microbes (bacteria, fungi, AMF,BGA AB etc.) which, when applied on the seed, plant surface or soil, colonizes the environmental of roots and stimulate plant growth by various ways like, increased supply of primary nutrients for the host plant	Biological nitrogen fixation (BNF) Phosphate solubilizing microbes	[140, 141]
Bio-stimulator	Microorganisms that characterized by their ability to produce or synthesis phytohormones or other secondary metabolites that stimulate plant growth	Like indole acetic acid (IAA), gibberellic acid (GA), cytokinins and ethylene	[141, 142]
Bio-control agents	Microorganisms that protect plant against diseases by controlling phytopathogenic using different mechanisms	Synthesis of antibiotics, siderophores, HCN, hydrolytic enzymes,Acquired and Induced systemic resistance	[140, 141, 143]

**Adapted from [144].*

Table 4.
Some mechanism of action of PGPRs that enhance the plant growth.*

indirect techniques [6, 145]. Concerning direct mechanisms, it involved the synthesis of compounds that expedite the uptake of crucial nutrients and micronutrients from the soil and accumulation of plant growth regulators, such as phosphorus and potassium solubilization, iron and zinc sequestration, siderophore and plant hormone accumulation, and atmospheric nitrogen fixation. Regarding the indirect techniques, it occurs through the accumulation of HCN and antifungal components, hostile activity regarding pathogenic organisms, and resistance to unfavorable stress conditions. Moreover, the bacteria can promote systemic resistance in plants via the accumulation of certain metabolites that provide extracellular signals and stimulate a series of internal processes. Ultimately, these signals are recognized by different plant cells responsible on the promotion of the defense system.

In addition to bacteria, fungi especially mycorrhizae are considered pivotal plant growth stimulators. Mycorrhizae are mainly divided into mycorrhizal fungi (MF) and vesicular-arbuscular mycorrhizal (VAM) fungi. These types of fungi are either still connected externally with the host plant (ectomycorrhizae) or they may organize endosymbiotic associations (VAM). They form extended networking of fungal mycelium, so, maximize nutrient uptake via roots. In this connection, [146] concluded that the endophyte root fungal of *Piriformos poraindica* promoted salt and drought tolerance in Chinese cabbage and barley, respectively. These stimulatory effects were achieved by promoting the concentration and activity of antioxidants and stimulating many other processes [147]. The possibility of microbial connections with the plants has several aspects. It starts with the induction of local or systemic stress mitigation techniques in plants to resist unfavorable stress conditions. Then, they assist plants to protect their growth, proliferation, and development via fixation, mobilization and/or accumulation of nutrients, hormones and organic phytostimulant components. These multipronged roles of microorganisms or their populations demonstrate their strength, achievable and critical options for different alleviation techniques for abiotic stress in plant crops.

Various suggested techniques explain the effect of microbes in mitigation of abiotic stress. Soil-dwelling microbes can be classified into genera *Achromobacter, Aeromonas, Azospirillum, Azotobacter, Bacillus, Enterobacter, Klebsiella, Pseudomonas,* and *Variovora which* exhibited enhancement of plant growth under different stress conditions [60, 75, 89, 122, 125, 148]. Several publications concerned with the role of microbes for alleviating abiotic stresses indicate the importance of microbes in this field (**Tables 5–9**). All soil-inhabiting bacteria are organized as plant growth promoters (PGP) if they are able to promote plant growth even under different unfavorable physicochemical conditions. There are several tools by which microbes promote plant growth as indole acetic acid (IAA), which is synthesized in the shoot apical meristem and gathered in the active root apical meristems. The auxins have growth-promoting roles in plant-involved cell elongation, consequently root growth induction and lateral root formation. In contrast, the high auxins concentrations, promote retardant effects on root growth [60, 186]. The same result was recorded as a result of high ethylene synthesis [186]. Results also concluded that the rhizosphere colonizing bacteria promote plant growth via phytohormones production [187]. Generally, agricultural practices observed that the PGPRs not only assist in alleviation of environmental stresses, but also increase the yield of several crop plants including barley, maize, rice, and soybean [174, 188, 189]. In this regard, *Pseudomonas* sp. PMDzncd2003 enforces salt tolerance on rice germinates under salt stress. It also has a high ability to root colonizing parallel to the ability to accumulate exo-polysaccharides (EPS) that

Crop plants	**Microorganisms**	**Effect/Mechanism**	**References**
Maize (*Zea mays*)	*Azospirillum lipoferum*	Increase accumulation of TSS, FAAs, and proline Enhance the growth parameters	[149]
	Bacillus Spp.	Increased accumulation of proline, TSS, FAAs, Decrease electrolyte leakage, reduce the activity of antioxidants enzyme (CAL, GPX peroxidase)	[59]
Soybean	*Pseudomonas putida* H-2–3	Decrease the level of AB and SA, Increase the accumulation of JA. Modulated antioxidants by declining SOD, flavonoids, and RSA	[150]
Wheat (*Triticum aestivum*)	*Bacillus amyloliquefaciens* 5113 *Azospirillum brasilense* NO40	Bacterial-mediated plant attenuated transcript level and improves homeostasis.	[23]
	A. brasilense NO40, *R. leguminosarum* (LR-30), *R. phaseoli* (MR-2) *Mesorhizobium ciceri* (CR-30,39),	Improved the growth, biomass, and drought tolerance index throughout the production of CAL, EPS, and IAA	[151]
Lavandula dentate	*Bacillus thuringiensis*	IAA induced higher proline and K-content improved nutritional, physiological, and metabolic activities, Decreased the activity of: GR and APX	[152]
Cicer arietinum L.	*Pseudomonas putida* MTCC5279 (RA)	Increase: osmolyte accumulation, ROS scavenging ability, and stress-responsive gene expressions	[24]

Crop plants	Microorganisms	Effect/Mechanism	References
Lettuce	*Azospirillum sp.*	Increased chlorophyll and ascorbic acid content, Promote air-part biomass, better overall visual quality, hue, chroma and antioxidant capacity, and a lower browning intensity	[153]
Arabidopsis	*Azospirilum brasilense sp* 245	Improved plants seed yield, plants survival, proline levels, and relative leaf water content; Decreased stomatal conductance, malondialdehyde, and relative soil water content	[154]
	Phyllobacterium brassicacearum strain STM196	Enhanced ABA content resulted in: Decreased leaf transpiration, Delay in reproductive development, Increased biomass and water use efficiency	[155]
Brassica oxyrrhina	*Pseudomonas libanensis* TR1 and *Pseudomonas reactans* Ph3R3	Enhanced plant growth, leaf relative water, and pigment content Decreased concentrations of proline and malondialdehyde in leaves	[133, 156]
Rice (*Oryza sativa* L.)	*Trichoderma harzianum*	Stimulate root growth independent of water status, Delay drought response	[12]
Medicago truncatula	Sinorhizobium medicae	Improve root nodulation and nutrient acquisition during drought stress	[157]
Wheat	*Bacillus* spp., *Enterobacter* spp., *Moraxella* spp., *Pseudomonas* spp.	Auxin synthesis	[158]
Vigna mungo L. *Pisum sativum* L	Consortium (*Ochrobactrum pseudogrignonense*, *Pseudomonas* sp., *Bacillus subtilis*)	Enhance the production of ACC deaminase, RO scavenging enzymes, and osmolytes	[159]
Wheat	*Pantoea agglomerans*	Improving soil aggregation through (EPS)	[160]
Arabiodopsis	*Paenibacillus polymyxa*	Induction of stress-resistant gene ERD 15	[161]
Sunflower	*Rhizobium sp.*	Enhancing Soil aggregation through EPS	[83]
Wheat	*Azospirillum sp.*	Improved Water relations	[92]
Pea	*Variovorax paradoxus*	production of ACC-deaminase	[162]
Arabiodopsis	*Paraphaeosphaeria quadriseptata*)	Induction of HSP	[163]
Rice	*Brome mosaic virus*	-	[164]
Common bean	*P. polymyxa and Rhizobium tropici*	Change in hormone balance and stomatal conductance	[90]
Pea	*Pseudomonas sp.*	Decreased ethylene production	[88]
Lettuce	*Pseudomonas mendocina, Glomus intraradices*	Improved antioxidant status	[86]
Sunflower	*Pseudomonas putida* P45	Improved soil aggregation due to EPS production	[84, 165]

Crop plants	Microorganisms	Effect/Mechanism	References
Trifolium	*Bacillus megaterium, Glomus sp.*	IAA and proline production	[87]
Tomato	*Achromobacter piechaudii*	Synthesis of ACC-deaminase	[79]
Sorghum	AM Fungi	Improved Water relation	[166]

Table 5.
List of some microorganisms that have the ability for mitigating drought stress condition through different mechanisms.

Crop plants	Microorganisms	Effect/Mechanism	References
Groundnut *(Arachis hypogaea* L.)	*Brachybacterium saurashtrense* (JG-06), *Brevibacterium casei* (JG-08), *Haererohalobacter* (JG-11)	Higher of K+/Na+ ratio, Ca2+, P, and N content. Shoot and root contain a higher concentration of auxin	[12, 167]
Mung bean (*Vigna radiate*)	*Rhizobium and Pseudomonas*	Improving growth, nodulation and yield of mung bean under natural and salt-affected conditions throughout ACC-deaminase production	[168]
Barley and oats	*Acinetobacter spp.* , *Pseudomonas sp.*	Production of enzyme ACC deaminase, lower ethylene and IAA promote plant growth	[169]
Wheat	*Azospirillum sp.*	Increased shoot dry weight and grain yield. Plants accumulate some organic solutes as. Proline, TSS and inorganic ions to maintain osmotic adjustment	
	Pseudomonas sp. Serratia sp.	Have ACC deaminase activity, Reduce ethylene level and enhance plant growth and yield	[170]
Maize (*Zeya Mays*)	*Pseudomonas and Enterobacter*	Enhance N, P, and K uptake and increase K+/Na+ ratios, Decrease triple response	[171]
Rice GJ-17	*Pseudomonas pseudoalcaligenes Bacillus pumilus*	Reduced: the toxicity of ROS, the activity of lipid peroxidation, and SOD activity	[172]
Rice	*Bacillus amyloliquefaciens* NBRISN13 (SN13)	Modulating differential transcription in a set of at least 14 genes	[173]
Barley (*Hordeum vulgare* L.)	Hartmannibacter diazotrophicus E19	Increased: root and shoot dry weight. Enhance ACC-deaminase activity lower ethylene content	[174]
lettuce seeds	*Azospirillum*	Stimulated ascorbic acid content, antioxidant capacity, higher biomass, and a lower browning intensity	[153]
Brassica napus (canola) and Maize	*Pseudomonas putida* UW4	Modulation of plant protein differential expression and ACC deaminase activity	[175]
Oryza sativa **L.** *Arabidopsis thaliana*	*Curtobacterium albidum*	Modulation of osmolytes and antioxidative enzymes, and induction of systemic tolerance	[176]

Crop plants	Microorganisms	Effect/Mechanism	References
Alfafa	*Enterobacter* sp.	Synthesis of 2-keto-4-methylthiobutyric acid (KMBA)	[177]
Barley	*Piriformaspora indica*	Increased antioxidative capacity	[178]
Wheat	*B. amylolequifaciens* *B. insolitus, Microbacterium sp.* *P. syringae*	Restricted Na^{+} influx	[76]
Groundnut	*Pseudomonas fluorescens*	production of ACC-deaminase	[77]
Rice	Scytonema	Production of GA and extracellular products	[179]
Tomato	*Achromobacter piechaudii*	Production of ACC-deaminase	[79]
Sorghum	AM Fungi	Amended Water relation	[166]

Table 6.
List of some microorganisms that have the ability for mitigating salinity stress condition through different mechanisms.

Temperature	Crop plants	Microorganisms	Mechanisms	Reference
Heat	Wheat	*Pseudomonas putida*	Phytohormone, HCN, ammonia, siderophore and P-solubilization, and accumulation of metabolites like proline, sugars, starch, amino acids, and proteins	[180]
Heat	Sorghum	*Pseudomonas sp.* AMK-P6	Induction of heat shock proteins and improved plant biochemical status	[66]
Low	Grapevine	*Burkholderia phytofirmans* PsJN	Synthesis of ACC-deaminase	[96]
Low	Canola	*P. putida*	Synthesis of ACC-deaminase	[95]

Table 7.
Some microorganisms that have the ability for mitigating temperature stress conditions through different mechanisms.

promote salinity tolerant [190]. Also, inoculation of rice with *Bacillus pumilus* mitigates salinity and high boron stresses [191]. The reported technique for cell protection under stress conditions was high antioxidant enzyme activity accompanied by the presence of bacterial inoculant. More studies are needed to investigate the communication between plant and bacterial colonizers at the molecular level.

Finally, [192] have proved the duty of *Trichoderma harzianum* on alleviation of stress in different rice genotypes through adjustment of dehydrin, malonialdehyde and aquaporin, and genes parallel to several physiological traits. Rhizobacteria-promoted resistance to water deficit and resilience (RIDER) by altering the phytohormone levels, enzyme activities, defense-related proteins incorporation, antioxidant levels, and epoxypolysaccharide accumulation for plants. These strategies help plants to mitigate unfavorable conditions [122, 125]. Using stress tolerant microorganisms is a promising tool in improving the productivity of crop plants grown in stress-susceptible areas. Application of *Trichoderma harzianum* improved oil content in NaCl affected Indian

Water stress	plants	Microbes	Mechanisms	References
Flooding	Dragonblood (*Pterocarpus officinalis*)	AM fungi & *Bradyrhizobium*	Development of adv. roots, aerenchyma, and hyper trophied lenticels	[181]
Flooding	Tomato	*Pseudomonas putida, Enterobacter cloacae*	Synthesis of ACC-deaminase	[94]

Table 8.
List of some microorganisms that have the ability for mitigating flooding stress conditions through different mechanisms.

Heavy metal S	Crop plants	Microorganisms	Mechanisms	References
Cadmium and Iron toxicity	*Hibiscus cannabinus*	Enterobacter sp.	Metal immobilization Production:, IAA siderophore)	[182]
Lead toxicity	Sunflower	*Pseudomonas gessardii, Pseudomonas fluorescens*	Lead uptake (increase in APX, CAL, SOD, GR, and proline contents)	[183]
Arsenic toxicity	Rice	*Achromobacter* sp.	ACCD (Arsenic uptake)	[184]
Ni & Cd Toxicity	Tomato	*Methylobacterium oryzae, Burkholderia sp*	Reduced uptake and translocation	[100]
Metal toxicity	Chickpea	PGPR	Sequestration of metal ions	[185]

Table 9.
List of some microorganisms that have the ability for mitigating heavy metal stress conditions through different mechanisms

mustard (*Brassica juncea*) via increasing the uptake of essential nutrients, promoting the accumulation of antioxidants and osmolytes, and decreasing NaCl uptake [67]. In addition to, up-regulation of monodehydroascorbate reductase in treated plants. It also alleviates salinity stress via accumulation of ACC-deaminase [193]. Moreover, inoculation of barley and oats, with *Acinetobacter* sp. and *Pseudomonas* sp. enhance the accumulation of IAA and ACC deaminase under saline soil [169].

7. Conclusion

Agriculture is the backbone of developed and particularly developing countries, with more than 60% of the population of the developing countries depending on it for their livelihood. Increasing food production to fulfill the needs of an increasing world population becomes of a major concern. Despite the necessity of doubling agricultural production, in terms of quantity and quality, to cope with the worsening increase in the global population and to meet the increasing humanitarian needs, the agricultural sector faces many abiotic and biotic stress conditions. Additionally, the great climate changes resulting from global warming lead to an increase in the negative impact of these stressors. Throughout this literature study, it is well established that the abiotic stress conditions (salinity, drought, high and low temperature, alkalinity, and organic and inorganic pollution have great side effects on plants (decreasing in plant growth and productivity, physiological changes, alteration in osmotic balance and ion cytotoxicity). Moreover, the side effects of abiotic stress conditions have been expected to be increased because of the bad or nonsustainable agricultural practices, water scarcity and reduced arable land, soil degradation, human activity, and the climate change (global warming of the planet). Hence, it has become a necessity to reduce the different causes behind the increasing abiotic stress conditions. On one hand, these can be achieved through good and sustainable agricultural practices such as agricultural rotation system, integrated crop management, integrated nutrient management, and integrated pest management re-mapping of agricultural map in the light of climate change, soil fertility, *etc*. On the other hand, in order to increase crop productivity, it becomes necessary to develop low-cost technologies for abiotic stress management. Soil microorganisms, surviving in the soil under extreme conditions, have shown high properties, which, if exploited can serve agriculture by increasing and maintaining crop productivity. Our literature study has indicated the paramount importance of these beneficial microorganisms in the mitigation of the negative consequences resulting from different abiotic stress conditions. Where, it is well established that beneficial soil microorganisms can promote growth and increase productivity through different mechanisms such as increasing the availability of essential nutrient elements and enhancement of their uptake, phyto-hormones production, ACC-deaminase production, biological control agents' production, *etc.* Even though, more efforts should be given in this field like that, isolation and characterization worldwide benefit microbes from different biological niches and under various harsh conditions. Further researches will be required concerning the optimization of the mass production of these microorganisms, the best carrier that allow increasing the shelf life of beneficial microorganisms and par consequence increasing its storage ability, also, the better ways for its field application. The application of these beneficial microorganisms is still limited and how to increase their application rate should be taken into account.

Author details

Talaat El Sebai[1*] and Maha Abdallah[2]

1 Agricultural Microbiology Department, Agricultural and Biologyl Research Institute, National Research Centre, Giza, Egypt

2 Botany Department, Agricultural and Biologyl Research Institute, National Research Centre, Giza, Egypt

*Address all correspondence to: tn.elsebai@nrc.sci.eg; talaatelsebai@gmail.com

References

[1] FAO. High Level Expert Forum - How to Feed the World in 2050. Rome: Economic and Social Development Department; 2009

[2] Borlaug N, Dowswell CR. Feeding a world of ten billion people: A 21st century challenge. In: Tuberosa T, Phillips RL, Gale M, editors. Proceedings of in the Wake of the Double Helix: From the Green Revolution to the Gene Revolution. Bologna, Italy; 2005. pp. 3-24

[3] Reynolds M, Tuberosa R. Translational research impacting on crop productivity in drought-prone environments. Current Opinion in Plant Biology. 2008;**11**:171-179

[4] Egamberdieva D, Wirth SJ, Alqarawi AA, Allah EF, Hashem A. Phytohormones and beneficial microbes: Essential components for plants to balance stress and fitness. Frontiers in Microbiology. 2017;**8**:2104. DOI: 10.3389/fmicb.2017.02104

[5] Kloepper JW, Lifshitz R, Zablotowicz RM. Free living bacterial inocula for enhancing crop productivity. Trends in Biotechnology. 1989;**7**:39-44

[6] Hayat R, Ali S, Amara U, Khalid R, Ahmed I. Soil beneficial bacteria and their role in plant growth promotion: A review. Annals of Microbiology. 2010;**60**:579-598

[7] Gray EJ, Smith DL. Intracellular and extracellular PGPR: Commonalities and distinctions in the plant bacterium signaling processes. Soil Biology and Biochemistry. 2005;**37**:395-412

[8] Backer R, Rokem JS, Ilangumaran G, Lamont J, Praslickova D, Ricci E, et al. Plant growth-promoting rhizobacteria: Context, mechanisms of action, and roadmap to commercialization of biostimulants for sustainable agriculture. Frontiers in Plant Science. 2018;**9**:1473-1490

[9] Gontia-Mishra I, Sasidharan S, Tiwari S. Recent developments in use of 1- aminocyclopropane-1-carboxylate (ACC) deaminase for conferring tolerance to biotic and abiotic stress. Biotechnology Letters. 2014;**36**:889-898

[10] Meena KK, Sorty AM, Bitla UM, Choudhary K, Gupta P, Pareek A, et al. Abiotic stress responses and microbe-mediated mitigation in plants: The omics strategies. Frontiers in Plant Science. 2017;**8**:172. DOI: 10.3389/fpls.2017.00172

[11] Masood S, Seiji Y, Shinwari ZK, Anwar R. Mapping quantitative trait loci (QTLs) for salt tolerance in rice (*Oryza sativa*) using RFLPs. Pakistan Journal of Botany. 2005;**36**(4):825-834

[12] Shukla N, Awasthi RP, Rawat L, Kumar J. Biochemical and physiological responses of rice (*Oryza sativa* L.) as influenced by *Trichoderma harzianum* under drought stress. Plant Physiology and Biochemistry. 2012a;**54**:78-88

[13] Enebe MC, Babalola OO. The influence of plant growth-promoting rhizobacteria in plant tolerance to abiotic stress: A survival strategy. Applied Microbiology and Biotechnology. 2018;**102**(18):7821-7835

[14] Julkowska MM, Testerink C. Tuning plant signaling and growth to survive salt. Trends in Plant Science. 2015;**20**:586-594

[15] Hu Y, Schmidhalter U. Drought and salinity: A comparison of their effects

on mineral nutrition of plants. Journal of Plant Nutrition and Soil Science. 2005;**168**:541-549

[16] Kumar A, Verma JP. Does plant—Microbe interaction confer stress tolerance in plants: A review? Microbiological Research. 2018;**207**:41-52

[17] Suzuki N, Bassil E, Hamilton JS, Inupakutika MA, Zandalinas SI, Tripathy D, et al. ABA is required for plant acclimation to a combination of salt and heat stress. PLoS One. 2016;**11**(1):e0147625

[18] Foyer CH, Noctor G. Oxygen processing in photosynthesis:Regulation and signaling. New Phytology. 2000;**146**:359-388

[19] Dubey RS. Photosynthesis in plants under stressful conditions. In: Pessarakli M et al., editors. Handbook of Photosynthesis. CRC Press; 2005. pp. 717-718

[20] Raza SH, Athar HR, Ashraf M. Influence of exogenously applied glycinebetaine on photosynthetic capacity of differently adapted wheat cultivars under salt stress. Pakistan Journal of Botany. 2006;**38**:341-351

[21] Jaleel CA, Kishore kumar P, Manivannan A (2008) Salt stress mitigation by calcium chloride in Phyllanthusamarus. Acta Botanica Croatica. 67:53-62

[22] Beck EH, Fettig S, Knake C, Hartig K, Bhattarai T. Specific and unspecific responses of plants to cold and drought stress. Journal of Biosciences. 2007;**32**(3):501-510

[23] Kasim WA, Osman ME, Omar MN, Abd El-Daim IA, Bejai S, Meijer J. Control of drought stress in wheat using plant-growth-promoting bacteria. Journal of Plant Growth Regulation. 2013;**32**:122-130

[24] Tiwari S, Lata C, Chauhan PS, Nautiyal CS. *Pseudomonas putida* attunes morphophysiological: Biochemical and molecular responses in *Cicer arietinum* L. during drought stress and recovery. Plant Physiology. 2016;**99**:108-117

[25] Beard JB. Turfgrass: Science and Culture. In: Pessarakli M, editor. Handbook of Plant and Crop Physiology. 3rd ed. USA: Taylor & Francis Publishing Group; 1973. pp. 453-485

[26] DiPaola JM. Syringing effects on the canopy temperatures of bentgrass greens. Agronomy Journal. 1984;**76**:951-953

[27] Fry JD, Huang B. Applied Turfgrass Science and Physiology. New Jersey, USA: John Wiley & Sons Hoboken; 2004. p. 2004

[28] Calatayud A, Gorbe E, Roca D, Martinez PF. Effect of two nutrient solution temperatures on nitrate uptake, nitrate reductase activity, NH4+ concentration and chlorophyll a fluorescence in rose plants. Environmental and Experimental Botany. 2008;**64**:65-74

[29] Chen L, Yin H, Xu J, Liu X. Enhanced antioxidative responses of a salt-resistant wheat cultivar facilitate its adaptation to salt stress. African Journal of Biotechnology. 2011;**10**(74):16887-16896

[30] Hewitt EJ. The essential nutrient element:Requirement and interaction in plants. Plant Physiology. 1963;**3**:137-360

[31] Forde BG. Nitrate transporters in plants:Structure, function and regulation. Biochimica et Biophysica Acta. 2000;**1465**:219-235

[32] Uhde-Stone C, Gilbert G, Jhonson J, et al. Acclimation of white

lupin to phosphate deficiency involves enhanced expression of gens related to organic acid metabolism. Plant and Soil. 2003;**248**:99-116

[33] Rangnekar P. Effect of calcium deficiency in the carbon metabolisms in photosynthesis and respiration in tomato leaf. Plant and Soil. 1975;**42**:565-583

[34] Singh OS, Sharma VK. Alternation in growth and metabolism of potato plants by calcium deficiency. Plant and Soil. 1972;**36**:363-369

[35] Rangnekar P. Effect of calcium deficiency on the translocation and utilization of C14–photosynthate in tomato plants. Plant and Soil. 1974;**41**:589-600

[36] Hermans CGN, Johnson RJ, Strasser N, Verbruggen N. Physiological characterization of magnesium deficiency in sugar beet: Acclimation to low magnesium differentially affects photosystems I and II. Planta. 2004;**220**:344-355

[37] Hermans C, Verbruggen N. Physiological characterization of Mg deficiency in *Arabidopsis thaliana*. Journal of Experimental Botany. 2005;**56**:2153-2161

[38] Lobb WR, Reynolds DG. Further investigations in the use of Sulphur in North Otago. New Zealand Journal of Agriculture. 1956;**92**:17-25

[39] Horiguchi T. Effects of nitrogen, phosphorus and manganese deficiencies on the formation of anthocyanin and other phenolic compounds in plants. Soil Science and Plant Nutrition. 1989;**60**:226-232

[40] Ferrari J, Sluijsmans CMJ. Mottling and magnesium deficiency in oats and their dependence on various factors. Plant and Soil. 1955;**6**:262-299

[41] Wallace T. In the Diagnosis of Mineral Deficiencies in Plants by Visual Symptoms. London: lIM Stationery Office; 1961

[42] Robson AD, Snowball K. Nutrient deficiency and toxicity symptoms. In: Reuter DJ, editor. Plant Analysis. An Interpretation Manual. Australia: Inkata Press; 1986. pp. 13-19

[43] Cartwright B, Hallsworth EG. Effect of copper deficiency on root nodules of subterranean clover. Plant and Soil. 1970;**33**:685-698

[44] Bohnsack C, Wand Albel LS. Early effects of boron deficiency on indoleacetic acid oxidase levels of squash root tips. Plant Physiology. 1977;**59**:1047-1050

[45] Kouchi H. Rapid cessation of mitosis and elongation of root tip cells of *Vida faha* as affected by boron deficiency. Soil Science and Plant Nutrition. 1977;**23**:113-119

[46] Kramer U, Cotter-Howells JD, Charnock JM, Baker AJM, Smith JA. Free histidine as a metal chelator in plants that accumulate nickel. Nature. 1996;**379**:635-638

[47] Troitskaya EA, Maevskaya AN. Temp GA (1975) hydroxyproline content in cell walls of plants with different boron requirements. Soviet Plant Physiology. 1975;**22**:854-857

[48] Skok J. The role of boron in the plant cell. In: Lamb CA et al., editors. Trace Elements. New York, USA: Academic Press; 1958. pp. 227-243

[49] Johanson CM, Stout PR, Broyer TC, Carlton AB. Comparative chlorine requirements of different plant species. Plant and Soil. 1957;**8**:337-353

[50] Terry N. Photosynthesis, growth and the role of chloride. Plant Physiology. 1977;**60**:69-75

[51] de Zelicourt A, Al-Yousif M, Hirt H. Rhizosphere microbes as essential partners for plant stress tolerance. Molecular Plant. 2013;**6**:242-245

[52] Nadeem SM, Ahmad M, Zahir ZA, Javaid A, Ashraf M. The role of mycorrhizae and plant growth promoting rhizobacteria (PGPR) in improving crop productivity under stressful environments. Biotechnology Advances. 2014;**32**:429-448. DOI: 10.1016/j.biotechadv.2013.12.005

[53] Souza RD, Ambrosini A, Passaglia LMP. Plant growth-promoting bacteria as inoculants in agricultural soils. Genetics and Molecular Biology. 2015;**38**:401-419. DOI: 10.1590/S1415-475738420150053

[54] Gopalakrishnan S, Sathya A, Vijayabharathi R, Varshney RK, Gowda CL, Krishnamurthy L. Plant growth promoting rhizobia: Challenges and opportunities. Biotech. 2015;**5**:355-377

[55] Omar MNA, Osman MEH, Kasim WA, Abd El-Daim IA. Improvement of salt tolerance mechanisms of barley cultivated under salt stress using *Azospirillum brasiliense*. Tasks Vegetation Science. 2009;**44**:133-147. DOI: 10.1007/978-1-4020-9065-3_15

[56] Sahoo RK, Ansari MW, Dangar TK, Mohanty S, Tuteja N. Phenotypic and molecular characterisation of efficient nitrogen-fixing *Azotobacter* strains from rice fields for crop improvement. Protoplasma. 2014a;**251**:511-523. DOI: 10.1007/s00709-013-0547-2

[57] Sahoo RK, Ansari MW, Pradhan M, Dangar TK, Mohanty S, Tuteja N. A novel *Azotobacter vinellandii* (SRIAz3) functions in salinity stress tolerance in rice. Plant Signal Behaviour. 2014b;**9**:e29377. DOI: 10.4161/psb.29377

[58] Tiwari S, Singh P, Tiwari R, Meena KK, Yandigeri M, Singh DP, et al. Salt-tolerant rhizobacteria-mediated induced tolerance in wheat (*Triticum aestivum*) and chemical diversity in rhizosphere enhance plant growth. Biology and Fertility of Soils. 2011;**47**:907-916. DOI: 10.1007/s00374-011-0598-5

[59] Vardharajula S, Ali SZ, Grover M, Reddy G, Bandi V. Drought tolerant plant growth promoting *Bacillus* spp.: Effect on growth, osmolytes, and antioxidant status of maize under drought stress. Journal of Plant International. 2011;**6**:1-14

[60] Sorty AM, Meena KK, Choudhary K, Bitla UM, Minhas PS, Krishnani KK. Effect of plant growth promoting bacteria associated with halophytic weed (*Psoralea corylifolia* L.) on germination and seedling growth of wheat under saline conditions. Applied Biochemistry and Biotechnology. 2016;**180**:872-882. DOI: 10.1007/s12010-016-2139-z

[61] Panlada T, Pongdet P, Aphakorn L, Rujirek N-N, Nantakorn B, Neung T. Alleviation of the effect of environmental stresses using co-inoculation of mungbean by *Bradyrhizobium* and rhizobacteria containing stress-induced ACC deaminase enzyme. Soil Science and Plant Nutrition. 2013;**59**:559-571. DOI: 10.1080/00380768.2013.804391

[62] Oliveira CA, Alves VMC, Marriel IE, Gomes EA, Scotti MR, Carneiro NP, Guimaraes CT, Schaffert RE Sa, NMH. Phosphate solubilizing microorganisms isolated from rhizosphere of maize cultivated in an oxisol of the Brazilian Cerrado biome. Soil Biology and

Biochemistry. 2009;**41**:1782-1787. DOI: 10.1016/j.soilbio.2008. 01.012

[63] Meena KK, Kumar M, Kalyuzhnaya MG, Yandigeri MS, Singh DP, Saxena AK, et al. Epiphytic pink-pigmented methylotrophic bacteria enhance germination and seedling growth of wheat (*Triticum aestivum*) by producing phytohormone. Antonie Van Leeuwenhoek. 2012;**101**:777-786. DOI: 10.1007/s10482-011-9692-9

[64] Remans R, Ramaekers L, Shelkens S, Hernandez G, Garcia A, Reyes JL, et al. Effect of *Rhizobium*, *Azospirillum* co-inoculation on nitrogen fixation and yield of two contrasting *Phaseolus vulgaris* L. genotypes cultivated across different environments in Cuba. Plant and Soil. 2008;**312**:25-37. DOI: 10.1007/s11104-008-9606-4

[65] Egamberdiyeva D, Höflich G. Influence of growth-promoting on the growth of wheat in different soils and temperatures. Soil Biology and Biochemistry. 2003;**35**:973-978. DOI: 10.1016/S0038-0717(03)00158-5

[66] Ali SZ, Sandhya V, Grover M, Kishore N, Rao LV, Venkateswarlu B. *Pseudomonas* sp. strain AKM-P6 enhances tolerance of sorghum seedlings to elevated temperatures. Biological Fertility Soil. 2009;**46**:45-55

[67] Ahmad P, Hashem A, Abd-Allah EF, Alqarawi AA, John R, Egamberdieva D, et al. Role of *Trichoderma harzianum* in mitigating NaCl stress in Indian mustard (*Brassica juncea* L) through antioxidative defense system. Frontiers in Plant Science. 2015;**6**:868. DOI: 10.3389/fpls.2015.00868

[68] Singh DP, Prabha R, Yandigeri MS, Arora DK. Cyanobacteria mediated phenyl propanoids and phytohormones in rice (*Oryza sativa*) enhance plant growth and stress tolerance. Antonie Van Leeuwenhoek. 2011;**100**:557-568. DOI: 10.1007/s10482-011-9611-0

[69] Palaniyandi SA, Damodharan K, Yang SH, Suh JW. *Streptomyces* sp. strain PGPA39 alleviates salt stress and promotes growth of 'Micro tom' tomato plants. Journal of Applied Microbiology. 2014;**117**:766-773. DOI: 10.1111/jam.12563

[70] Pinedo I, Ledger T, Greve M, Poupin MJ. *Burkholderia phytofirmans* PsJN induces long-term metabolic and transcriptional changes involved in *Arabidopsis thaliana* salt tolerance. Frontiers in Plant Science. 2015;**6**:466. DOI: 10.3389/flps.2015.00466

[71] Naveed M, Mitter B, Reichenauer TG, Wieczorek K, Sessitsch A. Increased drought stress resilience of maize through endophytic colonization by *Burkholderia phytofirmans* PsJN and *Enterobacter* sp FD17. Environmental and Experimental Botany. 2014a;**97**:30-39. DOI: 10.1016/j.envexpbot.2013.09.014

[72] Naveed M, Hussain MB, Zahir ZA, Mitter B, Sessitsch A. Drought stress amelioration in wheat through inoculation with *Burkholderia phytofirmans* strain PsJN. Plant Growth Regulation. 2014b;**73**:121-131. DOI: 10.1007/s10725-013-9874-8

[73] Hamaoui B, Abbadi JM, Burdman S, Rashid A, Sarig S, Okon Y. Effects of inoculation with *Azospirillum brasilense* on chickpeas (*Cicer arietinum*) and faba beans (*Vicia faba*) under different growth conditions. Agronomie. 2001;**21**:553-560

[74] Barassi CA, Ayrault G, Creus CM, Sueldo RJ, Sobrero MT. Seed inoculation with *Azospirillum* mitigates NaCl effects on lettuce. Scientia Horticulturae. 2006;**109**:8-14

[75] Dardanelli MS, Fernández de Córdoba FJ, Rosario Espuny M, Rodríguez Carvajal MA, Soria Díaz ME, Gil Serrano AM, et al. Effect of *Azospirillum brasilense* coinoculated with *Rhizobium* on *Phaseolus vulgaris* flavonoids and nod factor production under salt stress. Soil Biology and Biochemistry. 2008;**40**:2713-2721. DOI: 10.1016/j.soilbio.2008.06.016

[76] Ashraf M, Berge SH, Mahmood OT. Inoculating wheat seedling with exopolysaccharide-producing bacteria restricts sodium uptake and stimulates plant growth under salt stress. Biology Fertility Soils. 2004;**40**:157-162

[77] Saravanakumar D, Samiyappan R. Effects of 1-aminocyclopropane-1-carboxylic acid (ACC) deaminase from *Pseudomonas fluorescence* against saline stress under in vitro and field conditions in groundnut (*Arachis hypogeal*) plants. Journal of Applied Microbiology. 2007;**102**:1283-1292

[78] Zhang H, Kim MS, Sun Y, Dowd SE, Shi H, Pare PW. Soil bacteria confer plant salt tolerance by tissue-specific regulation of the sodium transporter HKT1. Molecular Plant-Microbe Interactions. 2008;**21**:737-744

[79] Mayak S, Tirosh T, Glick BR. Plant growth-promoting bacteria confer resistance in tomato plants to salt stress. Plant Physiology and Biochemistry. 2004a;**42**:565-572

[80] Mayak S, Tirosh T, Glick BR. Plant growth promoting bacteria that confer resistance to water stress in tomato and pepper. Plant Science. 2004b;**166**:525-530

[81] Casanovas EM, Barassi CA, Sueldo RJ. *Azospirillum* inoculation mitigates water stress effects in maize seedlings. Cereal Research Communications. 2002;**30**:343-350

[82] Sandhya V, Ali SZ, Grover M, Reddy G, Venkateswarlu B. Effect of plant growth promoting *Pseudomonas* spp. on compatible solutes, antioxidant status and plant growth of maize under drought stress. Plant Growth Regulation. 2010;**62**:21-30

[83] Alami Y, Achouak W, Marol C, Heulin T. Rhizosphere soil aggregation and plant growth promotion of sunflowers by exopolysaccharide producing *Rhizobium* sp. strain isolated from sunflower roots. Applied and Environmental Microbiology. 2000;**66**:3393-3398

[84] Sandhya V, Ali SZ, Grover M, Kishore N, Venkateswarlu B. *Pseudomonas* sp. strain P45 protects sunflowers seedlings from drought stress through improved soil structure. Journal of Oilseed Research. 2009a;**26**:600-601

[85] Arkhipova TN, Prinsen E, Veselov SU, Martinenko EV, Melentiev AI, Kudoyarova GR. Cytokinin producing bacteria enhance plant growth in drying soil. Plant and Soil. 2007;**292**:305-315. DOI: 10.1007/s11104-007-9233-5

[86] Kohler J, Hernández JA, Caravaca F, Roldán A. Plant growth promoting rhizobacteria and arbuscular mycorrhizal fungi modify alleviation biochemical mechanisms in water stressed plants. Functional Plant Biology. 2008;**35**:141-151

[87] Marulanda A, Porcel R, Barea JM, Azcon R. Drought tolerance and antioxidant activities in lavender plants colonized by native drought tolerant or drought sensitive *Glomus* species. Microbial Ecology. 2007;**54**(3):543-552

[88] Arshad M, Sharoona B, Mahmood T. Inoculation with

Pseudomonas spp. containing ACC deaminase partially eliminate the effects of drought stress on growth, yield and ripening of pea (*Pisum sativum* L.). Pedosphere. 2008;**18**:611-620

[89] Belimov AA, Dodd IC, Hontzeas N, Theobald JC, Safronova VI, Davies WJ. Rhizosphere bacteria containing 1-aminocyclopropane-1-carboxylate deaminase increase yield of plants grown in drying soil via both local and systemic hormone signalling. The New Phytologist. 2009;**181**:413-423

[90] Figueiredo MVB, Burity HA, Martinez CR, Chanway CP. Alleviation of drought stress in common bean (*Phaseolus vulgaris* L.) by co-inoculation with *Paenibacillus polymyxa* and *Rhizobium tropici*. Applied Soil Ecology. 2008;**40**:182-188

[91] Liddycoat SM, Greenberg BM, Wolyn DJ. The effect of plant growth-promoting rhizobacteria on asparagus seedling and germinating seeds subjected to water stress under greenhouse conditions. Canadian Journal of Microbiology. 2009;**55**:388-394

[92] Creus CM, Sueldo RJ, Barassi CA. Water relations and yield in *Azospirillum*-inoculated wheat exposed to drought in the field. Canadian Journal of Botany. 2004;**82**(2):273-281

[93] Chakraborty U, Chakraborty BN, Chakraborty AP, Dey PL. Water stress amelioration and plant growth promotion in wheat plants by osmotic stress tolerant bacteria. World Journal of Microbiology and Biotechnology. 2013;**29**:789-803

[94] Grichko VP, Glick BR. Amelioration of flooding stress by ACC deaminase containing plant growth promoting bacteria. Canadian Journal of Microbiology. 2001;**47**:77-80

[95] Chang WS, van de Mortel M, Nielsen L, de Guzman GN, Li X, Halverson LJ. Alginate production by *Pseudomonas putida* creates a hydrated microenvironment and contributes to biofilm architecture and stress tolerance under water-limiting conditions. Journal of Bacteriology. 2007;**189**:8290-8299

[96] Ait Bakra E, Nowak J, Clement C. Enhancement of chilling resistance of inoculated grapevine plantlets with a plant growth promoting rhizobacterium, Burkholderia phytofirmans strain PsJN. Applied and Environmental Microbiology. 2006;**72**(11):7246-7252

[97] Mastretta C, Taghavi S, van der Lelie D, Mengoni A, Galardi F, Gonnelli C, et al. Endophytic bacteria from seeds of *Nicotiana tabacum* can reduce cadmium phytotoxicity. International Journal of Phytoremediation. 2009;**11**:251-267

[98] Pishchik VN, Provorov NA, Vorobyov NI, Chizevskaya EP, Safronova VI, Tuev AN, et al. Interactions between plants and associated bacteria in soils contaminated with heavy metals. Microbiology. 2009;**78**:785-793

[99] Sheng M, Tang M, Chan H, Yang B, Zhang F, Huang Y. Influence of arbuscular mycorrhizae on photosynthesis and water status of maize plants under salt stress. Mycorrhiza. 2008;**18**:287-296

[100] Madhaiyan M, Poonguzhali S, Sa T. Metal tolerating methylotrophic bacteria reduces nickel and cadmium toxicity and promotes plant growth of tomato (*Lycopersicon esculentum* L.). Chemosphere. 2007;**69**:220-228

[101] Terré S, Asch F, Padham J, Sikora RA, Becker M. Influence of root zone bacteria on root iron plaque formation in rice subjected to iron

toxicity. In: Tielkes E, editor. Utilisation of Diversity in Land Use Systems: Sustainable and Organic Approaches to Meet Human Needs. Witzenhausen, Germany: Tropentag; 2007. p. 446

[102] Fan QJ, Liu JH. Colonization with arbuscular mycorrhizal fungus affects growth, drought tolerance and expression of stress-responsive genes in *Poncirus trifoliata*. Acta Physiologiae Plantarum. 2011;**33**:1533-1542

[103] Garmendia I, Goicoechea N, Aguirreolea J. Moderate drought influences the effect of arbuscular mycorrhizal fungi as biocontrol agents against *verticillium*-induced wilt in pepper. Mycorrhiza. 2005;**15**:345-356

[104] Pinior A, Grunewaldt-Stöcker G, von Alten H, Strasser RJ. Mycorrhizal impact on drought stress tolerance of rose plants probed by chlorophyll a fluorescence, proline content and visual scoring. Mycorrhiza. 2005;**15**:596-605

[105] Alguacil MM, Kohler J, Caravaca F, Roldán A. Differential effects of *Pseudomonas mendocina* and *Glomus intraradices* on lettuce plants physiological response and aquaporin PIP2 gene expression under elevated atmospheric CO2 and drought. Microbial Ecology. 2009;**58**:942-951

[106] Sohrabi Y, Heidari G, Weisany W, Golezani KG, Mohammadi K. Changes of antioxidative enzymes, lipid peroxidation and chlorophyll content in chickpea types colonized by different *Glomus* species under drought stress. Symbiosis. 2012;**56**:5-18

[107] Abbaspour H. Investigation of the effects of vesicular arbuscular mycorrhiza on mineral nutrition and growth of *Carthamus tinctorius* under salt stress conditions. Russian Journal of Plant Physiology. 2010;**57**:526-531

[108] Estrada BE, Aroca R, Barea JM, Ruiz-Lozano JM. Native arbuscular mycorrhizal fungi isolated from a saline habitat improved maize antioxidant systems and plant tolerance to salinity. Plant Science. 2013;**201**:43-51

[109] Evelin H, Giri B, Kapoor R. Ultrastructural evidence for AMF mediated salt stress mitigation in *Trigonella foenum-graecum*. Mycorrhiza. 2013;**23**:71-86

[110] Fan L, Dalpé Y, Fang C, Dubé C, Khanizadeh S. influence of arbuscular mycorrhizae on biomass and root morphology of selected strawberry cultivars under salt stress. Botany. 2011;**89**:397-403. DOI: 10.1139/B11-028

[111] Campanelli A, Ruta C, Mastro GD, Morone-Fortunato I. The role of arbuscular mycorrhizal fungi in alleviating salt stress in *Medicago sativa* L. var. icon. Symbiosis. 2013;**59**:65-76

[112] Mergulhão ACES, Burity HA, Tabosa JN, Figueiredo MVB, Maia LC. Influence of NaCl on *Brachiaria humidicola* inoculated or not with *Glomus etunicatum*. Investigacion Agraria, Produccion y Proteccion Vegetales. 2002;**17**:220-227

[113] Kaushik JC, Mandal BS. The role of mycorrhiza in stress management for seedling growth of *Dalbergia sissoo* and *Acacia nilotica*. Bulletin of the NIE. 2005;**15**:133-137

[114] Wu QS, Li GH, Zou YN. Roles of arbuscular mycorrhizal fungi on growth and nutrient acquisition of peach (*Prunus persica* L. Batsch) seedlings. Journal of Animal &; Plant Sciences. 2011;**21**:746-750

[115] Wu QS. Mycorrhizal efficacy of trifoliate orange seedlings on alleviating temperature stress. Plant, Soil and Environment. 2011;**57**:459-464

[116] Matsubara Y, Hirano I, Sassa D, Koshikawa K. Alleviation of high temperature stress in strawberry (*Fragaria ananassa*) plants infected with arbuscular mycorrhizal fungi. Environment Control in Biology. 2004;**42**:105-111

[117] Abdel Latef AA. Influence of arbuscular mycorrhizal fungi and copper on growth, accumulation of osmolyte, mineral nutrition and antioxidant enzyme activity of pepper (*Capsicum annuum* L.). Mycorrhiza. 2011;**21**:495-503

[118] Abdelmoneim TS, Almagrabi OA. Improved tolerance of maize plants to heavy metals stress by inoculation with arbuscular mycorrhizal fungi. Archives des Sciences. 2013;**66**:155-167

[119] Hildebrandt U, Regvar M, Bothe H. Arbuscular mycorrhiza and heavy metal tolerance. Phytochemistry. 2007;**68**:139-146

[120] Pawlowska TE, Charvat I. Heavy metal stress and developmental patterns in arbuscular mycorrhizal fungi. Applied and Environmental Microbiology. 2004;**70**:6643-6649

[121] de Andrade SAL, da Silveira APD. Mycorrhiza influence on maize development under Cd stress and P supply. Brazilian Journal of Plant Physiology. 2008;**20**:39-50

[122] Kaushal M, Wani SP. Rhizobacterial plant interactions: Strategies ensuring plant growth promotion under drought and salinity stress. Agriculture, Ecosystems and Environment. 2016a;**231**:68-78

[123] Heydarian Z, Yu M, Gruber M, Glick BR, Zhou R, Hegedus DD. Inoculation of soil with plant growth promoting bacteria producing 1-aminocyclopropane-1-carboxylate deaminase or expression of the corresponding acdS gene in transgenic plants increases salinity tolerance in *Camelina sativa*. Frontiers in Microbiology. 2016;**7**:1966

[124] Ghosh D, Sen S, Mohapatra S. Drought-mitigating *Pseudomonas putida* GAP-P45 modulates proline turnover and oxidative status in *Arabidopsis thaliana* under water stress. Annales de Microbiologie. 2018;**68**:1-16

[125] Kaushal M, Wani SP. Plant-growth-promoting rhizobacteria: Drought stress alleviators to ameliorate crop production in drylands. Annales de Microbiologie. 2016b;**66**(1):35-42

[126] Vacheron J, Desbrosses G, Bouffaud ML, Touraine B, Moënne-Loccoz Y, Muller D, et al. Plant growth promoting rhizobacteria and root system functioning. Frontiers in Plant Science. 2013;**4**:356

[127] Alam MA, Kaliyamoorthy S, Zaidi PH, Dinesh A, Vinayan MT, Nath UK. Dissecting heat stress tolerance in tropical maize (*Zea mays* L.). Food Crops Research. 2017;**204**:110-119

[128] Shameer S, Prasad TNVKV. Plant growth promoting rhizobacteria for sustainable agricultural practices with special reference to biotic and abiotic stresses. Plant Growth Regulation. 2018;**84**(3):603-615

[129] Yadav AN, Sachan SG, Verma P, Saxena AK. Prospecting cold deserts of north western Himalayas for microbial diversity and plant growth promoting attributes. Journal of Bioscience and Bioengineering. 2015;**119**(6):683-693

[130] Verma P, Yadav AN, Khannam KS, Mishra S, Kumar S, Saxena AK, et al. Appraisal of diversity and functional

attributes of thermotolerant wheat associated bacteria from the peninsular zone of India. Saudi. Journal of Biological Science. 2019;**26**(7):1882-1895

[131] Feng NX, Yu J, Zhao HM, Cheng YT, Mo CH, Cai QY, et al. Efficient phytoremediation of organic contaminants in soils using plant–endophyte partnerships. Science Total Environment. 2017;**583**:352-368

[132] Glick BR. Using soil bacteria to facilitate phytoremediation. Biotechnology Advances. 2010;**28**(3): 367-374

[133] Ma Y, Rajkumar M, Zhang C, Freitas H. Inoculation of *Brassica oxyrrhina* with plant growth promoting bacteria for the improvement of heavy metal phytoremediation under drought conditions. Journal of Hazardous Materials. 2016a;**320**:36-44

[134] Ullah A, Heng S, Munis MFH, Fahad S, Yang X. Phytoremediation of heavy metals assisted by plant growth promoting (PGP) bacteria: A review. Environmental and Experimental Botany. 2015;**117**:28-40

[135] Gadd GM. Metals, minerals and microbes: Geomicrobiology and bioremediation. Microbiology. 2010;**156**(3):609-643

[136] Yong X, Chen Y, Liu W, Xu L, Zhou J, Wang S, et al. Enhanced cadmium resistance and accumulation in *Pseudomonas putida* KT 2440 expressing the phytochelatin synthase gene of *S chizosaccharomyces* pombe. Letters in Applied Microbiology. 2014;**58**(3):255-261

[137] Złoch M, Thiem D, Gadzała-Kopciuch R, Hrynkiewicz K. Synthesis of siderophores by plant-associated metallotolerant bacteria under exposure to Cd2+. Chemosphere. 2016;**156**:312-325

[138] Abou-Shanab RA, El-Sheekh MM, Sadowsky MJ. Role of rhizobacteria in phytoremediation of metal-impacted sites. In: Emerging and Eco-Friendly Approaches for Waste Management. Singapore: Springer; 2019. pp. 299-328

[139] Burges A, Alkorta I, Epelde L, Garbisu C. From phytoremediation of soil contaminants to phytomanagement of ecosystem services in metal contaminated sites. International Journal of Phytoremedicine. 2018;**20**(4):384-397

[140] Vessey JK. Plant growth promoting rhizobacteria as biofertilizers. Plant and Soil. 2003;**255**:571-586

[141] Somers E, Vanderleyden J, Srinivasan M. Rhizosphere bacterial signalling: A love parade beneath our feet. Critical Review Microbiology. 2004;**30**:205-240

[142] Lugtenberg BJ, Chin-A-Woeng TF, Bloemberg GV. Microbe–plant interactions: Principles and mechanisms. Antonie Leeuwenhoek. 2002;**81**:373-383

[143] Chandler D, Davidson G, Grant WP, Greaves J, Tatchell GM. Microbial biopesticides for integrated crop management: An assessment of environmental and regulatory sustainability. Trends Food Science Technology. 2008;**19**:275-283

[144] Bhattacharyya PN, Jha DK. Plant growth-promoting rhizobacteria (PGPR): Emergence in agriculture world. Journal of Microbiology and Biotechnology. 2012;**28**:1327-1350

[145] Braud A, Jezequel K, Bazot S, Lebeau T. Enhanced phytoextraction of an agricultural Cr-, Hg- and Pb-contaminated soil by bioaugmentation with siderophore producing bacteria. Chemosphere. 2009;**74**:280-286

[146] Baltruschat H, Fodor J, Harrach BD, Niemczyk E, Barna B, Gullner G, et al. Salt tolerance of barley induced by the root endophyte *Piriformospora indica* is associated with a strong increase in anti-oxidants. The New Phytologist. 2008;**180**:501-510. DOI: 10.1111/j.1469-8137.2008.02583.x

[147] Franken P. The plant strengthening root endophyte *Piriformospora indica*: Potential application and the biology behind. Applied Microbiology and Biotechnology. 2012;**96**:1455-1464. DOI: 10.1007/s00253-012-4506-1

[148] Ortiz N, Armadaa E, Duqu E, Roldánc A, Azcóna R. Contribution of arbuscular mycorrhizal fungi and/or bacteria to enhancing plant drought tolerance under natural soil conditions: Effectiveness of autochthonous or allochthonous strains. Journal of Plant Physiology. 2015;**174**:87-96

[149] Bano QUDSIA et al. Effect of *Azospirillum* inoculation on maize (*Zea mays* L.) under drought stress. Pakistan Journal of Botany. 2013;**45**:13-20

[150] Kang SM, Kang SM, Radhakrishnan R, Khan A, Kim MJ, Park JM, et al. Gibberellin secreting rhizobacterium, *Pseudomonas putida* H-2-3 modulates the hormonal and stress physiology of soybean to improve the plant growth under saline and drought conditions. Plant Physiology and Biochemistry. 2014;**84**:115-124

[151] Hussain MB et al. Can catalase and exopolysaccharides producing rhizobia ameliorate drought stress in wheat? International Journal of Agriculture and Biology. 2014;**16**:3-13

[152] Armada E, Roldán A, Azcon R. Differential activity of autochthonous bacteria in controlling drought stress in native Lavandula and Salvia plants species under drought conditions in the natural arid soil. Microbial Ecology. 2014;**67**:410-420

[153] Fasciglione G, Casanovas EM, Quillehauquy V, Yommi AK, Goni MG, Roura SI, et al. *Azospirillum* inoculation effects on growth, product quality and storage life of lettuce plants grown under salt stress. Scientia Horticulturae. 2015;**195**:154-162

[154] Cohen AC, Bottinia R, Pontina M, Berlia FJ, Moreno D, Boccanlandro H, et al. *Azospirillum brasilense* ameliorates the response of *Arabidopsis thaliana* to drought mainly via enhancement of ABA levels. Physiologia Plantarum. 2015;**153**:79-90

[155] Bresson J, Varoquaux F, Bontpart T, Touraine B, Vile D. The PGPR strain *Phyllobacterium brassicacearum* STM196 induces a reproductive delay and physiological changes that result in improved drought tolerance in Arabidopsis. The New Phytologist. 2013;**200**:558-569

[156] Ma Y, Rajkumar M, ZhangC FH. The beneficial role of bacterial endophytes in heavy metal phytoremediation. Journal of Environmental Management. 2016b;**174**:14-25

[157] Staudinger C, Mehmeti-Tershani V, Gil-Quintana E, Gonzalez EM, Hofhansl F, Bachmann G, et al. Evidence for a rhizobia-induced drought stress response strategy in Medicago truncatula. Journal of Proteomics. 2016;**136**:202-213

[158] Raheem A, Shaposhnikov A, Belimov AA, Dodd IC, Ali B. Auxin production by rhizobacteria was associated with improved yield of wheat (*Triticum aestivum* L.) under drought stress. Archives of Agron and Soil Science. 2018;**64**(4):574-587

[159] Saikia J, Sarma RK, Dhandia R, Yadav A, Bharali R, Gupta VK, et al. Alleviation of drought stress in pulse crops with ACC deaminase producing rhizobacteria isolated from acidic soil of Northeast India. Scientific Reports. 2018;**8**(1):3560

[160] Amellal N, Burtin G, Bartoli F, Heulin T. Colonization of wheat rhizosphere by EPS producing Pantoea agglomerans and its effect on soil aggregation. Applied and Environmental Microbiology. 1998;**64**:3740-3747

[161] Timmusk S, Wagner EGH. The plant growth-promoting rhizobacterium *Paenibacillus polymyxa* induces changes in *Arabidopsis thaliana* gene expression, a possible connection between biotic and abiotic stress responses. Molecular Plant-Microbe Interactions. 1999;**12**:951-959. DOI: 10.1094/MPMI.1999. 12.11.951

[162] Dodd IC, Belimov AA, Sobeih WY, Safronova VI, Grierson D, Davies WJ. Will modifying plant ethylene status improve plant productivity in water-limited environments? In: 4th International Crop Science Congress. 2005

[163] McLellan CA, Turbyville TJ, Wijeratne K, Kerschen A, Vierling E, Queitsch C, et al. A rhizosphere fungus enhances *Arabidopsis* thermo tolerance through production of an HSP90 inhibitor. Plant Physiology. 2007;**145**:174-182

[164] Marquez LM, Redman RS, Rodriguez RJ, Roosinck MJ. A virus in a fungus in a plant: Three-way symbiosis required for thermal tolerance. Science. 2007;**315**(5811):513-515

[165] Sandhya V, Ali SZ, Grover M, Reddy G, Venkateswarlu B. Alleviation of drought stress effects in sunflower seedlings by exopolysaccharides producing *Pseudomonas putida* strain P45. Biology and Fertility Soil. 2009b;**46**:17-26

[166] Cho K, Toler H, Lee J, Ownley B, Stutz JC, Moore JL, et al. Mycorrhizal symbiosis and response of sorghum plants to combined drought and salinity stresses. Journal of Plant Physiology. 2006;**163**:517-528

[167] Shukla PS, Agarwal PK, Jha B. Improved salinity tolerance of *Arachisy pogaea* L. by the interaction of halotolerant plant-growth-promoting rhizobacteria.J. Plant Growth Regulation. 2012b;**31**:195-206

[168] Ahmad M, Zahir ZA, Asghar HN, Asghar M. Inducing salt tolerance in mung bean through co-inoculation with rhizobia and plant-growth-promoting rhizobacteria containing 1-aminocyclopropane-1-carboxylate deaminase. Canadian Journal of Microbiology. 2011;**57**:578-589

[169] Chang P, Gerhardt KE, Huang X-D, Yu X-M, Glick BR, Gerwing PD, et al. Plant growth promoting bacteria facilitate the growth of barley and oats in salt impacted soil: Implications for phytoremediation of saline soils. International Journal of Phytoremediation. 2014;**16**:1133-1147. DOI: 10.1080/15226514.2013. 821447

[170] Zahir ZA, Ghani U, Naveed M, Nadeem SM, Asghar HN. Comparative effectiveness of *Pseudomonas and Serratia* sp. containing ACC-deaminase for improving growth and yield of wheat (*Triticum aestivum* L.) under salt-stressed conditions. Archives of Microbiology. 2009;**191**:415-424

[171] Nadeem SM, Zahir ZA, Naveed M, Arshad M. Rhizobacteria containing ACC-deaminase confers salt tolerance in maize grown on salt-affected fields. Canadian Journal of Microbiology. 2009;**55**:1302-1309

[172] Jha Y, Subramanian RB. PGPR regulate caspase-like activity, programmed cell death, and antioxidant enzyme activity in paddy under salinity. Physiological Molecular Biology Plant. 2014;**20**:201-207

[173] Nautiyal CS, Srivastava S, Chauhan PS, Seem K, Mishra A, Sopory SK. Plant growth-promoting bacteria *Bacillus amyloliquefaciens* NBRISN13 modulates gene expression profile of leaf and rhizosphere community in rice during salt stress. Plant Physiology and Biochemistry. 2013;**66**:1-9

[174] Suarez C, Cardinale M, Ratering S, Steffens D, Jung S, Montoya AMZ, et al. Plant growth-promoting effects of *Hartmannibacter diazotrophicus* on summer barley (*Hordeum vulgare* L.) under salt stress. Applied Soil Ecology. 2015;**95**:23-30. DOI: 10.1016/j.apsoil.2015.04.017

[175] Cheng Z, Woody OZ, McConkey BJ, Glick BR. Combined effects of the plant growth-promoting bacterium *Pseudomonas putida* UW4 and salinity stress on the *Brassica napus proteome*. Applied Soil Ecology. 2012;**61**:255-263

[176] Vimal SR, Patel VK, Singh JS. Plant growth promoting *Curtobacterium albidum* strain SRV4: An agriculturally important microbe to alleviate salinity stress in paddy plants. Ecological Indicators. 2019;**105**(2019):553-562

[177] de Zélicourt A, Synek L, Saad MM, Alzubaidy H, Jalal R, Xie Y, et al. Ethylene induced plant stress tolerance by *Enterobacter* sp. SA187 is mediated by 2-keto-4-methylthiobutyric acid production. PLoS Genetics. 2018;**14**(3):e1007273

[178] Waller F, Achatz B, Baltruschat H, Fodor J, Becker K, Fischer M, et al. The endophytic fungus Piriformis indica reprograms barley to salt-stress tolerance, disease resistance, and higher yield. Proceedings of the National Academy of Sciences. 2005;**102**:13386-13391

[179] Rodriguez AA, Stella AM, Storni MM, Zulpa G, Zaccaro MC. Effect of cyanobacterial extracellular products and gibberellic acid on salinity tolerance in Oryza sativa L. Saline System. 2006;**2**:7. DOI: 10.1186/1746-1448-2-7

[180] Ali SZ, Sandhya V, Grover M, Linga VR, Bandi V. Effect of inoculation with a thermotolerant plant growth promoting *Pseudomonas putida* strain AKMP7 on growth of wheat (*Triticum* spp.) under heat stress. Journal of Plant Interactions. 2011;**6**(4):239-246

[181] Fougnies L, Renciot S, Muller F, Plenchette C, Prin Y, de Faria SM, et al. Arbuscular mycorrhizal colonization and nodulation improve tolerance in *Pterocarpus officinalis* Jacq. Seedlings. Mycorrhiza. 2007;**17**:159-166

[182] Chen Y, Yang W, Chao Y, Wang S, Tang YT, Qiu RL. Metal-tolerant *Enterobacter* sp. strain EG16 enhanced phytoremediation using *Hibiscus cannabinus* via siderophore-mediated plant growth promotion under metal contamination. Plant and Soil. 2017;**413**(1-2):203-216

[183] Saleem M, Asghar HN, Zahir ZA, Shahid M. Impact of lead tolerant plant growth promoting rhizobacteria on growth, physiology, antioxidant activities, yield and lead content in sunflower in lead contaminated soil. Chemosphere. 2018;**195**:606-614

[184] Corsini A, Colombo M, Gardana C, Zecchin S, Simonetti P, Cavalca L. Characterization of As (III) oxidizing *Achromobacter* sp. strain N2: Effects on arsenic toxicity and

translocation in rice. Annales de Microbiologie. 2018;**68**(5):295-304

[185] Gupta DK, Rai UN, Sinha S, Tripathi RD, Nautiyal BD, Rai P, et al. Role of *Rhizobium* (CA-1) inoculation in increasing growth and metal accumulation in *Cicer arietinum* L. growing under fly-ash stress condition. Bulletin of Environmental Contamination and Toxicology. 2004;**73**:424-431

[186] Jackson MB. Regulation of water relationships in flooded plants by ABA from leaves, roots and xylem sap. In: Davis WJ, editor. Abscisic Acid. Physiology and Biochemistry. Oxford: Bios Scientific); 1991. pp. 217-226

[187] Belimov AA, Dodd IC, Safronova VI, Hontzeas N, Davies WJ. *Pseudomonas brassicacearum* strain Am3 containing 1-aminocyclopropane-1-carboxylate deaminase can show both pathogenic and growth-promoting properties in its interaction with tomato. Journal of Experimental Botany 2007;**58**:1485-1495. DOI: 10. 1093/jxb/erm010

[188] Tapias DR, Galvan AM, Diaz SP, Obando M, Rivera D, Bonilla R. Effect of inoculation with plant growth-promoting bacteria (PGPB) on amelioration of saline stress in maize (*Zea mays*). Applied Soil Ecology. 2012;**61**:264-272. DOI: 10.1016/j.apsoil.2012.01.006

[189] Sharma A, Shankhdha D, Shankhdhar SC. Enhancing grain iron content of rice by the application of plant growth promoting rhizobacteria. Plant, Soil and Environment. 2013;**59**:89-94

[190] Sen S, Chandrasekhar CN. Effect of PGPR on growth promotion of rice (*Oryza sativa* L.) under salt stress. Asian Journal of Plant Science and Research. 2014;**4**:62-67

[191] Khan A, Sirajuddin Zhao XQ, Javed MT, Khan KS, Bano A, Shen RF, et al. *Bacillus pumilus* enhances tolerance in rice (*Oryza sativa* L.) to combined stresses of NaCl and high boron due to limited uptake of NaCl. Environmental and Experimental Botany. 2016;**124**:120-129. DOI: 10.1016/j.envexpbot.2015. 12.011

[192] Pandey V, Ansari MW, Tula S, Yadav S, Sahoo RK, Shukla N, et al. Dose-dependent response of *Trichoderma harzianum* in improving drought tolerance in rice genotypes. Planta. 2016;**243**:1251-1264. DOI: 10.1007/s00425-016-2482-x

[193] Brotman Y, Landau U, Cuadros-Inostroza Á, Takayuki T, Fernie AR, Chet I, et al. *Trichoderma*-plant root colonization: Escaping early plant defense responses and activation of the antioxidant machinery for saline stress tolerance. PLoS Pathogens. 2013;**9**:e1003221. DOI: 10.1371/journal.ppat.1003221

Chapter 13

Techniques against Distinct Abiotic Stress of Rice

Ananya Prova and Md. Saeed Sultan

Abstract

Plants cannot physically escape environmental stresses because they are sessile organisms, which can stunt their growth. As a result, plants have had to evolve distinct strategies to deal with abiotic stress. Indeed, responding to and eventually adapting to abiotic stress may be a driving force in speciation. Because of the complexity of stress, multiple sensors, rather than a single sensor, are more likely to be responsible for stress perception. Stress-induced gene issues can be divided into two categories: those involved in stress tolerance and those involved in signal transduction. Stress-tolerance genes help plants cope with stress in both short- and long-term responses. These can include the synthesis of chaperones and enzymes for osmolyte biosynthesis. And, as with cold stress, detoxification causes a change in the composition of membrane lipids. Gene products can also function as transcription regulators, controlling groups of stress-related genes, or as components in the production of regulatory molecules. It has been shown that multiple signaling pathways can be activated during stress, resulting in similar responses to different triggers.

Keywords: stemless, abiotic stress, resistance, membrane lipid, enzymes, defense, tolerance, toxicity

1. Introduction

Land plants have evolved to thrive in harsh environments since their inception. To address the various environmental constraints that affect their growth and development as stemless organisms, plants have evolved a variety of complex and efficient molecular and physiological mechanisms [1]. They are physically and chemically hostile to cold and heat, for example. Stress can be caused by a lack of or an excess of water, high salt levels, heavy metals, ultraviolet (UV) radiation, and other factors. These pressures, known as biological stress, pose a significant threat to agriculture and ecosystems, resulting in significant crop losses [2, 3]. Plants must adapt to their ever-changing environment because they lack stems. During development, plants are subjected to a variety of environmental influences, which can limit productivity. A yield gap is observed when the crop is grown under suboptimal environmental conditions, so the average yield achieved is much lower than the maximum potential yield of a particular crop [4]. Yield gaps for the three major grains: wheat, rice, and corn account for 40, 75, and 30% of the world's major growing regions, respectively [5]. The primary causes of crop yield gaps can be classified as some abiotic factors such as temperature, water, or minerals, or (ii) biological factors such as bacteria, fungi, or insect invasion [6, 7].

Rice cultivation and productivity are water-intensive, making them highly vulnerable to drought and flooding. Due to drought impairment, Asia's most pertinent rice-producing belts most pertinent rice-producing belts in Asia produce only 40% of total production efficiency [8, 9]. The overwhelming majority of strain-related studies in rice, as in other plant species, have primarily focused on single stresses, either abiotic or biotic. Despite undeniable progress in the overall field, this approach provides an oversimplified and unrealistic picture. Furthermore, changes in the global environment (GEC) endanger crop production. Droughts, floods, soil acidification, soil salt content, frigid and hot temperatures, and other harmful environmental conditions are all caused by GEC. All of these stressors have an impact on crop yield and quality, either directly or indirectly. Furthermore, with the world's population expected to reach 9 billion by 2050, crop production must be increased to feed an additional 2 billion people over the next 40 years. Rice (23%), wheat (17%), and corn account for roughly half of total human calories (10%).

Despite their immobility, plants must simply respond to and endure a variety of environmental and biotic stresses in the field. A biotic and abiotic factor contends both cause crop yield losses [10]. As a result, many crop improvement programs prioritize the development of stress-tolerant plant varieties [11–13]. Plants respond differently to different or concurrent stresses, and breeding for single stress (e.g., drought, salinity, pathogen) rather than multiple stresses (e.g., abiotic or biotic) may be risky. It is worth noting that increasing tolerance to one type of stress may reduce tolerance to another [14, 15]. Climate change is increasing the frequency of extreme weather events, and plants are subjected to a variety of stresses in the field, including additional pressure from plant diseases [16].

As a result, recognizing the similarities and differences among stress response pathways is essential for optimizing targeted crop improvement. Stomatal closure, reduced photosynthesis, increased reactive oxygen scavenging activity, reduced leaf growth, and increased root length are all indications of plant responses to abiotic stresses [17]. Plant pathogens, for example, cause stomata to close, reducing photosynthesis [18, 19]. The production of toxic compounds such as phytoalexins and reactive oxygen species, as well as the induction of localized cell death, are other pathogen-induced plant responses [20]. Many of these responses are governed by phytohormones [21, 22].

Hormones like abscisic acid (ABA) and jasmonic (JA) play a key role in regulating inanimate stress tolerance. For pathogen immunity, plants primarily rely on salicylic acid (SA), JA, and ethylene signaling. The abiotic stress response is regulated by many transcription factor (TF) families, both ABA-dependent and ABA-independent. ABA-induced basic leucine zipper (bZIP) transcription factors are among the aforementioned [23, 24]. These TFs are responsible for stoma closure, dehydration resistance gene expression, and other adaptive physiological responses [25–27]. ABA, on the other hand, frequently increases plant sensitivity to biological interactions [28–32] and frequently interacts detrimentally with SA [28–35]. As a result, plants must have evolved the ability to detect and respond to multiple environmental cues in varying combinations. Developing stress-tolerant plants and testing their performance against individually imposed stresses, according to this viewpoint, may be insufficient. Furthermore, plants exposed to multiple stresses activate a single response that is not simply additive but results from synergistic and antagonistic interactions, resulting in unpredictable effects when each stress is taken into account [11].

The presence of secondary stress, according to this theory, can either aggravate the negative consequences of primary stress or, on the other hand, contribute to a better response.

2. Defenses against abiotic stresses in general

2.1 Cuticle

The cuticle is a fine translucent lipid structure that seals the aerial surfaces of land plants' organs on the outside. The thin hydrophobic layer is essentially a cutin matrix filled with cuticular waxes and encapsulated with them. As the plant's primary interface with the environment, the cuticle cuticle, as the plant's primary interface with the environment, is critical for controlling liquid and gas fluxes, defending against pathogen and bug attacks, and resisting abiotic stresses. The ability of land plants to deploy an outer shield made of simple molecules is a brilliant innovation that is critical to their success in terrestrial colonization [36–38]. The cell wall, the second barrier that actively remodels in response to abiotic stresses [39, 40], is far more complex and poorly understood to order to counteract this [41, 42]. The cuticle is macromolecular polyester of C16 or C18 oxygenated fatty acids (FAs), which are exclusively produced by epidermal cells. Waxes, on the other hand, are a high-end blend of C24 to C34 FA derivatives such as alcohols, aldehydes, alkanes, esters, and ketones. The biosynthetic pathways of these organisms are nearly complete and well documented [10, 37, 43–45]. In a nutshell, C24 and C34 are constituted from acetyl coenzyme A (CoA) in plastids via de novo FA synthesis, adding the addition of two carbons in each recurring cycle. Until C16/C18 products emerge, they are transported to the endoplasmic reticulum (ER), they are oxidized and incorporated to form cutin precursors (monoacylglycerols) or elongated modified to include wax components. The alcohol-forming (or acyl-reduction) pathway for primary alcohols and esters and the alkane-forming (or decarbonylation) pathway for aldehydes, alkanes, secondary alcohols, and ketones, are two distinct modification pathways. These materials must be transported from the ER to the cytomembrane (PM), where cutin monomers polymerize and wax members crystallize, to form the apoplastic cuticle on the outer surface. Membrane vesicle trafficking [46] is one of the systems involved in intracellular shipments to the ATP-binding cassette (ABC) transporters that channel the PM [38, 43, 47].

2.2 Unsaturated fatty acids

C16/C18 FAs are not only important components of the cuticle, but also of membranes, which serve as basic biological barriers. Phospholipids and glycolipids with a glycerol core and two FA-derived "tails" are the primary components of botanic membranes. Membrane properties are greatly influenced by FAs. Their degree of unsaturation, in particular, is an important determinant of membrane fluidity. The UFA chain will kink at a cis-double bond, which will act as a steric hindrance in the intermolecular package, causing the intermolecular package to become more fluid [48, 49]. Membrane fluidity is vulnerable to abiotic stresses, particularly extreme temperatures. Cold-driven rigidification and heat-driven fluidization can cause biomembrane dysfunction, as exemplified by protein deactivation and ion leakage [50]. Cytoskeleton destabilization is also a direct consequence [51, 52].

C18 UFAs are used as a raw material in the production of a variety of aliphatic compounds in plants, including membrane glycerolipids, TAG, cutin/suberin, jasmonates, and nitroalkenes (NO2-FAs). All of these products, as well as C18 UFAs, help plants; defend themselves against biotic and abiotic stresses. Multiple mechanisms implicate C18 UFAs in stress defense, either directly and indirectly. Biomembranes are a functional platform for many cellular processes, including substance exchange, signal transduction, and many metabolic reactions, in addition to being a structural barrier for cells and intracellular organelles [53]. The signaling of Ca2+, a versatile second messenger involved in virtually every stress response in plants, is based on membrane isolation and transportation. With the help of its channels, such as the PM cyclic nucleotide-gated channels (CNGCs) and the tonoplast TWO PORE CHANNEL 1 (TPC1), the sharp influx of the cation ignites ca2+ signaling into the cytosol [51, 54, 55]. Furthermore, efflux through Ca2+-ATPases and Ca2+/H+ exchangers has quickly rinsed it [56, 57]. Other ion transporters, such as the K+ rectifier ARABIDOPSIS K+ TRANSPORTER 1 (AKT1) [54] and the Na+/H+ antiporter SALT OVERLY SENSITIVE 1 (SOS1) [26, 34], preserve a sufficient K+/Na+ ratio in the cytoplasm, which is necessary for salt tolerance [58, 59]. The electrochemical gradient created by transmembrane proton pumps, such as PM H+-ATPase, vacuolar H+-ATPase (V-ATPase), and vacuolar H+-translocating inorganic pyrophosphatase (V-PPase), energizes these secondary transporters [60–62]. The PM H+-ATPase, in particular, is a critical site that responds to salt and other stresses such as cold and heavy metals, as well as active transport across the PM [63].

Membrane fluidity is vulnerable to a variety of stresses, including extreme temperatures. Cold-induced rigidification and thermal fluidization, on the other hand, are harmful to membrane function, causing protein deactivation, electrolyte leakage, and perhaps even cytoskeleton destabilization [51, 52]. As a thermodynamic property, membrane fluidity could be used as a sensor in heat flux signaling. Interestingly, dimethylsulfoxide (DMSO) and benzyl alcohol (BA) can both mimic the effects of cold and heat at 25°C. Plants are poikilothermic organisms, emphasizing the significance of membrane remodeling, as well as the threat of climate change.

2.3 Scavengers of reactive species

The endless generation of noxious RS, particularly reactive oxygen species (ROS) such as superoxide (O2), oxide (H2O2), hydroxyl (OH), and singlet oxygen (O2), as well as reactive carbonyl species (RCS) such as malondialdehyde and methylglyoxal, is an inherent paradox in aerobic organisms' normal metabolism (MG; CH3COCHO). The two types of RS are inextricably linked. RCS can be caused by ROS-induced lipid peroxidation, while ROS are frequently raised by RCS activities. Almost all abiotic stresses can cause a surge of ROS and RCS, transforming their scavengers into particular defenses. Nonetheless, ROS and MG have been shown to perform a signaling role at low levels, which is tactically exploited to aid stress perception and elicitor retortion [64, 65]. As a result, it's critical to maintain the delicate RS homeostasis, which must be taken into account when trying to manipulate RS scavengers for multi-stress tolerance.

2.4 Reactive oxygen species

The active transport of chloroplasts imposes a greater burden of ROS on plant cells. When these small chemicals are overproduced, they attack a variety of biomolecules such as carbohydrates, lipids, proteins, and nucleic acids, causing oxidative catastrophes such as increased photoinhibition inhibition and membrane damage, which can be measured by the amount of MDA generated per oxidation of UFA [64, 66, 67]. MDA is a potential RCS that attacks under acidic conditions, forming covalent adducts known as advanced lipoxidation end products (ALEs), usually causes protein dysfunction and its consequences. Plants have evolved sophisticated ROS scavenging systems that employ both non-enzymatic and enzymatic methods. Many metabolites, such as betalain, carotenoids, flavonoids, and vitamin E, have antioxidant properties [68, 69]. Superoxide dismutase (SOD), catalase (CAT), and various peroxidases are examples of special enzymes (POD). SOD converts oxygen to hydrogen peroxide (H2O2), which is then reduced to water by CAT and POD. Dehydroascorbic acid reductase (DHAR), monodehydroascorbic acid reductase (MDHAR), and glutathione reductase are all core component of the ascorbate glutathione (ASAGSH) cycle, which is required for ascorbic acid peroxidase (APX) (GR). The intervention of detoxifying enzymes can undoubtedly be used to achieve multiple stress tolerance. Transgenic plants have used APX, an important enzyme that makes sure the removal of H2O2, to combat drought, salt, and intense light [70–72]. It also revealed glutathione peroxidase (GPX) activity, as well as cold air, heat, ultraviolet light, and heavy metals [73–75].

2.5 Carbonyl reactive species

Methylglyoxal, a substantial type of RCS, is attracting more attention in stressful situations. Because of the non-enzymatic dephosphorylation of two intermediates, glyceraldehyde3-phosphate, and dihydroxyacetone phosphate, glycolysis is the primary source of this cytotoxin in plant cells. Even before MG increases to a dangerous level, it can harm a variety of biomolecules, particularly because of its aldehyde group. Plant defenses against abiotic stresses ALEs, MG can help accelerate the photoreduction of O2 to O2 in chloroplasts and consume GSH via spontaneous combination into hemithioacetal, actually results in a vicious cycle and eventual cell death.

2.6 Chaperones at the molecular level

Heat shock proteins (HSPs) are molecular chaperones that are articulated either induced or constitutively to assisted protein folding, assembly, transport, and degradation. HSPs' anti-stress role isn't limited to their definition. This large family of proteins is a universal rescue system used by nearly all living organisms to combat all dangerous factors can trigger protein damage. They work to prevent denatured proteins from accumulating, assist in their refolding, or present them to lysosomes or proteasomes for proteolysis, restoring cellular homeostasis [76, 77]. Furthermore, some unusual hydrophilic proteins, such as members of the late embryogenesis abundant (LEA) and cold-regulated (COR) families, may act as chaperones to protect proteins and membranes from stress injury [78, 79]. The five conserved HSP classes based on molecular weight are HSP100/Clp, HSP90, HSP70/DnaK, HSP60/Chaperonin, and small HSP (smHSP). The most widely conserved among species is HSP70, which consists of an N-terminal ATPase domain and a C-terminal substrate-binding domain. In response to stress, smHSPs accumulate quickly and are more likely to seize non-native proteins and transfer them to ATP-dependent chaperones like the HSP70 system for re-naturation [80, 81]. According to a new study, Arabidopsis transformed with HSP16.4 from pepper (*Capsicum annum*) was less susceptible to drought, heat, and their combination, and ROS scavenging enzymes were more active under stressful conditions [82].

2.7 Compatible solutes

Conformance solutes are small organic compounds with electrical neutrality, high solubility, and low toxicity that can cause some problems when present in high concentrations in cells. Qualified molecules include sugar and amino acid derivatives, as well as their derivatives, such as ash, trehalose, inoside, mannit, proline (PRO), and glycine betaine (GB). Except for the protein and membrane RS and stabilizer, these metabolites begin under stressful conditions for dehydration and can be started. A compatible dissolved compatible lysis solute is a small organic connection with electrical neutrality, high solubility, and low toxicity that really can cause a variety of problems in a significant cell concentration. Qualified molecules include sugar and amino acid derivatives, as well as their derivatives, such as ash, trehalose, inoside, mannit, proline (PRO), and glycine betaine (GB).

3. Plant responses by signal transduction

Plant response to environmental changes has been related to changes in signaling molecules [e.g., sugars, hormones, calcium, reactive oxygen species (ROS), nitrous oxide (NO)] [54, 83], along with large-scale genomic restructuring, including transposon activation [84, 85], and rapid changes in gene expression patterns (e.g., genes encoding transcription factors [86, 87]. A percentage of transcription factors (TFs) from different crops have been discovered to play critical roles in abiotic stress responses. The ability of transcriptional regulators to act as master regulators has been hailed as a long-term solution for modifying complex traits in crop plants [85, 88]. Several transcription factor families, including AP2/ERF, bZIP, Zn-finger, NAC, MYB, and WRKY, have been implicated in abiotic stresses in past few decades [88–90]. Three major methodologies have been used to identify TFs associated with abiotic stress responses in rice: comparative genomics—abiotic stress-responsive genes from Arabidopsis and maize have been used to identify rice orthologs; forward genetics—genes related to traits like drought or hypoxia tolerance were identified through association mapping; genome-wide expression profiles—transcriptome analysis using microarrays was used to identify novel abiotic stress-response genes. Plant stress responses have also been interconnected to chromatin remodeling and nuclear organization. Salinity and heat-shock stresses, for example, caused decondensation of interphase ribosomal chromatin in rice and wheat [91, 92]. Heterochromatin maintenance mechanisms may repress transcription in normal circumstances, but they may fail to cause chromatin remodeling and novel gene expression profile in stressful situations [93, 94].

Changes in signaling pathways molecules for example, B. sugar, hormone, calcium, reactive oxygen species (ROS), basic oxide (NO)] are linked to plant response to the environmental changes. Scharf et al. [83] only Not even with the most thorough genomic reconstruction, which included transposon activation (for example, gene encoding transcription factor) (e.g., transcription factor) [75, 87]. Many transcription factors (TFS) derived from various plants were found to play a key role in their stress responses. TFS's ability was viewed as a long-term solution for reconfiguring different dynamics in crops, as well as a long-term solution as a master regulatory authority [88]. TF families such as AP2 / ERF, BZIP, Zn-finger, NAC, MYB, BZIP, Zn-finger, NAC, MyB, WRY, and others have been implicated in the rise in stress resistance in recent decades [13, 88–90]. In rice, three effective interventions were used to identify TFS. By assigning associations, the forward genetics gene associated with characteristics like drought or hypoxia resistance was discovered. Microarrays were used to identify new and assistant voltage attractive genes using genomic effect expression profile transcript analysis. The plant's response to stress also involves chromatin remodeling and nuclear organization [95]. Salt and heat shock stress, for example, caused decondensation of interphase ribosomal chromatin in rice and wheat [91, 92]. Heterochromatin maintenance mechanisms can suppress transcription under normal conditions, but under stress, these mechanisms can disrupt, leading to chromatin remodeling and new genetic patterns [93, 94].

4. Abiotic stress responses: epigenetic mechanisms and gene expression regulation

The effects of stress on genomic epigenetic marks, which affect gene expression regulation, are referred to as environmental epigenetics [96, 97]. Epigenetic memory is achieved by interacting with a variety of molecular mechanisms, including DNA methylation, post-translational modification of the nucleosome core histone protein's N-terminal region, and chromatin remodeling [98, 99]. Many proteins, known as transcription factor-interacting proteins (TFIPs), have been found to regulate epigenetic responses to environmental stress, but only a few have been found in rice. Rice underlying genetic factors were discovered primarily through comparative genomics. Plant plasticity responses to unpredictable abiotic stresses rely heavily on epigenetic mechanisms.

4.1 DNA methylation and abiotic stress

Cytosine methylation is a conserved epigenetic mark that plays a role in genome defense against endogenous transposable elements and viral DNA, as well as gene regulation regulation throughout plant development. Methyltransferases catalyze the addition of a methyl group to cytosine residues (MTases). Furthermore, in plants, this can happen in both asymmetric (CHH) and symmetric (CG and CHG) situations. DOMAINS REARRANGED METHYLTRANSFERASE TFS AND EPIGENETIC MECHANISMS IN ABIOTIC STRESS RESPONSES 847 (DRM), METHYLASE 1 (MET1), CHROMOMETHYLTRANSFERASE (CMT), and DNA methyltransferase homolog 2 (Dnmt2) are four key families of plant MTases that seem to have distinct functions in de novo and/or maintenance methylation [100]. While methylated cytosines are replaced with unmethylated ones during DNA replication, active demethylation occurs without DNA replication throughout a base excision repair mechanism mediated by DNA glycosylases [101]. Other frameworks, such as the RNA-directed DNA methylation (RdDM) pathway, mediated by siRNAs [102], and chromatin remodeling factors, also impact DNA methylation [78, 103]. Overall, these regulatory pathways provide a dynamic platform for establishing DNA methylation patterns, which may be critical for epigenomic plasticity and rapidly respond to developmental cues and environmental stress. The technique of DNA methylation's implication on transcription is still unidentified. Several lines of evidence suggest that cytosine methylation has a broad array of functions that are likely individualized for different genes [100, 104]. Methylated cytosines may attract methyl binding proteins, which in spin may attract histone modifiers and chromatin remodeling proteins, resulting in a complex that can disrupt transcription factor linkage [36, 104].

On either hand, high-resolution DNA methylation mapping has demonstrated some common aspects related to the H3K9me3 and H3K27me3 are infused in genes of euchromatic regions [105]. Biotinylation and sumoylation, two other histone modifications, have been interconnected to gene repression [106]. Large and powerful histone lysine acetylation has been linked toward a more open chromatin structure and thus enhanced transcription, whereas weak acetylation has been linked to chromatin compaction and gene silencing [107]. Histone acetyltransferases (HAT) and histone deacetylases enforce histone lysine acetylation (HDAC). In plants, there have been four major classes of HDAC encoding genes [108], also with HD2 class being the only one that exists [109, 110]. In hybrid rice, the OsHDT1 gene is involved in regulation of gene expression [111]. There are at least 19 HDAC genetic traits in the nucleotide sequence, and most of them are differentially regulated by different abiotic stress conditions [12, 112]. Most rice HDAC genes were exceptionally responsive to drought or salt stresses, primarily through transcriptional repression, as according microarray data [112]. As a result, abiotic stresses may start regulating the transcription of chromatin modifier enzymes. Down regulation of HDAC may be required in this case to allow the induction of stress-responsive genes [12, 113].

The highest density of methylated cytosines is reported in transcriptionally inactive heterochromatic regions, which contain countless transposable elements (TEs) and repetitive sequences. Lower but still significant cytosine methylation levels were observed in euchromatic regions. Surprisingly, DNA methylation related to active genes was more abundant in transcribed regions than in promoters in both Arabidopsis [114] and rice. The magnitude of methylation within the gene body was negatively correlated with transcript elongation performance in Arabidopsis [114]. It's reasonable to assume that rice has a similar principle. Abiotic stresses may end up causing changes in DNA methylation levels, which may be posted a link to chromatin remodeling and stress-responsive gene transcription regulation. Genome-wide analyses in several plant species reveal a global methylation readjustment in response to stress, owing primarily to demethylation [115–117]. Because conserved patterns were observed between different genotypes or tissues, the AFLP-based methylation-sensitive approach (MSAP) demonstrated that some of these improvements (methylation/demethylation) are site-specific. This method is best suited to CG methylation analysis. Other studies, on the other hand, have found that stress induces transcriptional induction of silent loci without a loss of DNA methylation, but instead a decrease in nucleosome occupancy [118, 119]. Elevated expression of the AtHKT1 gene, which encodes for a vacuolar Na +/H+ transporter, was also linked to lower DNA methylation in the Arabidopsis met1-3 mutant. The methylation pattern of a putative small RNA target region in the AtHKT1 promoter is required for the differential expression of this gene in roots and leaves, which may influence salt sensitivity and response [120]. Two *Laguncularia racemosa* species that grow in salt marsh and riverside habitats had different global DNA methylation patterns [121]. It's possible that epigenetic variation plays a role in helping plants adapt to different environments under natural conditions. It's possible that epigenetic variation plays a role in helping plants adjust to different environments under natural circumstances. Several MTases have been defined in rice, with such microarray data indicating that some are found to be elevated preferentially even during the commencement of floral organs [81, 122].

Furthermore, during the booting and heading stages, increased levels of methylation in rice leaves were detected (as ascertained by MSAP) than during the tillering stage [116]. As a result, drought-induced demethylation levels were higher at the tillering stage than it is at the booting and heading stages [116]. Changes in DNA methylation may differentially modulate response of plants to abiotic stress across the whole of development, according to these studies.

4.2 Abiotic stress and histone modifications

Nuclear DNA is packed and organized in eukaryotic cells in affiliations with a histone protein core-forming nucleosome, is also one of the chromatin's structural units. Combinations of histone variants and covalent modifications of histone tails, also including acetylation, methylation, phosphorylation, ubiquitination, biotinylation, or SUMOylation, resulted in changes in nucleosome structure. Assemble an integrated histone code that has been linked to gene expression regulation [106, 123, 124]. Depending on which lysine is methylated and how many methyl groups are added, the methylation process of lysine residues on histone H3 has been linked to transcription activation or repression [96, 105, 106]. Histone H3 lysine four trimethylation (H3K4me3), for example, has been linked to euchromatin and gene activation in maize, revealing inactive gene sequences not found in transposons [125]. Histone H3 lysine 9 dimethylation (H3K9me2) has been accompanied with transposons in Arabidopsis as an indication for heterochromatin and repressed transcription [105, 126]. H3K9me3 and H3K27me3 are, on the other hand, abundant in euchromatic genes [105]. Biotinylation and sumoylation, two other histone modifications, have been linked to gene repression [106]. In terms of histone lysine acetylation, strong acetylation has been linked toward a more relaxed chromatin structure that promotes transcription, so even though weak acetylation has been linked to chromatin densification and gene silencing [107, 127].

Histone acetyltransferases (HAT) and histone deacetylases enforce histone lysine acetylation (HDAC). In plants, there are four major classes of HDAC encoding genes [128, 129], with the HD2 class being the only one [109]. In hybrid rice, the OsHDT1 gene is involved in genetic variations [111]. There are at least 19 HDAC genes in the rice genome, and most of them are differentially regulated by different abiotic stress conditions [12, 112]. Most rice HDAC genes were exceptionally responsive to drought or salt stresses, pretty much exclusively through transcriptional repression, thus according microarray data [112]. As a result, abiotic stresses may regulate the transcription of chromatin modifier enzymes. Down regulation of HDAC may be taken into account to allow the induction of stress-responsive genes [12, 130]. Gene expression regulation has been coupled to histone modifications and DNA methylation crosstalks. In Arabidopsis, for example, the loss of DNA methylation in the ddm1 mutants was associated with low levels of dimethylation of histone H3 at lysine 9 (H3K9me2) [127, 131, 132]. A SUVH [Su(var)3–9 homologs] protein that plays in H3K9 methylation has been discovered to directly bind to methylated DNA, revealing the existence of a self-reinforcing feedback loop for DNA and histone methylation preservation in this species [133]. Recently, several rice SUVH genes with a deduced role in heterochromatin formation were revealed [134]. Through DNA methylation and H3K9me3, some of these have been shown how to resolve retrotransposon repression [74, 135].

Furthermore, most protein-coding genes with methylated DNA in rice are associated with H3K4me2 and/or H3K4me3, and when H3K4me3 is present, the repressive impact of DNA methylation on gene expression is reduced. Due to the rapid technological advances such as chromatin immunoprecipitation (ChIP) and genome-wide sequencing, the consequence of abiotic stresses on the genome-wide landscape of histone modifications is beginning to be deciphered (ChIP-Seq). Submergence stress in rice resulted in a decline in H3K4me2 levels, an increase in H3K4me3 levels, and a gradual increase in H3 acetylation at the ALCOHOL DEHYDROGENASE 1 (ADH1) and PYRUVATE DECARBOXYLASE 1 (PDC1) genes' 5- and 3-coding regions [136]. These changes were associated with increased ADH1 and PDC1 expression as a result of stress [136]. Cold stress, on the other hand, lessened H3K27me3 in the promoters of two cold-responsive genes, COLD REGULATED 15A (COR15A) and GALACTINOL SYNTHASE 3 (ATGOLS3), whereas salt stress increased H3K9ac, H3K14ac, and H3K4me3, while depleting H3K9me2 at stress-responsive genes [60]. These observations show the importance of histone code plasticity in transcriptional regulation during plant responses to various abiotic stresses when taken together.

5. Drought tolerance's molecular mechanism

Drought stress could severely limit rice production, leading to significant financial losses. It has become a more serious issue as the world's temperature rises. In light of current and projected global food demand, it is critical to prioritize increasing crop productivity on drought-prone rainfed lands. Drought-tolerant rice varieties are intended to address the assembly target in rainfed areas, and genetic improvement of rice for drought tolerance should also be a high priority theme of research in the next two decades. Breeding for drought tolerance could be an interesting challenge. The present study would be severely hampered by the complex nature and multigenic control of drought-tolerant traits.

Environmental drought impulses are intercepted by membrane sensors, which are still heavily portrayed. The signals are then transmitted via various signal transduction pathways, resulting in the outflow of drought-responsive attributes with effective gene functions and drought tolerance [98, 137]. Drought is a complex phenomenon, making it difficult to comprehend [73, 138]. As a result, hybridization and selection strategies could not provide precise drought tolerance results. Using DNA markers in molecular studies, at the other hand, can append the procedure by providing precise outcomes. These molecular markers are also useful for identifying drought-tolerant germplasm in a mass and using it to improve crops. Many studies have been conducted in order to identify some qualitative trait loci (QTL) linked to various traits [139, 140]. DNA studies based on marker-based phenotyping were the very first methodologies used to distinguish genes associated in rice drought resilience. Despite the progress, only a few traits have been officially approved for drought resistance [138, 141]. Molecular breeding can improve crop varieties, and yield assortments, produce productive, safe harvests, and also have high agronomic credibility in this way.

6. Rice drought tolerance genes and transgenic approaches

Many remarkable genes are highly expressed in rice after exposure to drought varieties, with approximately 5000 genes upregulated and 6000 genes downregulated [142, 143]. These genes can be divided into three categories: membrane transport, signaling, and transcriptional regulation [140, 144]. Many important genes/transcription factors are expressed differently in rice and are used to create transgenic plants for drought strains [73, 140]. The majority of the genes regulated by drought are ABA-independent, as are the ABA-independent regulatory requirements that manage rice's drought tolerance mechanisms [131, 145]. OsJAZ1 has also been shown to mitigate drought tolerance in rice by impairing ABA signaling, which synchronizes plant responses to expansion and success under drought stress [12]. Osmoregulation and late embryogenesis abundant (LEA) proteins, which confer terminal drought tolerance in rice, are also linked with a number of genes [137, 140]. In transgenic rice, the gene DRO1 causes root elongation and deeper rooting. In rice under water deficit conditions, other genes like as OsPYL/RCAR5 and EcNAC67 induced leaf water content, delayed leaf rolling, improved growth parameters mass, and stomatal regulation [146, 147]. Over expression of OsDREB2B, CYP735A, and OsDREB1F [82, 148] pronounced the DREB2-like gene OsDRAP1 conferring drought tolerance in rice accelerated root morphological diversifications in rice under drought strain. Increased grain yield in rice under drought is critical, and it can be achieved by using transgenic approaches to start introducing genes like OsNAC5 [112], OsbZIP71 [51], OsWRKY47 [149], OsbZIP46 [150], and OsNAC10 into the crops. The WRKY genes play an important role in plant improvement by responding to drought strains and can be used to create drought-tolerant transgenic plants [151, 152]. Several genes were investigated using transgenic approaches to confer drought tolerance in rice grown in a research lab or glasshouse conditions. However, those genes must be investigated further.

7. The role of micro RNA in rice drought tolerance

Micro RNAs (miRNAs) are small noncoding regulatory RNAs that modulate gene expression during abiotic stress, as has been acknowledged [151, 153]. These 20–24 nucleotide long proteins control gene expression at the post-translational level [140]. Several miRNAs have also been found to alter gene expression in rice by up- up-and down-regulation, which confers drought tolerance [134, 154, 155]. Arabidopsis [156] was the first to reveal the expression of miR393, miR319, and miR397 in response to drought, and rice control transcriptional factors OsAUX1 and OsTIR1 confer tiller number increment, early flowering, and auxin increased sensitivity [157].

Rice does have 30 miRNAs, 11 of which are down-regulated and eight of which are up-regulated under drought stress [126]. Under drought stress, MiR160 and MiR167 regulate the expression of the ARFs gene, that further regulates early auxin response [140]. Through ROS homeostatic genes, DST- amiRNA enhances drought resistance by increasing stomatal closure and decreasing stomatal density [155, 158]. Over expression of the UDP-glucose-4-epimerase gene, facilitated by OSA-miR169-3p and Osa-miR166e-3p, regulated root development and cell wall biogenesis, along with carbohydrate metabolism [154, 159]. Ten miRNAs (miR531, miR827, miR8175, miR977, miR6300, miR1861, miR440, miR9773, miR3982, and miR1876) were recently discovered to be regulated under drought stress and confer tolerance attributes in traditional rice land races [160]. Drought tolerance can be accomplished by gene manipulation of these miRNAs. As a consequence, miRNAs regulate many drought tolerance responses, potentially enhancing the development of drought-tolerant rice genotypes.

Natural rice genotypic variation could be investigated to seek novel genotypes with a drought-tolerant trait of interest and a gene/locus affiliated with them. Drought-tolerant rice varieties can be developed and use these novel genotypes in traditional breeding programs using marker-assisted selection. The breeding program's aim is to create high-yield lines with improved performance parameters, as well as to commercialize the cultivars. Numerous researchers have investigated the progeny of drought-tolerant genotypes in the past [139, 161, 162], but the overall performance has been far lower than expected due to the difficulty in finding suitable donors with a higher tolerance level, as well as the environment-specific nature of the genotypes. The majority of marker-assisted breeding approaches for improving drought-tolerant rice varieties have always been carried out at the International Rice Research Institute in the last decade [163]. In India [145, 160], the Philippines [159, 161], and Malaysia, several works on marker-assisted progression of popular varieties were carried out [164]. Several QTLs for drought tolerance in rice have been incorporated into leading cultivars using marker-assisted breeding techniques [162]. By using a marker-assisted backcrossing approach, they were able to successfully incorporate QTLs such as qDTY9.1, qDTY2.2, qDTY10.1, and qDTY4.1 in the high yielding IR64 variety [27]. They also developed the drought-tolerant elite Malaysian rice cultivar MR219 by pyramiding three QTLs, qDTY2.2, qDTY3.1, and qDTY12.1. They created TDK1 rice varieties with three QTLs for high yield in drought conditions (qDTY3.1, qDTY6.1, and qDTY6.2). Drought has only received attention as a constraint, and no effective methods for developing drought-tolerant rice varieties have yet been successful. Farmers prefer to grow high-yielding cultivars with better grain quality but are drought-prone, or traditional drought-tolerant varieties with low yield. As a result, more effort will be needed in the future to develop unique rice varieties that can produce high yields in drought and acclimate to a range of adverse climatic conditions.

8. Salt tolerance

Salt stress may cause progress in multiple physiological and metabolic pathways depending on the severity and duration of the stress, leading to a reduction in rice productivity [50, 120, 165, 166]. To estimate the phenotypic coefficient of variation (PCV), genotypic variance (GCV), broad-sense heritability, and genetic Advance, genetic characterization of salt tolerance-related traits is required (GA). Assume there is sufficient variation in the germplasm for salt tolerance-related attributes with greater heritability and genetic advance. As a result, by utilizing salt-tolerant landraces/germplasm in breeding programs, it may be possible to improve the personality characteristics associated to salt tolerance in rice. The low Na-K ratio was controlled by both additive and dominance gene effects, according to a genetic component analysis (GCA) study [89, 167]. The results of the amalgamating ability analysis show that both general combing ability (GCA) and specific combining ability (SCA) effects are important in understanding salt tolerance genetics. They also revealed that selection for common heritable traits like the Na-K ratio could have been built in subsequent generations under controlled conditions to reduce environmental effects. Additive gene action is linked to narrow-sense heritability [168, 169], and additive gene action may enhance or fix the action of the desired combination of genes. As a result, early generation preference for salinity tolerance is possible. Recognizing the gene action in rice that generates salt tolerance will support future breeding efforts [112, 170].

9. Submergence

Submergence is among the most important abiotic stresses in rice-growing areas prone to flash floods [171]. Submergence tolerance is a necessary trait for rice in rain-fed lowland conditions (Oryza sativa). A significant gene known as Sub1 is mainly accountable for this trait. Indica cultivar FR13A is a highly tolerant rice variety that can withstand complete submersion for up to two weeks. Near the centromere of chromosome 9, they have a substantial quantitative trait locus known as submergence1 (Sub1) [157, 172, 173]. Background genetic information for submergence tolerance was well documented out of some research using QTL mapping and map-based cloning techniques [172, 174, 175]. Because of the great specificity of contemporary rice varieties, salt stress is a significant constraint in many rice-producing areas. One of the most severe abiotic stresses restricts rice growth and development of plants, resulting in yield reduction of more than 50% [176, 177]. Salinity tolerance is multifaceted, involving a range of biological mechanisms such as sodium exclusion from root system. Salinity is estimated to affect over 150 million hectares of current and potential rice land in tropical and subtropical regions of the world [176, 178]. Despite the fact that rice is the source of nutrition for half of the world's population, it is more susceptible to salt stress than other cereals [179, 180].

If rice plants are immersed in water for even more than five days resulting from environmental or abiotic stress, they become the deepest submergence-tolerant contributors and are widely used by rice breeders. The pyramiding of submergence and salinity tolerance is especially important in coastal areas where floodwaters are frequently saline. On chromosome 1, a major salinity QTL has recently been introduced and characterized [79, 170]. Although many QTLs may be necessary to accomplish adequate salinity tolerance in the field, additional QTLs for vegetative growth and reproductive-stage salinity tolerance may be compelled to provide salinity stress defense during in the rainy season. Using molecular marker technologies to stack multiple tolerance genes/QTLs into single rice varieties provides breeders with a once-in-a-lifetime chance to advance tolerant cultivars more faster for specific environments [172, 181]. There are various types of biotic stresses. The Sub1 gene, which is managed to find on chromosome 9 of rice, is well-known for conferring submergence tolerance intolerant rice cultivar FR13A and its progenies [157, 173, 175].

10. Modern breeding techniques

Using a variety of innovative tools, genomic assisted breeding (GAB) is routinely used to improve the genetics of salt-tolerant rice. Genomic breeding, forward breeding, rapid breeding, and haplotype-based breeding are all examples of genomic breeding [115, 182]. 5G breeding methods are used to improve genotype productivity by improving genome sequence availability (genome assembly), characterization of germplasm at the genomic and morpho-agronomic level, genomic detection and understanding function, genomic breeding, and genome editing [118, 183, 184]. These could be used to improve efficiency and accuracy of breeding for complex abiotic stress tolerance traits. Through SNP-based speed breeding, SNP-assisted introgression of the hst1 (Salt-tolerant 1) gene enhanced salt tolerance in a high-yielding rice variety [41]. The emphasis in contemporary breeding is on data-driven parent selection. Genetic technique, trait categorization through diagnostic trait markers, genomic screening, and breeding value estimation are all applied to local and exotic germplasms. Native germplasm with low yield potential may contain traits of interest (ToI) like salinity and submergence tolerance, aroma, and resistance to disease. Then, in order to develop pre-breeding materials, ToI is first transferred to a privileged background with a higher yield. The elite line with the desired traits is then used to accomplish the product profile for breeding [185].

11. Underpinning defense systems is a regulatory network

The five general defense mechanisms are coordinated by a delicate regulatory network composed of numerous signaling molecules and gene regulation indicators in the face of abiotic stresses. We'll look at some of the more well-known ones here. Stress hormones (ABA), reactive oxygen species (ROS), hydrogen sulfide (H2S), nitrogen oxides (NO), polyamines (PAs), phytochrome B (PHYB), and calcium interplay with others at various levels, synergistically or antagonistically, to establish a specific directive for downstream effectors, especially transcription factors (TFs), to alter gene expression and protein/enzyme activities in a specific pattern, thereby launching a proper response. Hormones of Stress Phytohormones like ABA, ethylene (ET), jasmonic acid (JA), and salicylic acid (SA) are important organizers of systemic stress defense, and they work together in the complex hormonal signalosome [3]. S Notably, melatonin, a universal multi-regulatory molecule across all life forms, is increasingly recognized as a potent stimulator against stress in the plant. One notable aspect of this yet-to-be-licensed phytohormone is that it operates as if a commander of other phytohormones [28, 186]. Nevertheless, ABA is the main stress hormone, which not only extensively interplays with other phytohormones but with all following signaling molecules. Particularly, components of all biochemical defenses remarked above can be mobilized by ABA, including cuticular waxes [187, 188], HSPs [95], Pro [120], antioxidants [189, 190], and RS detoxifying enzymes [137, 191]. Stress stimuli can rapidly trigger de novo synthesis of ABA from oxidative cleavage of β-carotene, with 9-cisepoxycarotenoiddioxygenase (NCED) is the rate-limiting enzyme.

ABA can also end up causing organic changes in order to cope with stressful situations. Probably one of the best is the closing of stomata, tiny pores formed by paired guard cells that allow gas exchange and thus minimize water loss from transpiration and thus mitigate dehydration. The activity of ion channels and aquaporins is modulated to achieve this movement. As a result, the outflow of K+ and anions pulls water out through osmosis, causing guard cell shrinkage, which is ensured by actin filament reshuffling [48, 192, 193]. Another unique feature is seed dormancy, which enables seeds to avoid existing stresses and wait for ideal germination conditions [194–197]. Endogenous ABA elevation and exogenous ABA addition both help plants cope with a wide assortment of stresses. For in-field applications, the development of ABA analogs with greater security is promising. Overloaded ABA signaling, such as that caused by over expressed NCED or constitutively active PYLs, can cause vegetative growth retardation and grain yield reduction [81, 198], while foliar ABA spraying can cause leaf senescence in rice (*Oryza sativa*) and maize (*Zea mays*). As a result, it's critical to gain a better understanding of ABA homeostasis, its wide range of biological effects, and crosstalk with other pathways in order to create crop stress tolerance strategies that don't sacrifice economic traits. ROS are continuously produced in plant cells as byproducts of aerobic metabolism in chloroplasts, mitochondria, and peroxisomes, among other cytoplasmic organelles. However, as noted previously, it is not only toxins that can only be removed but also signaling molecules that are required for a variety of physiological processes, including stress resistance. A ROS signal is shaped by a variety of factors, including dose, duration, source, and type

[199, 200]. The extensive crosstalk between H2O2 and other signaling molecules has been reviewed, including ABA, ET, JA, SA, NO, and Ca2+ [136]. Ca2+ influx, in particular, is a notable event in H2O2 signaling, which modulates H2O2 levels by activating producing (e.g., RBOHs) or scavenging enzymes. Because of the degradation of PAs, H2O2 is strongly intertwined to them [189]. Numerous enzymes in the apoplast can produce ROS on their own in response to increases stimuli. Respiratory burst oxidase homologs are the most common (RBOHs). These PM-localized NADPH oxidases are activated by Ca2+ binding to the EF-hand motifs in the N-terminal cytosolic region, in combination with phosphorylation by receptor-like cytoplasmic protein kinases, for example (RLCKs). The MAPK signalosome plays a key role in H2O2 intracellular signaling, which is triggered by metabolic disturbance and/or apoplastic discharge. It is a pertinent stress signaling divergent node. MAPK kinase (MAPKKK), MAPK kinase (MAPKK), and MAPK make up each phosphorylation cascade. Many MAPKs and cascades have been discovered to decode the H2O2 signal in different ways, but how the precision is determined is still unspecified. The MAPK pathway, on the other hand, functions upstream of ROS by modulating the activities of RBOHs, either positively or negatively [201, 202]. Two MAPKs, MPK3 and MPK6, are aligned in defensive response in Arabidopsis, and can eventually elevate the levels of defending factors such as GSTs and HSPs. H2O2 can activate them though the ANP1, a MAPKKK, and other kinases such as oxidative signal-inducible 1 (OXI1), which is required for full activation of MPK3/6, and NUCLEOTIDE DIPHOSPHATE KINASE 2 (NDPK2), which can interact with MPK3/6 and potentiate its activities. NPK1 (tobacco ANP1 ortholog) and AtNDPK2 imparted tolerance to a variety of stresses, which would include salt and extreme temperatures, on transgenic plants. MKK2, a MAPKK in another anti-stress cascade, MEKK1-MKK2-MPK4/6, did this very same [44, 203].

12. Nitric oxide and hydrogen sulfide

At low concentrations, toxic secondary gaseous molecules, H2S and NO, show impressive powers in providing protection against a wide range of stresses, just like H2O2. Many anti-stress mechanisms are shared by the two. Both, for example, can reduce salt toxicity by increasing Na+ exclusion by unlocking SALT OVERLY SENSITIVE 1 (SOS1), a PM Na+/H+ antiporter [155, 204]. The ability to combat oxidative stress should be their most prominent role, as both of them not only act as antioxidants in their own right but can also repress ROS production and activate ROS elimination. H2S can be assimilated into GSH as a source of sulfur, resulting in an increase in this important R.S. scavenger [20, 205]. The first recognized gasotransmitter, carbon monoxide (CO), is also an elicitor. CO research on this topic, however, is still in its infancy [114]. During various stresses, the three R.S., H2O2, H2S, and NO, are typically available and together demonstrate intricate interactions depending on the context. It can ablate NO accumulation and aid stomata opening with the antagonist's face [176]. NO, on the other hand, is a mediator of H2S in the promotion of adventitious root development [103, 206], which can boost O2 uptake and thus reduce hypoxia stress caused by waterlogging [33, 173]. H2S, on the other hand, is a NO mediator in maize heat tolerance and bermudagrass (Cynodon dactylon) cadmium resistance [116]. Furthermore, with the addition of another R.S. player, MG, the situation will become even more complicated. Their chemical reactions add to the complexity, blocking one another while also forming new compounds with physiological implications, such as peroxynitrite (ONOO) formation by NO and O2 nitrosothiol formation by NO and H2S. There is even a competition between M.G. and GSH participants. Thiol modification, such as oxidation by H2O2, sulfhydration by H2S, nitrosylation by NO, glycation by M.G., and glutathionylation by GSH, can directly modulate protein function [152, 176]. Even though GSH is a derivative of H2S but a cocktail mixer of the other three, it provides additional pathway for their crosstalk. Besides this, the MAPK pathway is likely to be a point of convergence for the four signaling RS. Two gasotransmitters originate from various sources and are natural products of botanic metabolism. Cysteine desulfhydrases (DES) and sulfite reductase (SIR) for H2S, along with nitrate reductase (NR) and a nitric oxide synthase (NOS)-like a pathway for NO, though a truthful NOS has yet to be discovered in plants. PAs have long been acknowledged for their protective role in plant response to a variety of stresses [75, 165, 189]. Indeed, protein expression of every PA biosynthetic enzyme, such as arginine decarboxylase (ADC), spermidine synthase (SPDS), and S-adenosylmethionine synthase (SAMS), simultaneously breakthroughs stress tolerance mechanisms in various plant species, seeking to make exogenous PA application unnecessary [8, 75, 207]. Dissecting the mechanisms underlying PAs' anti-stress effects is complicated. Because of their polycationic nature, RS-scavenging property, and signaling function, it's conceivable that these multifaceted substances contribute to stress defense in a multitude of ways. Protonated PAs, for instance, not only take an active part in ion homeostasis at the physiological PH, but they can also bind to negatively charged molecules such as membrane lipids and integral proteins, which helps mitigate stress-induced membrane damage. PAs not only convey with ABA in stress signaling, but they can also induce rapid NO production. PAs are linked to other stress-related metabolites, such as Pro and ET, which are linked to PA anabolism and H2O2. Furthermore, GABA is produced as a result of PA catabolism [8, 75]. PAs also fall into the Janus category [168], which must be taken into account in their practical implementation due to the obvious specific link with H2O2. Phytochromes PHYB, a modest family of chromophore-containing proteins that serve as photoreceptors to perceive red (R) and far-red (FR) light, is emerging as a negative regulator in stress tolerance. PHYB's signaling activity undergoes reversible photoconversion, that also involves R activation and FR deactivation in response to protein aggregation. The Pr (R-absorbing) form of nascent PHYB is inactive. Dimeric PHYB will translocate into the nucleus once converted to the bioactive Pfr (FR-absorbing) form, where it can interact with and trigger the proteasomal degradation of phytochrome interacting factors (PIFs), a subfamily of basic helix-loop-helix (bHLH) TFs, to remodel the expression profile of thousands of light-responsive genes, guiding photomorphogenesis [131, 208–210].

PHYB was discovered recently to be a thermosensor [46, 71]. Warm ambient temperatures can effectively induce elongation development, which is phenocopied by shade avoidance and is controlled by the PHY-PIF cascade. Indeed, thermal (or dark) reversion, which is independent of light but sensitive to temperature, can end up causing to spontaneously revert to warm temperatures, especially at night, can relieve PIF4 repression by fastly unplugging PHYB and increasing PIF4 transcription, resulting in thermo morphogenesis. Because active PHYB was discovered to interact with PIF-binding sites (G-boxes) at PIF4-targeted promoters, it was tried to suggest that it could play a co-repressor or competitor role in gene regulation with PIF4 [211].

In case of light and temperature-induced growth, another cascade downstream of PHYB involves the RING E3 ligase CONSTITUTIVE PHOTOMORPHOG for degradation to depress the growth genes. ENIC 1 (COP1), and the TF ELONGATED HYPOCOTYL 5 (HY5), with COP1 ubiquitinating. Notably, COP1 can indirectly potentiate the activity of PIF4, thereby connecting the two branches [211, 212]. Plant

Defenses against diverse abiotic stresses, ROS and GOLS are essential for raffinose synthesis via binding upon oligomerization to the heat shock elements (HSEs) located in their promoter regions. Therefore, HSFs are capable of launching three general defensive systems. Surprisingly, plants have a powerful and relatively variable number of HSFs. Nonetheless, genetic manipulation of HSFs remains a viable option for conferring multiplex sensitivity to plants [83, 89]. It's also worth noting that HSFA6b and HSFA3 allow ABA to play a legitimate role in HS response. The former is stimulated directly by AREB1, whereas the latter is activated downstream of DREB2A, a destination that both AREB1 and HSFA6b share [95]. Another member of the class 2 DREB family, DREB2C, is also an HSFA3 activator [96]. It can also work against salt toxicity by trying to target chaperones like COR15A and DESICCATION-RESPONSIVE PROTEIN 29A (RD29A) [173, 213]. Interestingly, DREB2C from *Ammopiptanthus mongolicus* and evergreen broadleaf shrubs living in the desert was newly reported to up-regulate 11-pyrroline-5-carboxylate synthetase initiates Pro biosynthesis, as well as FADs that catalyze 18:3 production, thereby promoting Arabidopsis endurance to drought, freezing, and heat [188]. A safe conclusion can be drawn that this single TF governs all four cellular general defenses. Furthermore, a computational analysis of the Arabidopsis DREB2C promoter revealed a variety of elements that are sensitive to ABA (ABRE), MeJA, SA (TCA), heat (HSE), low thermal conductivity (LTR), and stress (TC rich) [206, 214], implying that TF is a central point in stress signaling. It is an ABA-inducible transcription factor that can delay seed germination by exerting good feedback on ABA biosynthesis by trans-activating NCED9. The enzyme genes in Arabidopsis are primarily controlled by TFs from two families: the AP2/ERF superfamily's SHINE 1 (SHN1), -2, -3, and DEWAX, as well as the R2R3-MYB family's MYB16, MYB30, and MYB106. MYB96, on the other hand, is a repressor of cutin synthesis, whereas MYB41 is an inducer of wax manufacturing [187]. MYB96, in specific, is a key player in ABA signaling, which helps regulate the whole of wax metabolism. Not only elongation and modification enzymes, but also ABC transporters and nsLTPs are directly or indirectly targeted by at least one isoform gene [103, 215]. MYB96 transgenesis continued to improve Arabidopsis drought and freezing tolerance, but it also caused significant dwarfism [99, 215]. Wax production 1 (WXP1), an ERF member from *Medicago truncatula,* may be a better candidate, as it was the one responsible for the previously mentioned observation that higher n-alkane and predominant alcohol contents results in improved viability under drought and freezing without interfering with transgenic Arabidopsis growth [99].

12.1 E3-Ubiquitin ligases, water stress responses

At some point during the plant growth cycle, climate change is threatening more than 20 million ha of rain-fed lowland rice (12 percent of total rice area and about 20% of global production) [216–218]. As extreme events become more familiar, water scarcity is expected to worsen, with yield losses of up to 81 percent [161, 216]. Drought has a particularly negative impact on seedlings (2 to 3 weeks old) and reproductive tiers (pollen-development stage) [207, 211], delaying flowering and lowering yields [186, 219]. Drought leads to inefficient water use, stomatal closure, photosynthesis impairment, and decelerated cell division and expansion [186]. As a proactive approach to cope with water deficit, plants directly influence divergent genetic and metabolic methodologies, such as cellular osmotic potential, stomatal aperture, impaired antioxidant, phytohormones, and chlorophyll content. This really is likely to result in the adaptation and maintenance of their physiological activity under drought conditions [220]. As a result, it's critical to comprehend the molecular mechanisms underpinning rice drought response, particularly the role of ubiquitination, and to apply these skills to the development of drought-tolerant crop varieties. Over the last two decades, several rice E3-ubiquitin ligases and their interacting proteins have been linked to drought response in plants. To fully comprehend their interactome and function, however, more research is required.

12.2 E3-Ubiquitin ligases in salinity responses

High salinity, drought, and ABA induce Oryza sativa salt-induced RING finger protein 2 (OsSIRP2), which encodes a RING-type E3-ubiquitin ligase that binds specifically to the nucleus of rice protoplasts for both control and high-salinity conditions. OsSIRP2 been shown to confer tolerance to salinity and osmotic stresses in Arabidopsis when overexpressed [221]. OsSIRP2 had been shown to interact with the rice transketolase 1 (OsTKL1) in the cytoplasm, causing it to be targeted for degradation by the UPS. OsTKL1 is a member of the transketolase family, which is involved in the Calvin cycle's oxidative pentose phosphate pathway. It is observed in the chloroplast and is considered necessary for the regeneration of ribulose 1,5-bisphosphate [189, 191]. In tobacco, reduced TKL1 activity causes photosynthesis to be inhibited [165]. The enzymatic activity of transketolase is also compelled for the stress-induced manufacturing of cytosolic NADPH, which is an important component of a plant's defense against ROS-induced damage [214]. It is critical to perform functional characterization of these two proteins in rice to stronger understand the mechanism of OsSIRP2 and the physiological implying of the OsSIRP2–OsTKL1 interaction in salt (and drought) stress responses, which include photosynthesis performance. It's crucial to figure out if and how OsSIRP2's deleterious regulation of OsTKL1 promotes stress tolerance. Finally, because OsSIRP2 did not change specificity under salt stress, the translocation of OsSIRP2 from the nucleus to the cytoplasm to ubiquitinate OsTKL1 raises a question of the underlying mechanism driving this export.

12.3 Response to low temperature

OsPUB2 and OsPUB3 are homologous U-box type E3-ubiquitin ligases that have recently been identified as positive regulators of rice's cold stress response [222]. Minimum temperature, drought, and salt stress upregulate OsPUB2, whereas OsPUB3 expression is unaffected by any of the aforementioned stresses. Overexpression of OsPUB2 or OsPUB3 in rice plants, on the other hand, confers a cold-tolerance phenotype in the result of enhanced survival rates, total chlorophyll, and diminished ion leakage. Furthermore, gene expression analysis reveals that under both control and cold conditions, overexpression of the two OsPUB genes is linked to upregulation of cold stress-inducible genes also including glutamate decarboxylase (GAD), WRKY77, and multidrug resistance protein 4 (MRP4). Furthermore, both were crafted more stable by the cold. Both E3-ligases are found in small cytosolic punctate bodies in Nicotiana benthamiana leaf protoplasts' subcellular localization. However, it remains to see whether those two homologous E3-ubiquitin ligases collaborate to confer cold tolerance to rice plants and, if so, which target protein(s) are used to achieve this tolerance. Furthermore, the OsPUB2 mutants should really be studied for phenotypic analysis under a variety of stresses. The RING-type upregulation of osmotically responsive gene 1 (OsHOS1) is another E3-ubiquitin ligase that modulates rice plant response to cold stress [204]. OsHOS1 proteins bind to the nucleus's Inducer of CBF

Expression 1 (OsICE1) and direct it to be broken down by the UPS. Stress-responsive transcription factor dehydration responsive element (DRE)-binding protein 1A (OsDREB1A) transcript thresholds and protein levels of OsICE1, a master integrator of cold stress, are higher in OsHOS1-silenced (RNAi) lines.

13. Conclusion

Plants are tough and have evolved strategies accordingly to a variety of environments over the course of their evolution. As a result, understanding the molecular mechanisms that underlie stress tolerance is essential for improving crop stress tolerance as the impact of abiotic stresses grows as a result of global climate change. Distinctive cutting-edge/modern breeding strategies are aggregated in the holistic breeding approach. Genotypes could help the farmers cope with rising temperatures, increase varietal turnover, and help meet the challenges of abiotic stress-prone ecosystems by increasing productivity and ensuring food security. In addition, rice cultivation areas in slightly elevated abiotic stress-prone areas under which salt stress is extremely crucial for rice production during pre- monsoon season will be decided to expand in various rice-growing countries, including Bangladesh. Furthermore, the stress associated with HNT must be highlighted because this stress could pose a threat to food security in areas where rice is a staple food.

Conflict of interest

No conflict of interest.

Author details

Ananya Prova[1*] and Md. Saeed Sultan[2]

1 Department of Plant Pathology, EXIM Bank Agricultural University Bangladesh (EBAUB), Chapainawabganj, Bangladesh

2 Department of Plant Pathology, Bangabandhu Sheikh Mujibur Rahman Agricultural University (BSMRAU), Gazipur, Bangladesh

*Address all correspondence to: provaananya16@gmail.com

References

[1] Ahmad P, Prasad MNV, editors. Environmental Adaptations and Stress Tolerance of Plants in the Era of Climate Change. Berlin, Germany: Springer Science & Business Media; 2011

[2] Shepherd T, Wynne Griffiths D. The effects of stress on plant cuticular waxes. The New Phytologist. 2006;**171**:469-499. DOI: 10.1111/j.1469-8137.2006. 01826.x

[3] Shrivastava P, Kumar R. Soil salinity: A serious environmental issue and plant growth promoting bacteria as one of the tools for its alleviation. Saudi Journal of Biological Science. 2015;**22**:123-131. DOI: 10.1016/j.sjbs.2014.12.001

[4] Kim KC, Lai Z, Fan B, Chen Z. Arabidopsis WRKY38 and WRKY62 transcription factors interact with histone deacetylase 19 in basal defense. Plant Cell. 2008;**20**:2357-2371

[5] Farooq M, Siddique KH, Rehman H, Aziz T, Lee D-J, Wahid A. Rice direct seeding: Experiences, challenges and opportunities. Soil and Tillage Research. 2011;**111**(2):87-98

[6] FAO. World Food and Agriculture. Rome: Food and Agriculture organization United Nations; 2020

[7] Fischer RA, Edmeades GO. Breeding and cereal yield progress. Crop Science. 2010;**50**:S-85

[8] Alcázar R, Altabella T, Marco F, Bortolotti C, Reymond M, Koncz C, et al. Polyamines: Molecules with regulatory functions in plant abiotic stress tolerance. Planta. 2010;**231**:1237-1249. DOI: 10.1007/s00425-010-1130-0

[9] Dixit S, Swamy BPM, Vikram P, Ahmed HU, Cruz MTS, Amante M, et al. Fine mapping of QTLs for rice grain yield under drought reveals sub-QTLs conferring a response to variable drought severities. Theoretical and Applied Genetics. 2012;**125**(1):155-169

[10] Banerjee A, Roychoudhury A. Abscisic-acid-dependent basic leucine zipper (bZIP) transcription factors in plant abiotic stress. Protoplasma. 2017;**254**(1):3-16

[11] Arnholdt-Schmitt B. Stress-induced cell reprogramming. A role for global genome regulation? Plant Physiology. 2004;**136**:2579-2586

[12] Feng ZT, Deng YQ, Zhang SC, Liang X, Yuan F, Hao JL, et al. KC accumulation in the cytoplasm and nucleus of the salt gland cells of Limonium bicolor accompanies increased rates of salt secretion under NaCl treatment using NanoSIMS. Plant Sciences. 2015;**238**:286-296. DOI: 10.1016/j.plantsci.2015.06.021

[13] Hirayama T, Shinozaki K. Research on plant abiotic stress responses in the post-genome era: Past, present and future. The Plant Journal. 2010;**61**:1041-1052

[14] Aroca R. Plant responses to drought stress: From morphological to molecular features. In: Aroca R, editor. Plant Responses to Drought Stress: From Morphological to Molecular Features. Berlin: Springer; 2013

[15] Lira-Medeiros CF, Parisod C, Fernandes RA, Mata CS, Cardoso MA, Ferreira PCG. Epigenetic variation in mangrove plants occurring in contrasting natural environment. PLoS One. 2010;**5**:e10326

[16] Fich EA, Segerson NA, Rose JK. The plant polyester cutin: Biosynthesis, structure, and biological roles.

Annual Review of Plant Biology. 2016;**67**:207-233. DOI: 10.1146/annurev-arplant-043015-111929

[17] Lang-Mladek C, Popova O, Kiok K, Berlinger M, Rakic B, Aufsatz W, et al. Transgenerational inheritance and resetting of stress-induced loss of epigenetic gene silencing in Arabidopsis. Molecular Plant. 2010;**3**:594-602

[18] Bannister AJ, Kouzarides T. Regulation of chromatin by histone modifications. Cell Research. 2011;**21**:381-395

[19] Lekklar C, Pongpanich M, Suriya-Arunroj D, Chinpongpanich A, Tsai H, Comai L, et al. Genome-wide association study for salinity tolerance at the flowering stage in a panel of rice accessions from Thailand. BMC Genomics. 2019;**20**:76. DOI: 10.1186/s12864-018-5317-2

[20] Siddiqui MH, Al-Whaibi MH, Basalah MO. Role of nitric oxide in tolerance of plants to abiotic stress. Protoplasma. 2011;**248**:447-455. DOI: 10.1007/s00709-010-0206-9

[21] Lobell DB, Bänziger M, Magorokosho C, Vivek B. Nonlinear heat effects on African maize as evidenced by historical yield trials. Nature Climate Change. 2011;**1**:42

[22] Prakash C, Sevanthi AM, Shanmugavadivel PS. Use of QTLs in developing abiotic stress tolerance in rice. In: Advances in Rice Research for Abiotic Stress Tolerance. New York: Elsevier; 2019. pp. 869-893

[23] Atkinson NJ, Urwin PE. The interaction of plant biotic and abiotic stresses: From genes to the field. Journal of Experimental Botany. 2012;**63**(10):3523-3543

[24] Sunkar R, Zhu JK. Novel and stress-regulated microRNAs and other small RNAs from Arabidopsis. Plant Cell. 2004;**16**(8):2001-2019

[25] Hasanuzzaman M, Hossain MA, Fujita M. Selenium-induced up-regulation of the antioxidant defense and methylglyoxal detoxification system reduces salinity-induced damage in rapeseed seedlings. Biological Trace Element Research. 2011;**143**(3):1704-1721

[26] Hwang JE, Lim CJ, Chen H, Je J, Song C, Lim CO. Overexpression of Arabidopsis dehydration-responsive elementbinding protein 2C confers tolerance to oxidative stress. Molecules and Cells. 2012;**33**:135-140. DOI: 10.1007/s10059-012-2188-2

[27] Qi YC, Wang FF, Zhang H, Liu WQ. Over expression of suadea salsa S-adenosylmethionine synthetase gene promotes salt tolerance in transgenic tobacco. Acta Physiologiae Plantarum. 2010;**32**:263-269. DOI: 10.1007/s11738-009- 0403-3

[28] Kanwar MK, Yu J, Zhou J. Phytomelatonin: Recent advances and future prospects. Journal of Pineal Research. 2018;**65**:e12526. DOI: 10.1111/jpi.12526

[29] Maiti R, Satya P. Research advances in major cereal crops for adaptation to abiotic stresses. GM Crops Food. 2014;**5**:259-279

[30] Silva VA et al. Reciprocal grafing between clones with contrasting drought tolerance suggests a key role of abscisic acid in coffee acclimation to drought stress. Plant Growth Regulation. 2018;**85**:221-229

[31] Singh S, Kumar A, Panda D, Modi MK, Sen P. Identification and characterization of drought responsive miRNAs from a drought tolerant rice genotype of Assam. Plant Gene. 2020;**21**:100213

[32] Sui N. Photoinhibition of Suaeda salsa to chilling stress is related to energy dissipation and water-water cycle. Photosynthetica. 2015;**53**:207-212. DOI: 10.1007/s11099-015-0080-y

[33] Chen TS, Yuan F, Song J, Wang BS. Nitric oxide participates in waterlogging tolerance through enhanced adventitious root formation in the euhalophyte Suaeda salsa. Functional Plant Biology. 2016;**43**:244-253. DOI: 10.1071/ fp15120

[34] Cheng C, Daigen M, Hirochika H. Epigenetic regulation of the rice retrotransposon Tos17. Molecular Genetics and Genomics. 2006;**276**:378-390

[35] Henkes S, Sonnewald U, Badur R, Flachmann R, Stitt M. A small decrease of plastid transketolase activity in antisense tobacco transformants has dramatic effects on photosynthesis and phenylpropanoid metabolism. Plant Cell. 2001;**13**:535-551. DOI: 10.1105/ tpc.13.3.535

[36] Farooq M, Wahid A, Lee DJ, Ito O, Siddique KH. Advances in drought resistance of rice. Critical Reviews in Plant Sciences. 2009;**28**(4):199-217

[37] Pighin JA, Zheng H, Balakshin LJ, Goodman IP, Western TL, Jetter R, et al. Plant cuticular lipid export requires an ABC transporter. Science. 2004;**306**:702-704. DOI: 10.1126/science.1102331

[38] Sun W, Van Montagu M, Verbruggen N. Small heat shock proteins and stress tolerance in plants. Biochimica et Biophysica Acta. 2002;**1577**:1-9. DOI: 10.1016/S0167-4781(02)00417-7

[39] Fang Y, Xiong L. General mechanisms of drought response and their application in drought resistance improvement in plants. Cellular and Molecular Life Sciences. 2015;**72**:673-689. DOI: 10.1007/s00018-014-1767-0

[40] Peskan-Berghöfer T et al. Sustained exposure to abscisic acid enhances the colonization potential of the mutualist fungus Piriformospora indica on Arabidopsis thaliana roots. The New Phytologist. 2015;**208**:873-886

[41] Johnson LM, Bostick M, Zhang X, Kraft E, Henderson I, Callis J, et al. The SRA methyl-cytosine-binding domain links DNA and histone methylation. Current Biology. 2007;**17**:379-384

[42] Sangwan V, Örvar BL, Beyerly J, Hirt H, Dhindsa RS. Opposite changes in membrane fluidity mimic cold and heat stress activation of distinct plant MAP kinase pathways. The Plant Journal. 2002;**31**:629-638. DOI: 10.1046/j.1365-313X.2002.01384.x

[43] FAO. FAOSTAT, Crops. Food and Agriculture Organization of the United Nations, Database on Crops. Rome: FAO; 2019

[44] Jalmi SK, Sinha AK. ROS mediated MAPK signaling in abiotic and biotic stress- striking similarities and differences. Frontiers in Plant Science. 2015;**6**:769. DOI: 10.3389/fpls.2015.00769

[45] McClintock B. The significance of responses of the genome to challenge. Science. 1984;**226**:792-801

[46] Legris M, Klose C, Burgie ES, Rojas CC, Neme M, Hiltbrunner A, et al. Phytochrome B integrates light and temperature signals in Arabidopsis. Science. 2016;**354**:897-900. DOI: 10.1126/ science.aaf5656

[47] Manivong P, Korinsak S, Korinsak S, Siangliw JL, Vanavichit A, Toojinda T. Marker-assisted selection to improve submergence tolerance, blast resistance and strong fragrance in glutinous rice. Thai Journal of Genetics. 2014;**7**:110-122

[48] Zhao SS, Jiang YX, Zhao Y, Huang SJ, Yuan M, Zhao YX, et al. CASEIN KINASE1-LIKE PROTEIN2 regulates actin filament stability and stomatal closure via phosphorylation of actin depolymerizing factor. Plant Cell. 2016;**28**:1422-1439. DOI: 10.1105/tpc.16.00078

[49] Zhao Y, Zhao SS, Mao TL, Qu XL, Cao WH, Zhang L, et al. The plant-specific actin binding protein SCAB1 stabilizes actin filaments and regulates stomatal movement in Arabidopsis. Plant Cell. 2011;**23**:2314-2330. DOI: 10.1105/tpc.111.086546

[50] Ghosh TK, Kaneko M, Akter K, Murai S, Komatsu K, Ishizaki K, et al. Abscisic acid-induced gene expression in the liverwort Marchantia polymorpha is mediated by evolutionarily conserved promoter elements. Physiologia Plantarum. 2016;**156**(4):407-420

[51] Kim H, Lee K, Hwang H, Bhatnagar N, Kim DY, Yoon IS, et al. Overexpression of PYL5 in rice enhances drought tolerance, inhibits growth, and modulates gene expression. Journal of Experimental Botany. 2014;**65**(2):453-464

[52] Oladosu Y, Rafii MY, Samuel C, Fatai A, Magaji U, Kareem I, et al. Drought resistance in rice from conventional to molecular breeding: A review. International Journal of Molecular Sciences. 2019;**20**(14):3519

[53] Zheng Y, Liao CC, Zhao SS, Wang CW, Guo Y. The glycosyltransferase QUA1 regulates chloroplast-associated calcium signaling during salt and drought stress in Arabidopsis. Plant & Cell Physiology. 2017;**58**:329-341. DOI: 10.1093/pcp/pcw192

[54] Mikami K, Murata N. Membrane fluidity and the perception of environmental signals in cyanobacteria and plants. Progress in Lipid Research. 2003;**42**:527-543. DOI: 10.1016/S0163-7827(03)00036-5

[55] Upadhyaya H, Panda SK. Drought stress responses and its management in rice. In: Advances in Rice Research for Abiotic Stress Tolerance. New York: Elsevier; 2019. pp. 177-200

[56] Fukuoka S, Saka N, Mizukami Y, Koga H, Yamanouchi U, Yoshioka Y, et al. Gene pyramiding enhances durable blast disease resistance in rice. Scientific Reports. 2015;**5**(1):1-7

[57] Garrett KA, Dendy SP, Frank EE, Rouse MN, Travers SE. Climate change effects on plant disease: Genomes to ecosystems. Annual Review of Phytopathology. 2006;**44**:489-509

[58] Chapagain S, Park YC, Kim JH, Jang CS. Oryza sativa salt-induced RING E3 ligase 2 (OsSIRP2) acts as a positive regulator of transketolase in plant response to salinity and osmotic stress. Planta. 2018;**247**:925-939. DOI: 10.1007/s00425-017-2838-x

[59] Faisal ARM, Biswas S, Zerin T, Rahman T, Seraj ZI. Down regulation of the DST transcription factor using artificial microRNA to increase yield, salt and drought tolerance in rice. American Journal of Plant Sciences. 2017;**8**(9):2219-2237

[60] Chan SW-L, Henderson IR, Jacobsen SE. Gardening the genome: DNA methylation in Arabidopsis thaliana. Nature Reviews. Genetics. 2005;**6**:351-360

[61] Strahl BD, Allis CD. The language of covalent histone modifications. Nature. 2000;**403**:41-45

[62] Takahashi S, Murata N. How do environmental stresses accelerate photoinhibition? Trends in Plant Science. 2008;**13**:178-182. DOI: 10.1016/j.tplants.2008.01.005

[63] Pollard M, Beisson F, Li Y, Ohlrogge JB. Building lipid barriers biosynthesis of cutin and suberin. Trends in Plant Science. 2008;**13**:236-246. DOI: 10.1016/j. tplants.2008.03.003

[64] Zhu J, Jeong JC, Zhu Y, Sokolchik I, Miyazaki S, Zhu JK, et al. Involvement of Arabidopsis HOS15 in histone deacetylation and cold tolerance. Proceedings of the National Academy Science USA. 2008;**105**:4945-4950

[65] Zilberman D, Gehring M, Tran RK, Ballinger T, Henikoff S. Genome-wide analysis of Arabidopsis thaliana DNA methylation uncovers an interdependence between methylation and transcription. Nature Genetics. 2007;**39**:61-69

[66] Roychoudhury SP, Basu S. Cross-talk between abscisic acid-dependent and abscisic acid-independent pathways during abiotic stress. Plant Cell Reports. 2013;**32**(7):985-1006

[67] Saibo NJ, Lourenco T, Oliveira MM. Transcription factors and regulation of photosynthetic and related metabolism under environmental stresses. Annals of Botany. 2009;**103**:609-623

[68] Fischer R, Byerlee D, Edmeades GO. Can technology deliver on the yield challenge to 2050. In: Expert Meeting on How to Feed the World in. 2009

[69] Tuteja N, Sopory SK. Chemical signaling under abiotic stress environment in plants. Plant Signaling & Behavior. 2008;**3**:525-536

[70] Bilgin DD, Zavala JA, Zhu JIN, Clough SJ, Ort DR, DeLucia EH. Biotic stress globally downregulates photosynthesis genes. Plant, Cell & Environment. 2010;**33**(10):1597-1613

[71] Jung JH, Domijan M, Klose C, Biswas S, Ezer D, Gao M, et al. Phytochromes function as thermosensors in Arabidopsis. Science. 2016;**354**:886-889. DOI: 10.1126/science.aaf6005

[72] Madlung A, Comai L. The effect of stress on genome regulation and structure. Annals of Botany. 2004;**94**:481-495

[73] Ikeda Y, Kinoshita T. DNA demethylation: A lesson from the garden. Chromosoma. 2009;**118**:37-41

[74] Melotto M, Underwood W, Koczan J, Nomura K, He SY. Plant stomata function in innate immunity against bacterial invasion. Cell. 2006;**126**:969-980

[75] Minocha R, Majumdar R, Minocha SC. Polyamines and abiotic stress in plants: A complex relationship. Frontiers in Plant Science. 2014;**5**:175. DOI: 10.3389/ fpls.2014.00175

[76] Iftekharuddaula KM, Ahmed HU, Ghosal S, Moni ZR, Amin A, Ali MS. Development of new submergence tolerant rice variety for Bangladesh using marker-assisted backcrossing. Rice Science. 2015;**22**:16-26. DOI: 10.1016/j. rsci.2015.05.003

[77] Shi H, Ye T, Chan Z. Nitric oxide-activated hydrogen sulfide is essential for cadmium stress response in bermudagrass (Cynodon dactylon (L). Pers.). Plant Physiology in Biochemistry. 2014;**74**:99-107. DOI: 10.1016/j.plaphy.2013.11.001

[78] Meyer P. DNA methylation systems and targets in plants. FEBS Letters. 2010;**585**:2008-2015

[79] Santos AP, Ferreira L, Maroco J, Oliveira MM. Abiotic stress and induced DNA hypomethylation cause interphase chromatin structural changes in rice rDNA loci. Cytogenetic and Genome Research. 2011;**132**:297-303

[80] Sah SK, Reddy KR, Li J. Abscisic acid and abiotic stress tolerance in

crop plants. Frontiers in Plant Science. 2016;7:571. DOI: 10.3389/fpls.2016.00571

[81] Sreenivasulu N, Harshavardhan VT, Govind G, Seiler C, Kohli A. Contrapuntal role of ABA: Does it mediate stress tolerance or plant growth retardation under long-term drought stress? Gene. 2012;**506**:265-273. DOI: 10.1016/j.gene.2012.06.076

[82] Han N, Shao Q, Bao HY, Wang BS. Cloning and characterization of a Ca2C/HC antiporter from halophyte Suaeda salsa L. Plant Molecular Biology Reporter. 2011;**29**:449-457. DOI: 10.1007/s11105-010-0244-7

[83] Scharf KD, Berberich T, Ebersberger I, Nover L. The plant heat stress transcription factor (Hsf) family: Structure, function and evolution. Biochimica et Biophysica Acta. 2012;**1819**:104-119. DOI: 10.1016/j.bbagrm.2011.10.002

[84] Cheah BH, Nadarajah K, Divate MD, Wickneswari R. Identification of four functionally important microRNA families with contrasting differential expression profiles between drought-tolerant and susceptible rice leaf at vegetative stage. BMC Genomics. 2015;**16**(1):692

[85] Kim T-H, Böhmer M, Hu H, Nishimura N, Schroeder JI. Guard cell signal transduction network: Advances in understanding abscisic acid, CO2, and Ca2+ signaling. Annual Review of Plant Biology. 2010;**61**:561-591

[86] Labra M, Ghiani A, Citterio S, Sgorbati S, Sala F, Vannini C, et al. Analysis of cytosine methylation pattern in response to water deficit in pea root tips. Plant Biology. 2002;**4**:694-699

[87] Sharma R, Mohan Singh RK, Malik G, Deveshwar P, Tyagi AK, Kapoor S, et al. Rice cytosine DNA methyltransferases—gene expression profiling during reproductive development and abiotic stress. The FEBS Journal. 2009;**276**:6301-6311

[88] Bin Rahman ANMR, Zhang JH. Flood and drought tolerance in rice: Opposite but may coexist. Food and Energy Security. 2016;**5**(2):76-88

[89] Guo H, Xiao T, Zhou H, Xie Y, Shen W. Hydrogen sulfide: A versatile regulator of environmental stress in plants. Acta Physiologiae Plantarum. 2016;**38**:16. DOI: 10.1007/s11738-015-2038-x

[90] Murphy DJ, Walker DA. The properties of transketolase from photosynthetic tissue. Planta. 1982;**155**:316-320. DOI: 10.1007/BF00429458

[91] Li K, Pang CH, Ding F, Sui N, Feng ZT, Wang BS. Over expression of Suaeda salsa stroma ascorbate peroxidase in Arabidopsis chloroplasts enhances salt tolerance of plants. South African Journal of Botany. 2012;**78**:235-245. DOI: 10.1016/j.sajb.2011.09.006

[92] Pandey V, Shukla A. Acclimation and tolerance strategies of rice under drought stress. Rice Science. 2015;**22**(4):147-161

[93] Almeida DM, Almadanim MC, Lourenço T, Abreu IA, Saibo NJM, Oliveira MM. Screening for abiotic stress tolerance in rice: Salt, cold, and drought. Methods in Molecular Biology. 2016;**1398**:155-182. DOI: 10.1007/978-1-4939-3356-3_14

[94] Kunst L, Samuels L. Plant cuticles shine: Advances in wax biosynthesis and export. Current Opinion in Plant Biology. 2009;**12**:721-727. DOI: 10.1016/j.pbi.2009.09.009

[95] Huang YC, Niu CY, Yang CR, Jinn TL. The heat stress factor HSFA6b connects ABA signaling and ABA-mediated heat responses. Plant Physiology.

2016;**172**:1182-1199. DOI: 10.1104/pp.16.00860

[96] Chen H, Hwang JE, Lim CJ, Kim DY, Lee SY, Lim CO. Arabidopsis DREB2C functions as a transcriptional activator of HsfA3 during the heat stress response. Biochemical and Biophysical Research Communications. 2010;**401**:238-244. DOI: 10.1016/j.bbrc.2010.09.038

[97] Lievens L, Pollier J, Goossens A, Beyaert R, Staal. Abscisic acid as pathogen efector and immune regulator. Frontiers in Plant Science. 2017;**8**:587

[98] Kumar S, Trivedi PK. Glutathione S-transferases: Role in combating abiotic stresses including arsenic detoxification in plants. Frontiers in Plant Science. 2018;**9**:751. DOI: 10.3389/fpls.2018.00751

[99] Tunc-Ozdemir M, Miller G, Song L, Kim J, Sodek A, Koussevitzky S, et al. Thiamin confers enhanced tolerance to oxidative stress in Arabidopsis. Plant Physiology. 2009;**151**:421-432. DOI: 10.1104/pp.109.140046

[100] Tsuji H, Saika H, Tsutsumi N, Hirai A, Nakazono M. Dynamic and reversible changes in histone H3-Lys4 methylation and H3 acetylation occurring at submergenceinducible genes in rice. Plant & Cell Physiology. 2006;**47**:995-1003

[101] Hasanuzzaman M, Nahar K, Hossain MS, Mahmud JA, Rahman A, Inafuku M. Coordinated actions of glyoxalase and antioxidant defense systems in conferring abiotic stress tolerance in plants. International Journal of Molecular Science. 2017;**18**:200

[102] Byun MY, Cui LH, Oh TK, Jung YJ, Lee A, Park KY, et al. Homologous U-box E3 ubiquitin ligases OsPUB2 and OsPUB3 are involved in the positive regulation of low temperature stress response in rice (*Oryza sativa* L.). Frontiers in Plant Science. 2017;**8**:16

[103] Zhang X, Wang L, Meng H, Wen H, Fan Y, Zhao J. Maize ABP9 enhances tolerance to multiple stresses in transgenic Arabidopsis by modulating ABA signaling and cellular levels of reactive oxygen species. Plant Molecular Biology. 2010;**75**:365-378

[104] Wang W-S, Pan Y-J, Zhao X-Q, Dwivedi D, Zhu L-H, Ali J, et al. Drought-induced site-specific DNA methylation and its association with drought tolerance in rice (Oryza sativa L.). Journal of Experimental Botany. 2011;**62**:1951-1960

[105] Thomashow MF. Plant cold acclimation: Freezing tolerance genes and regulatory mechanisms. Annual Review of Plant Physiology and Plant Molecular Biology. 1999;**50**:571-599. DOI: 10.1146/annurev.arplant.50.1.571

[106] Roychoudhury S, Paul A. Abscisic acid-inducible genes during salinity and drought stress. In: Berhardt LV, editor. Advances in Medicine and Biology. Vol. vol. 51. New York: Nova Science Publishers; 2012. pp. 1-78

[107] Wang D, Pan Y, Zhao X, Zhu L, Fu B, Li Z. Genome-wide temporal-spatial gene expression profiling of drought responsiveness in rice. BMC Genomics. 2011;**12**:149

[108] Lusser A, Brosch G, Loidl A, Haas H, Loidl P. Identification of maize histone deacetylase HD2 as an acidic nucleolar phosphoprotein. Science. 1997;**277**:88-91

[109] Kregel KC. Heat shock proteins: Modifying factors in physiological stress responses and acquired thermotolerance. Journal of Applied Physiology. 2002;**92**:2177-2186

[110] Liu W, Ji SX, Fang XL, Wang QG, Li Z, Yao FY, et al. Protein kinase LTRPK1 influences cold adaptation and microtubule stability in rice. Journal of Plant Growth Regulation. 2013;**32**:483-490. DOI: 10.1007/s00344-012-9314-4

[111] Joshi R, Wani SH, Singh B, Bohra A, Dar ZA, Lone AA, et al. Transcription factors and plants response to drought stress: Current understanding and future directions. Frontiers in Plant Science. 2016;**7**:1029

[112] Han N, Lan WJ, He X, Shao Q, Wang BS, Zhao XJ. Expression of a Suaeda salsa vacuolar HC/Ca2C transporter gene in Arabidopsis contributes to physiological changes in salinity. Plant Molecular Biology Reporter. 2012;**30**:470-477. DOI: 10.1007/s11105-011-0353-y

[113] Wang X, Elling AA, Li X, Li N, Peng Z, He G, et al. Genome-wide and organ-specific landscapes of epigenetic modifications and their relationships to mRNA and small RNA transcriptomes in maize. Plant Cell. 2009;**21**:1053-1069

[114] Wang L, Zhu J, Li X, Wang S, Wu J. Salt and drought stress and ABA responses related to bZIP genes from *V. radiata* and *V. angularis*. Gene. 2018;**651**:152-160

[115] Jayani RS, Ramanujam PL, Galande S. Studying histone modifications and their genomic functions by employing chromatin immunoprecipitation and immunoblotting. Methods in Cell Biology. 2010;**98**:35-56

[116] Shi Y, Yue X, An L. Integrated regulation triggered by a cryophyte &-3 desaturase gene confers multiple-stress tolerance in tobacco. Journal of Experimental Botany. 2018;**69**:2131-2148. DOI: 10.1093/jxb/ery050

[117] Vaidyanathan H, Sivakumar P, Chakrabarty R, Thomas G. Scavenging of reactive oxygen species in NaCl-stressed rice (*Oryza sativa* L.)-differential response in salt-tolerant and sensitive varieties. Biologia Plantarum;**165**:1411-1418

[118] Jiang CJ, Shimono M, Sugano S, Kojima M, Yazawa K, Yoshida R, et al. Abscisic acid interacts antagonistically with salicylic acid signaling pathway in rice–Magnaporthe grisea interaction. Molecular Plant-Microbe Interactions. 2010;**23**(6):791-798

[119] Mafakheri A, Siosemardeh A, Bahramnejad B, Struik P, Sohrabi Y. Efect of drought stress on yield, proline and chlorophyll contents in three chickpea cultivars. Australian Journal of Crop Science. 2010;**4**:580

[120] Ashraf M, Foolad MR. Roles of glycine betaine and proline in improving plant abiotic stress resistance. Environmental and Experimental Botany. 2007;**59**:206-216. DOI: 10.1016/j.envexpbot.2005.12.006

[121] Khozaei M, Fisk S, Lawson T, Gibon Y, Sulpice R, Stitt M, et al. Overexpression of plastid transketolase in tobacco results in a thiamine auxotrophic phenotype. Plant Cell. 2015;**27**:432-447. DOI: 10.1105/tpc.114.131011

[122] Pecinka A, Dinh HQ, Baubec T, Rosa M, Lettner N, Scheid OM. Epigenetic regulation of repetitive elements is attenuated by prolonged heat stress in Arabidopsis. Plant Cell. 2010;**22**:3118-3129

[123] Azarin KV, Usatov AV, Kolokolova NS, Usatova OA, Alabushev AV, Kostylev PI. Effects of salt stress on ion balance at vegetative stage in rice (*Oryza sativa* L.). OnLine Journal of Biological Sciences. 2016;**16**(1):76-81

[124] Hazel JR. Thermal adaptation in biological membranes: Is homeoviscous

adaptation the explanation? Annual Review of Physiology. 1995;**57**:19-42. DOI: 10.1146/annurev.ph.57.030195.000315

[125] Shigenaga AM, Argueso CT. Seminars in Cell and Development Biology. New York: Elsevier. pp. 174-189

[126] Vu HTT, Le DD, Ismail AM, Le HH. Marker-assisted backcrossing (MABC) for improved salinity tolerance in rice (*Oryza sativa* L.) to cope with climate change in Vietnam. AJCS. 2012;**6**:1649-1654

[127] Kim Y, Chung YS, Lee E, Tripathi P, Heo S, Kim KH. Root response to drought stress in rice (Oryza sativa L.). International Journal of Molecular Science. 2020;**21**(4):1513

[128] Fransz PF, De Jong JH. Chromatin dynamics in plants. Current Opinion in Plant Biology. 2002;**5**:560-567

[129] Lourenço T, Sapeta H, Figueiredo DD, Rodrigues M, Cordeiro A, Abreu IA, et al. Isolation and characterization of rice (*Oryza sativa* L.) E3-ubiquitin ligase OsHOS1 gene in the modulation of cold stress response. Plant Mol. O Biologico. 2013;**83**:351-363. DOI: 10.1007/s11103-013-0092-6

[130] Fernandes JC, Goulao LF, Amâncio S. Regulation of cell wall remodeling in grapevine (*Vitis vinifera* L.) callus under individual mineral stress deficiency. Journal of Plant Physiology. 2016;**190**:95-105. DOI: 10.1016/j.jplph.2015.10.007

[131] Franklin KA, Quail PH. Phytochrome functions in Arabidopsis development. Journal of Experimental Botany. 2010;**61**:11-24. DOI: 10.1093/jxb/erp304

[132] Wania SH, Kumar V, Shriram V, Sah SK. Phytohormones and their metabolic engineering for abiotic stress tolerance in crop plants. Crop Journal. 2016;**4**:162-176. DOI: 10.1016/j.cj.2016.01.010

[133] Hernandez JA, Jiménez A, Mullineaux P, Sevilia F. Tolerance of pea (*Pisum sativum* L.) to long-term salt stress is associated with induction of antioxidant defences. Plant, Cell & Environment. 2000;**23**(8):853-862

[134] Qin QL, Liu JG, Zhang Z, Peng RH, Xiong AS, Yao QH, et al. Isolation, optimization, and functional analysis of the cDNA encoding transcription factor OsDREB1B in *Oryza sativa* L. Molecular Breeding. 2007;**19**:329-340

[135] Chinnusamy V, Zhu JK. Epigenetic regulation of stress responses in plants. Current Opinion in Plant Biology. 2009;**12**:133-139

[136] Saxena I, Srikanth S, Chen Z. Cross talk between H2O2 and interacting signal molecules under plant stress response. Frontiers in Plant Science. 2016;7:570. DOI: 10.3389/fpls.2016.00570

[137] Chen M, Zhang WH, Lv ZW, Zhang SL, Hidema J, Shi FM, et al. Abscisic acid is involved in the response of Arabidopsis mutant sad2- 1 to ultraviolet-B radiation by enhancing antioxidant enzymes. South African Journal of Botany. 2013;**85**:79-86. DOI: 10.1016/j.sajb.2012.11.006

[138] Tenhaken R. Cell wall remodeling under abiotic stress. Frontiers in Plant Science. 2016;**5**:771. DOI: 10.3389/fpls.2014.00771

[139] Baek D, Jiang J, Chung JS, Wang B, Chen J, Xin Z, et al. Regulated AtHKT1 gene expression by a distal enhancer element and DNA methylation in the promoter plays an important role in salt tolerance. Plant & Cell Physiology. 2011;**52**:149-161

[140] Semenov MA, Shewry PR. Modelling predicts that heat stress, not drought, will increase vulnerability of wheat in Europe. Scientific Reports. 2011;**1**:66

[141] McFarlane HE, Watanabe Y, Yang W, Huang Y, Ohlrogge J, Samuels AL. Golgi- and TGN-mediated vesicle trafficking is required for wax secretion from epidermal cells. Plant Physiology. 2014;**164**:1250-1260. DOI: 10.1104/pp.113.234583

[142] Barik SR, Pandit E, Pradhan SK, Mohanty SP, Mohapatra T. Genetic mapping of morpho-physiological traits involved during reproductive stage drought tolerance in rice. PLoS One. 2019;**14**(12):e0214979

[143] Hoang TML, Tran TN, Nguyen TKT, Williams B, Wurm P, Bellairs S, et al. Improvement of salinity stress tolerance in rice: Challenges and opportunities. Agronomy. 2016;**6**:1-23. DOI: 10.3390/agronomy6040054

[144] Huang LY, Wang YX, Wang WS, Zhao XQ, Qin Q, Sun F, et al. Characterization of transcription factor gene OsDRAP1 conferring drought tolerance in rice. Frontiers in Plant Science. 2018;**9**:94

[145] Fu W, Wu K, Duan J. Sequence and expression analysis of histone deacetylases in rice. Biochemical and Biophysical Research Communications. 2007;**356**:843-850

[146] Hu Y, Qin F, Huang L, Sun Q, Li C, Zhao Y, et al. Rice histone deacetylase genes display specific expression patterns and developmental functions. Biochemical and Biophysical Research Communications. 2009;**388**:266-271

[147] Mittler R. Abiotic stress, the field environment and stress combination. Trends in Plant Science. 2006;**11**:15-19

[148] Huang LJ, Cheng GX, Khan A, Wei AM, Yu QH, Yang SB, et al. CaHSP16.4, a small heat shock protein gene in pepper, is involved in heat and drought tolerance. Protoplasma. 2019;**256**:39-51. DOI: 10.1007/s00709-018-1280-7

[149] Molla KA, Debnath AB, Ganie SA, Mondal TK. Identification and analysis of novel salt responsive candidate gene based SSRs (cgSSRs) from rice (*Oryza sativa* L.). BMC Plant Biology. 2015;**15**:122

[150] Sandhu N, Kumar A. Bridging the rice yield gaps under drought: QTLs genes and their use in breeding programs. Agronomie. 2017;**7**(2):27

[151] Van Ittersum MK, Cassman KG, Grassini P, Wolf J, Tittonell P, Hochman Z. Yield gap analysis with local to global relevance—A review. Field Crops Research. 2013;**143**:4-17

[152] Mostofa MG, Ghosh A, Li ZG, Siddiqui MN, Fujita M, Tran LP. Methylglyoxal—A signaling molecule in plant abiotic stress responses. Free Radical Biology & Medicine. 2018;**122**:96-109. DOI: 10.1016/j.freeradbiomed.2018.03.009

[153] Lobell DB, Cassman KG, Field CB. Crop yield gaps: Their importance, magnitudes, and causes. Annual Review of Environment and Resources. 2009;**34**(1):179

[154] Century K, Reuber TL, Ratcliffe OJ. Regulating the regulators: The future prospects for transcription-factor-based agricultural biotechnology products. Plant Physiology. 2008;**147**:20-29

[155] Deng YQ, Bao J, Yuan F, Liang X, Feng ZT, Wang BS. Exogenous hydrogen sulfide alleviates salt stress in wheat seedlings by decreasing NaC content.

Plant Growth Regulation. 2016;**79**:391-399. DOI: 10.1007/s10725-015-0143-x

[156] Sahebi M, Hanafi MM, Rafii MY, Mahmud TMM, Azizi P, Osman M, et al. Improvement of drought tolerance in rice (*Oryza sativa* L.): Genetics, genomic tools, and the WRKY gene family. Biomedical Research Institute. 2018;**2018**:3158474

[157] Singh AK, Gopalakrishnan S, Singh VP, Prabhu KV, Mohapatra T, Singh NK, et al. Marker assisted selection: A paradigm shift in Basmati breeding. Indian Journal of Genetics. 2011;**71**:120-128

[158] Huang X, Hou L, Meng J, You H, Li Z, Gong Z, et al. The antagonistic action of abscisic acid and cytokinin signaling mediates drought stress response in Arabidopsis. Molecular Plant. 2018;**11**(7):970-982

[159] Jongdee B, Pantuwan G, Fukai S, Fischer K. Improving drought tolerance in rainfed lowland rice: An example from Thailand. Agricultural Water Management. 2006;**80**(1-3):225-240

[160] Ren XL, Qi GN, Feng HQ, Zhao S, Zhao SS, Wang Y, et al. Calcineurin B-like protein CBL10 directly interacts with AKT1 and modulates KC homeostasis in Arabidopsis. The Plant Journal. 2013;**74**:258-266. DOI: 10.1111/tpj.12123

[161] de Torres Zabala M, Bennett MH, Truman WH, Grant MR. Antagonism between salicylic and abscisic acid reflects early host–pathogen conflict and moulds plant defence responses. The Plant Journal. 2009;**59**(3):375-386

[162] Roxas VP, Lodhi SA, Garrett DK, Mahan JR, Allen RD. Stress tolerance in transgenic tobacco seedlings that over express glutathione S-transferase/glutathione peroxidase. Plant & Cell Physiology. 2000;**41**:1229-1234. DOI: 10.1093/pcp/pcd051

[163] Nelson GC, Valin H, Sands RD, Havlík P, Ahammad H, Deryng D, et al. Climate change effects on agriculture: Economic responses to biophysical shocks. Proceedings of the National Academy Science USA. 2014;**111**(9):3274-3279

[164] Pang CH, Li K, Wang BS. Overexpression of SsCHLAPXs confers protection against oxidative stress induced by high light in transgenic Arabidopsis thaliana. Physiologia Plantarum. 2011;**143**:355-366. DOI: 10.1111/j.1399-3054.2011.01515.x

[165] Gill SS, Tuteja N. Polyamines and abiotic stress tolerance in plants. Plant Signaling & Behavior. 2010;**5**:26-33. DOI: 10.4161/psb.5.1.10291

[166] Gopala Krishnan S, Dwivedi P, Dhawan G, Nagarajan M, Viswanathan C, Bhowmick PK, et al. Marker assisted transfer of QTLs, qDTY1.1 into Basmati rice variety "Pusa Basmati 1" and qDTY3.1 into elite rice variety "Pusa 44" for enhancing grain yield under reproductive stage drought stress (IDT9-034). Inter Drought V, Hyderabad, HICC, 21-25 February 20172017. p. 206

[167] Chen Z, Newman I, Zhou M, Mendham N, Zhang G, Shabala S. Screening plants for salt tolerance by measuring KC flux: A case study for barley. Plant Cell Environment. 2005;**28**:1230-1246. DOI: 10.1111/j.1365-3040.2005.01364.x

[168] Gupta K, Sengupta A, Chakraborty M, Gupta B. Hydrogen peroxide and polyamines act as double edged swords in plant abiotic stress responses. Frontiers in Plant Science. 2016;7:1343. DOI: 10.3389/fpls.2016.01343

[169] Nguyen D, Rieu I, Mariani C, van Dam NM. How plants handle multiple stresses: Hormonal interactions

underlying responses to abiotic stress and insect herbivory. Plant Molecular Biology. 2016;**91**:727-740

[170] Qi YC, Liu WQ, Qiu LY, Zhang SM, Ma L, Zhang H. Over expression of glutathione S-transferase gene increases salt tolerance of Arabidopsis. Russian Journal of Plant Physiology. 2010;**57**:233-240. DOI: 10.1134/s102144371002010x

[171] Hasanuzzaman M, Nahar K, Fujita M. Extreme temperature responses, oxidative stress and antioxidant defense in plants. Abiotic Stress-plant Responses and Applications in Agriculture. 2013;**13**:169-205

[172] Pandey S. Economic costs of drought and rice farmers' coping mechanisms. International Rice Research Notes. 2009;**32**(1):5

[173] Song J, Shi GW, Gao B, Fan H, Wang BS. Waterlogging and salinity effects on two Suaeda salsa populations. Physiology Plant. 2011;**141**:343-351

[174] Ashraf M, Akram NA. Improving salinity tolerance of plants through conventional breeding and genetic engineering: An analytical comparison. Biotechnology Advances. 2009;**27**(6):744-752

[175] Le Gall H, Philippe F, Domon JM, Gillet F, Pelloux J, Rayon C. Cell wall metabolism in response to abiotic stress. Plants. 2015;**4**:112-166. DOI: 10.3390/plants4010112

[176] Lisjak M, Teklic T, Wilson ID, Whiteman M, Hancock JT. Hydrogen sulfide: Environmental factor or signalling molecule? Plant, Cell & Environment. 2013;**36**:1607-1616. DOI: 10.1111/pce.12073

[177] Xu K, Mackill DJ. A major locus for submergence tolerance mapped on rice chromosome 9. Molecular Breeding. 1996;**2**:219-224. DOI: 10.1007/BF00564199

[178] Dash PK, Rai R, Rai V, Pasupalak S. Drought induced signaling in rice: Delineating canonical and non canonical pathways. Frontiers in Chemistry. 2018;**6**:264

[179] Shamsudin NAA, Swamy BPM, Ratnam W, Cruz MTS, Raman A, Kumar A. Marker assisted pyramiding of drought yield QTLs into a popular Malaysian rice cultivar, MR219. BMC Genetics. 2016;**17**(1):30

[180] Shen XY, Wang ZL, Song XF, Xu JJ, Jiang CY, Zhao YX, et al. Transcriptomic profiling revealed an important role of cell wall remodeling andethylene signaling pathway during salt acclimation in Arabidopsis. Plant Molecular Biology. 2014;**86**:303-317. DOI: 10.1007/s11103-014-0230-9

[181] Xu K, Xu X, Ronald PC, Mackill DJ. A high-resolution linkage map in the vicinity of the rice submergence tolerance locus Sub1. Molecular & General Genetics. 2000;**263**:681-689. DOI: 10.1007/s004380051217

[182] Dramé KN, Manneh B, Ismail AM. Rice genetic improvement for abiotic stress tolerance in Africa. In: Wopereis MCS, Johnson DE, Ahmadi N, Tollens E, Jalloh A, editors. Realizing Africa's Rice Promise. Wallingford: CABI; 2013. DOI: 10.1079/9781845938123.0144

[183] Dissanayake PK, Wijeratne AW. Development of a varietial screening procedure for salt tolerance of Rice (*Oryza sativa* L.) varieties at germination stage. Journal of Agricultural Science. 2006;**2**:63-72

[184] Gendrel AV, Lippman Z, Yordan C, Colot V, Martienssen RA. Dependence of heterochromatic histone H3 methylation

patterns on the Arabidopsis gene DDM1. Science. 2002;**297**:1871-1873

[185] Septiningsih EM, Collard BCY, Heuer S, Bailey-Serres J, Ismail AM, Mackill DJ. Applying genomics tools for breeding submergence tolerance in rice. In: Varshney RK, Tuberosa R, editors. Translational Genomics for Crop Breeding, Vol. II: Abiotic Stress, Yield and Quality. Ames, IA; Chichester; Oxford: John Wiley & Sons; 2013. pp. 9-30

[186] Arnao MB, Hernández-Ruiz J. Melatonin and its relationship to plant hormones. Annals of Botany. 2018;**121**:195-207. DOI: 10.1093/aob/mcx114

[187] Lee SB, Suh MC. Advances in the understanding of cuticular waxes in Arabidopsis thaliana and crop species. Plant Cell Reports. 2015;**34**:557-572. DOI: 10.1007/s00299-015-1772-2

[188] Yin Y, Jiang X, Ren M, Xue M, Nan D, Wang Z, et al. AmDREB2C, from Ammopiptanthus mongolicus, enhances abiotic stress tolerance and regulates fatty acid composition in transgenic Arabidopsis. Plant Physiology and Biochemistry. 2018;**130**:517-528. DOI: 10.1016/j.plaphy.2018.08.002

[189] Liu S, Lv Z, Liu Y, Li L, Zhang L. Network analysis of ABAdependent and ABA-independent drought responsive genes in Arabidopsis thaliana. Genetics and Molecular Biology. 2018;**41**:624-637. DOI: 10.1590/1678-4685-GMB-2017- 2229

[190] Wojtaszek P. Oxidative burst: An early plant response to pathogen infection. The Biochemical Journal. 1997;**322**:681-692

[191] Hoque TS, Hossain MA, Mostofa MG, Burritt DJ, Fujita M, Tran LS. Methylglyoxal: An emerging signaling molecule in plant abiotic stress responses and tolerance. Frontiers in Plant Science. 2016;7:1341. DOI: 10.3389/fpls.2016. 01341

[192] Xiong L, Yang Y. Disease resistance and abiotic stress tolerance in rice are inversely modulated by an abscisic acid–inducible mitogen-activated protein kinase. Plant Cell. 2003;**15**:745-759

[193] Yoshida T et al. Four Arabidopsis AREB/ABF transcription factors function predominantly in gene expression downstream of SnRK2 kinases in abscisic acid signalling in response to osmotic stress. Plant, Cell & Environment. 2015;**38**:35-49

[194] Yamasaki K, Kigawa T, Inoue M, Tateno M, Yamasaki T, Yabuki T, et al. Solution structure of an Arabidopsis WRKY DNA binding domain. Plant Cell. 2005;**17**:944-956

[195] Yang MF, Song J, Wang BS. Organ-specific responses of vacuolar HC-ATPase in the shoots and roots of C3 halophyte Suaeda salsa to NaCl. Journal of Integrative Plant Biology. 2010;**52**:308-314. DOI: 10.1111/j.1744-7909.2010.00895.x

[196] Yasuda M et al. Antagonistic interaction between systemic acquired resistance and the abscisic acid–mediated abiotic stress response in Arabidopsis. Plant Cell. 2008;**20**:1678-1692

[197] Yeats TH, Rose JK. The formation and function of plant cuticles. Plant Physiology. 2013;**163**:5-20. DOI: 10.1104/pp.113.222737

[198] Xu J, Audenaert K, Hofe M, De Vleesschauwer D. Abscisic acid promotes susceptibility to the rice leaf blight pathogen Xanthomonas oryzae pv. oryzae by suppressing salicylic acid-mediated defenses. PLOS One. 2013;**8**:e67413

[199] Gaspar T, Franck T, Bisbis B, Kevers C, Jouve L, Hausman J, et al. Concepts in plant stress physiology. Application to plant tissue cultures. Plant Growth Regulation. 2002;**37**(3):263-285

[200] Gechev TS, Van Breusegem F, Stone JM, Denev I, Laloi C. Reactive oxygen species as signals that modulate plant stress responses and programmed cell death. BioEssays. 2006;**28**:1091-1101. DOI: 10.1002/bies.20493

[201] Yuan F, Lyu MJA, Leng BY, Zhu XG, Wang BS. The transcriptome of NaCl-treated Limonium bicolor leaves reveals the genes controlling salt secretion of salt gland. Plant Molecular Biology. 2016;**91**:241-256. DOI: 10.1007/s11103-016-0460-0

[202] Zain NAM, Ismail MR, Mahmood M, Puteh A, Ibrahim MH. Alleviation of water stress effects on mr220 rice by application of periodical water stress and potassium fertilization. Molecules. 2014;**19**:1795-1819. DOI: 10.3390/molecules19021795

[203] Wang W, Vinocur B, Altman A. Plant responses to drought, salinity and extreme temperatures: Towards genetic engineering for stress tolerance. Planta. 2003;**28**:1-14. DOI: 10.1007/s00425-003-1105-5

[204] Kong XQ, Wang T, Li WJ, Tang W, Zhang DM, Dong HZ. Exogenous nitric oxide delays salt-induced leaf senescence in cotton (*Gossypium hirsutum* L.). Acta Physiol. Plant. 38:61. Acta Physiology Plant. 2016;**38**:61

[205] Hancock JT, Whiteman M. Hydrogen sulfide and cell signaling: Team player or referee? Plant Physiol. The Biochemist. 2014;**78**:37-42. DOI: 10.1016/j.plaphy. 2014.02.012

[206] Zhang Y, Chen C, Jin XF, Xiong AS, Peng RH, Hong YH, et al. Expression of a rice DREB1 gene, OsDREB1D, enhances cold and high-salt tolerance in transgenic Arabidopsis. BMC Report. 2008;**42**:486-492

[207] Sarris A. Proceedings of FAO Rice Conference 2004. UK: FAO; 2004

[208] Zhong L, Xu Y-H, Wang J-B. DNA-methylation changes induced by salt stress in wheat Triticum aestivum. African Journal of Biotechnology. 2009;**8**:6201-6207

[209] Zhou DX. Regulatory mechanism of histone epigenetic modifications in plants. Epigenetics. 2009;**4**:15-18

[210] Zhou JJ, Liu QQ, Zhang F, Wang YY, Zhang SY, Cheng HM, et al. Overexpression of OsPIL15, a phytochrome-interacting factor-like protein gene, represses etiolated seedling growth in rice. Journal of Integrative Plant Biology. 2014;**56**:373-387. DOI: 10.1111/jipb.12137

[211] Delker C, van Zanten M, Quint M. Thermo sensing enlightened. Trends in Plant Science. 2017;**22**:185-187. DOI: 10.1016/j.tplants.2017.01.007

[212] Zargar A, Sadiq R, Naser B, Khan F. A review of drought indices. Environmental Reviews. 2011;**19**(1):333-349

[213] Zhang X. The epigenetic landscape of plants. Science. 2008;**320**:489-492

[214] Sazegari S, Niazi A, Ahmadi FS. A study on the regulatory network with promoter analysis for Arabidopsis DREB-genes. Bioinformation. 2015;**11**:101-106. DOI: 10.6026/97320630011101

[215] Pandey R, Muller A, Napoli CA, Selinger DA, Pikaard CS, Richards EJ, et al. Analysis of histone acetyltransferase and histone deacetylase families of Arabidopsis thaliana suggests functional diversification of chromatin modification among multicellular eukaryotes. Nucleic Acids Research. 2002;**30**:5036-5055

[216] Ahmadikhah A, Marufinia A. Effect of reduced plant height on drought tolerance in rice. 3 Biotech. 2016;**6**:221

[217] Ding Y, Wang X, Su L, Zhai J, Cao S, Zhang D, et al. SDG714, a histone H3K9 methyltransferase, is involved in Tos17 DNA methylation and transposition in rice. The Plant Cell. 2007;**19**(1):9-22

[218] Tanou G, Molassiotis A, Diamantidis G. Induction of reactive oxygen species and necrotic death-like destruction in strawberry leaves by salinity. Environmental and Experimental Botany. 2009;**65**:270-281

[219] Wang W, Vinocur B, Shoseyov O, Altman A. Role of plant heatshock proteins and molecular chaperones in the abiotic stress response. Trends in Plant Science. 2004;**9**:244-252. DOI: 10.1016/j.tplants.2004.03.006

[220] Ding Y, Dommel M, Mou Z. Abscisic acid promotes proteasome-mediated degradation of the transcription coactivator NPR 1 in Arabidopsis thaliana. The Plant Journal. 2016;**86**(1):20-34

[221] Cao S, Du XH, Li LH, Liu YD, Zhang L, Pan X, et al. Over expression of Populus tomentosa cytosolic ascorbate peroxidase enhances abiotic stress tolerance in tobacco plants. Russian Journal of Plant Physiology. 2017;**64**:224-234. DOI: 10.1134/s1021443717020029

[222] Bernard A, Joubès J. Arabidopsis cuticular waxes: Advances in synthesis, export and regulation. Progress in Lipid Research. 2013;**52**:110-129. DOI: 10.1016/j.plipres.2012.10.002

Section 4

Cross Cutting Mechanisms of Plant Responses to Stresses

Chapter 14

Interactive Effects of Salinity, Drought, and Heat Stresses on Physiological Process and Selection Criteria for Breeding Stress-Resistant Cotton

Volkan Mehmet Cinar, Serife Balci and Aydın Unay

Abstract

The cotton crop is adversely affected by the combination of salinity, drought, and heat stress during all growth stages in cultivated areas. The negative impacts of salinity together with water scarcity on osmotic stress dramatically increased the sensitivity of reproductive development. After membrane integrity and signaling networks are depressed under stress at the cell level, the metabolic and physiological processes are disrupted in the next stage. The restricted root growth, ion and water uptake, phloem, photosynthetic and respiratory capacity, incompatible hormonal balance, and reduction in yield due to lower boll retention are the most important symptoms. The seed treatments and foliar applications of osmoprotectant and fertilization appear to reduce multiple stress factors in possible climate change conditions. The osmotic adjustment, antioxidative ability, electrolyte leakage in the membrane, and chlorophyll fluorescence are evaluated as selection criteria for improving genotypes. Direct selection of plants with high yield under stress conditions may increase the success of cotton breeding. It is important to know the molecular approaches and gene functions responsible for abiotic stress. In this chapter, the effects of high temperature, salinity, and drought on cotton plants and characteristics associated with tolerance were focused on cotton improvement. The classical breeding methods and molecular approaches should be combined for breeding new cotton varieties.

Keywords: breeding, high temperature, saline condition, stress physiology, water deficiency

1. Introduction

Agricultural areas with saline soils are estimated at 1100 Mha in the World. These areas are classified as saline (60%), sodic (26%), and saline-sodic soils (14%). Areas suffering from salinity are mostly in the Middle East, Australia, North Africa, and Eurasia. However, 20 to 50 percent of irrigated soils in arid or semi-arid climates

Figure 1.
Cotton plants under saline soil (A), stunted cotton plants under drought conditions (B), and cotton plants under alkali soil (C).

are salt-affected [1]. A report produced by Cotton 2040 emphasized that cotton affected by heat stress and drought will reach 40% and 50%, respectively by 2040 [2]. Although heat stress is a problem along with drought [3], cotton-growing face many stress factors such as drought (**Figure 1B**), high temperature (**Figure 2C**), salinity (**Figure 1A**), alkalinity (**Figure 1C**), and heavy metal contamination, which are seen together in the same ecology. A single stressor can play a predominant or protective role depending on stress resources [4]. The long-term effect of climatic change resulted in higher pH values with drought, unfavorable soil organic C, total N and P, and available N [5], and soil salinity due to increasing the sea level [6].

Cotton's response to stress is different at plant growth and development stages. Germination and seedling growth are adversely affected by high salinity and low temperature at the early stages in cotton (*Gossypium hirsutum* L.) [7, 8]. Although progenitors of cotton have spread in adverse conditions, modern cotton varieties are non-resistant to stresses from the squaring stage to boll retention stages, in which yield formation occurs [9]. Water deficiency [10–13] and heat stress [13–15], especially during flowering and boll formation can reduce yield and fiber quality. In many

Figure 2.
A picture of a plastic tunnel under heat stress at 11.00 am–17.00 pm (A), non-affected plants by heat stress (B), and aborted/shed bolls affected by heat stress (C).

areas of the world where cotton is cultivated, as in the Aegean, Mediterranean, and Southeast Anatolia regions of Turkey, maximum temperatures during the reproductive stage in July and August are above 40°C with low humidity and precipitation. The boll components such as seed and boll weight cannot be affected by short-term temperature changes, whereas the effects of sudden temperature rises during anthesis on seed number per boll are very high [16].

This review presents the physiological mechanism of stress tolerance and principles of genotype improvement, both classically and by genetic engineering in cotton. Also, it discussed the effects of agronomic management such as seed and foliar treatments to alleviate drought, salt and heat stress.

2. Roles of osmolytes

The stress-induced protein breakdown and amino acid synthesis caused the accumulation of amino acids in cells under abiotic stress conditions. The overall accumulation of amino acids has two different symptoms: cell damage and the beneficial effect of specific amino acids such as proline during stress acclimation [17, 18]. Drought, salinity, and heavy metals caused an increase in the accumulation of proline in resistant plants. Although proline was not synthesized in tobacco exposed to heat stress [19], the combination of heat and drought stress [20], and drought stress [21] induced an increase in the proline content of some cotton cultivars. In contrast with this finding, genotypic differences were non-significant for proline content under drought conditions [22].

The first approach to stress resilience is to restrict ethylene synthesis, which triggers the abscission of leaves and all reproductive organs under stress conditions [23]. Boll retention and yield increased by application of ethylene inhibitors such as aminoethoxyvinylglycine (AVG) or by downregulation of genes responsible for ethylene synthesis [24]. Abscisic acid (ABA), another growth-inhibitory hormone, is an important regulator of abiotic stress tolerance. The main function of ABA is to stimulate stomatal closure and gene expression to respond to the drought, salinity, and excessive temperature in the adaptive mechanism of plants [25, 26]. Under stress conditions, plants can alter their metabolism such as the synthesis of compatible solutes. In cell metabolism, the different unfavorable conditions increased the concentration of γ-aminobutyric acid (GABA) through enhanced activity of enzymes involved in GABA biosynthesis [27]. Under alkali stress conditions, the accumulation of GABA and putrescine in young and old leaves of cotton increased [28]. Polyamines such as putrescine, spermidine, and spermine have a protective role against salinity, drought and heat stress, and putrescine and spermine could be evaluated as selection criteria for stress-tolerant genotype breeding in cotton [29]. Exogenously applied polyamines increased the stress tolerance and yield under drought and salinity stress conditions in cotton [30, 31].

The osmoprotectant solutes such as glycine betaine (GB) are evaluated in three different ways. Firstly, the increase in GB under stress conditions is analyzed as biochemical. Secondly, the selection of the plants with high GB levels is used for improving the stress-tolerant cotton genotypes in conventional breeding and genetic engineering. Thirdly, the success of seed treatment with GB and foliar application of GB were examined in stress conditions. The accumulation of GB increased in transgenic crops, whereas foliar application of GB or seed dressed are common applications in plants where a certain amount of synthesis is insufficient [32, 33]. Generally localized in the chloroplast, GB has an important role in protecting photosystem II, stabilizing membranes, and alleviating ROS (reactive oxygen species) damage [34] and chilling damage [35], enhancing tolerance to lead (Pb) [36]. Cottonseed coating with GB enhanced seed cotton yield by approximately 20% [37]. Also, combined foliar application with GB and salicylic acid (SA) increased the tolerance to salt stress due to an increase in the leaf gas exchange with positively correlated stomatal properties and stimulate antioxidant enzyme activity in cotton seedlings [38, 39].

3. Heat stress

On the cell basis, heat stress firstly induced structural changes in chloroplast protein [40] and plasma membrane [41, 42], and these changes stimulated cell elongation, expansion, and differentiation. ROS, as a signaling molecule, induced excessive MDA (malondialdehyde) synthesis by increasing lipid peroxidation, and membrane structure disintegrated under heat stress conditions [43–45]. Also, molecular chaperones and heat shock proteins (HSP) accumulate to protect the membrane integrity by the expression of the sHSP coding gene in leaves under drought and heat stress conditions [46]. Heat stress caused the accumulation of specific isoforms of activase in cotton leaves [47, 48]. Although the efficiency of Rubisco activase decreased under heat stress, an enzyme can develop an alternative function by relocating on a thylakoid membrane [49], and this contributes to the acclimation of photosynthesis during high temperatures in cotton [50]. In addition, the synthesis of phytohormones such as SA and jasmonic acid (JA) to respond to heat stress is enhanced by ROS and Ca^{2+} [51]. The association between heat stress and plant nutrition demonstrated that the fertilization containing some macro and microelements such as K, Zn, and B could be used to alleviate the harmful effects of higher temperatures. These elements play an important role in chlorophyll synthesis and delay senescence in cotton [52, 53]. Zn spray eliminates the adverse effects of heat stress in cotton [54].

The reproductive period is synchronized with many processes in cotton. Although square retention is less affected by heat stress [55], sexual reproduction is defined as both sensitive to high temperature and thermotolerant during flowering – boll retention depending on the time, length, and severity of stress in cotton [56–58]. Heat stress during this period adversely affected the development of both sexual organs and caused flower abnormalities such as small flower, elongated stigmas (**Figure 3A** and **B**) [55], gametophyte, pollen germination, and pollen tube growth [59, 60]. Abiotic stress limited fertilization by preventing pollen tube development [61]. The tapetum, the innermost layer of somatic cells in anther lobes, is responsible for microsporogenesis and secretion of enzymes for the release of microspores from tetrads [62]. Programmed cell death (PCD) induced tapetal degeneration by heat stress, resulting in male sterility [63]. The regulation of tapetal PCD and anther dehiscence were controlled by the GhCKI gene (*G. hirsutum* casein kinase I) in the heat stress tolerance of cotton [64].

The decrease in the photosynthetic capacity of cotton at a temperature above 32°C reduces the accumulation of sucrose [65]. The decrease in sucrose accumulation adversely affected the fiber quality by inhibiting cellulose synthesis in higher temperature conditions [66]. Hereby, sucrose transport from subtending leaf of boll to developing boll negatively affected, and boll number per plant and boll weight reduced [67]. Plant growth is mostly regulated by sucrose, which is a source of substrates' energy production and biosynthesis by decomposing into hexoses. Furthermore, the plant responses are determined by ROS scavenging capacity and

Figure 3.
Elongated stigma affected by heat stress (A), and normal flower structure (B).

the signal pathway of sucrose [68]. In many crops, drought increased carbohydrate deprivation and ABA levels, whereas the ability of reproductive sinks to use sucrose and starch was reduced [69].

Auxin, an essential hormone, is very important for stress tolerance [70] and plays a role as a coordinator of plant growth and development [71]. The inhibition of auxin synthesis by overexpressing miR157 caused the sensitivity to heat stress and anther sterility in the reproductive period of cotton [72]. The auxin signaling pathway is controlled by the suppression of the sucrose synthase gene, and the antagonistic relationship between auxin and sucrose regulated plant growth and development [73, 74]. The favorable balance between auxin and sucrose is indispensable for the response of anther to heat stress [75].

Shedding is the formation of the abscission layer by natural or stress between sympodial or monopodial branches and reproductive organs such as square, flower, and boll in cotton (**Figure 2B** and **C**) [76]. The amount of hormones and regulation among hormone-controlled shedding, as auxin inhibits, and abscisic acid (ABA) promotes. Shedding is induced by increasing ABA and decreasing auxin under drought conditions [77]. In the reproductive period, failed fertilization caused by non-available pollen in higher temperatures resulted in flower shedding.

4. Salinity stress

The physiological drought via osmotic stress and ion toxicity caused by Na^+, Cl^-, and SO_4^{2-} are two major forms of damage. Although cotton is classified as moderately tolerant to salt stress (7.7 dS m^{-1}), salinity caused a decrease in seed cotton yield [78], boll number per plant [79–81], and an increase in early maturity [82]. Seed germination and early seedling growth in cotton are the most sensitive stages [83]. Both forms of salt damage caused abnormal plant growth such as stunted root and shoot growth by reducing photosynthetic capacity (**Figure 1A**) [84]. The excessive salt accumulation under salinity caused cellular injury in transpiring leaves [85]. However, cotton was successfully cultivated as a monoculture crop in saline-alkali soil. Plant cells are affected by low amounts of available water, loss of membrane functions, and ionic toxicity under excess sodium (Na^+). As with most abiotic stress, salinity contains an osmotic component, and cellular dehydration causes and disrupts the internal balance (homeostasis). The plant's first reaction is to reduce the Na^+ level in the cytosol by restricting influx, increasing efflux, and accumulating Na^+ in the vacuole for maintaining the cell metabolism. Transporters such as antiporter, uniporter, and symporter localized in the membrane are responsible for reducing Na^+ [86–88]. In a study conducted by [89], the Na^+/H^+ antiporter gene (GhSOS1) was detected in the plasma membrane of cotton (*G. hirsutum* L.). This gene is a Salinity Overly Sensitive, which has an important role to synthesize protein in upregulating under stress conditions such as salinity and drought. The suberization and lignification may occur around endodermal cells to inhibit apoplastic absorption of toxic ions in cotton [90, 91].

Many researchers focused on a root-associated microorganism to alleviate salinity stress in cotton. The arbuscular mycorrhizal fungi (AMF) are capable of increasing P and Zn uptake and promoting leaf proline accumulation [92, 93] but it should be noted that *Glomus mosseae*, AMF species, isolated from saline soil found to be less successful in alleviating salt stress compared to that of non-saline soil [94]. The microorganisms with the PGPR effect such as *Pseudomonas fluorescens* have IAA producing ability involved in the synthesis of important compounds under salinity stress [95]. In addition, the higher rhizosphere colonization of PGPR induced moderate N application due to the signaling molecule role of nitric oxide (NO) in the denitrification process [96]. Also, melatonin, as an indole hormone, alleviated the adverse effects of salt stress to reduce ROS production and ion toxicity and increases proline content in cotton seedlings [97].

5. Drought stress

The increase in the tensile force of the xylem, due to drought, caused a greater probability of rupture in the water column and formation of gas emboli in the xylem [98], and embolism decreased carbon assimilation depending on the linkage between water and CO_2 exchange. Furthermore, stomatal closure and decrease in the stomatal area take place [99], and desiccation of all plants is induced in the further process (**Figure 1B**) [100]. Thus, the most negative impacts occurred in photosynthetic capacity and transpiration rate [101]. The environmentally induced PCD (programmed cell death) occurred in response to drought, and PCD increased ROS accumulation, DNA fragmentation, organelle degeneration, and cytoplasm shrinkage [102]. The tylose formation and xylem inhabitation by wilt pathogens in water-limiting conditions may cause drought sensitivity to stress the susceptibility [103]. Primarily, MAPK (Mitogen-activated protein kinase), and secondly, ROS play an important role in intracellular signaling [104, 105]. *GhMKK3* and *GhMPK2* from *G. hirsutum* increased root hair development and ROS production by regulating ethylene synthesis, respectively under drought conditions [106].

Arbuscular mycorrhizal fungi (AMF) have an important role to alleviate drought stress by spreading on the soil and water transport by hyphae [107]. Similarly, PGPR coated phosphorus exhibited high performance due to the increasing stomatal conductance, net photosynthetic capacity, and yield of cotton under osmotic stress [108]. As a result of studies investigating the relationship between nutrients and drought, it was suggested that high N concentrations may decrease the effects of drought through nitrogen metabolism, proline synthesis antioxidant capacity, and osmotic regulation in cotton [109, 110]. However, K application has osmotic regulation due to increasing the osmoprotectant and regulates N metabolism [53, 111, 112] and photo-assimilation and translocation process [113] in drought-stressed cotton plants. Also, supplemental Zn alleviated the negative effects of drought stress by increasing antioxidant capacity and decreasing MDA content in cotton [112, 114]. The nanoparticles such as nano-TiO_2 and nano-SiO_2 could alleviate drought stress because their foliar applications increase photosynthetic pigments, antioxidant capacity, and proline content in cotton [115].

6. Heritability and breeding for stress tolerance in cotton

The combination of multiple stress factors and the identification of the model plant is complex at the physiological and genetic levels [116]. Growth chamber, greenhouse, and field studies were conducted by many researchers to screen the cotton genotypes in different stress studies. Stress tolerant genotypes can be bred through a selection of promising single plants under stress conditions followed by testing in either stress and non-stress conditions or selection under more optimum conditions for effective selection criteria [117]. Previous studies recommended the selection of cotton plants with higher seedling vigor, enhanced early root development, and lower root/shoot ratio for drought tolerance, pollen carrying a dominant heat-tolerance allele for heat stress, and salinity barrier for salt stress [118].

In stress-tolerant cotton breeding, genetic stock in strains with the D genome rather than those with the A genome should be considered. Transcriptomic analyses indicated that thick cuticles and a double layer palisade layer of D-genome species such as *G. harknessii*, *G. armourianum,* and *G. turneri* are important germplasm resources for water deficiency. Similarly, *G. gossypioides* and *G. thurberi* with aggressive and deep root structures for drought tolerance and *G. aridum*, *G. davidsonii,* and *G. klotzschianum* for both drought and salt tolerance can be used in cotton breeding [84, 119–122]. In support of this knowledge, [123] revealed that QTLs (quantitative trait locus) responsible for salt tolerance are usually localized on the D-subgenome. At the same time, novel genes and alleles in wild relatives are important to overcome the abiotic stress tolerance caused by narrow gene pools in cultivated cotton species [124].

The selection of plants with thick cuticle and waxy surfaces is important to reduce solar radiation in the breeding of heat tolerance [125]. The intense absorbency of cotton cultivars increased sensitivity to heat stress [126]. The most important cultural management is to arrange the sowing time to avoid the higher temperature in the reproductive period. However, planting cotton before the recommended time faced the problem of low temperature during the early growing stage. Therefore, breeding of high-temperature tolerant cotton varieties has been suggested as the best method [127].

A pre-screening of cotton germplasm and evaluation of hybrids by constructing a polythene tunnel at the reproductive stage is a very common method in cotton (**Figure 2A**) [128]. However, some researchers preferred delayed planting set of cotton plants to test genotypes in heat tolerance breeding [129–132]. In addition, some researchers have used tetrazolium chloride for identifying heat tolerance in both vegetative tissues and pollen viability in cotton [133–135]. Various studies have emphasized the success of wild cotton species, their stacking progenitor alleles, and *Gossypium tomentosum* (heat-resistant species) for stress tolerance [13, 136, 137]. The pollen characteristics, germination ability, and tube length have been screened to determine the tolerant and susceptible genotypes under higher temperatures in cotton [13, 138]. On the other hand, the increase in fiber wax content of susceptible cotton genotypes under heat and drought stress indicated that acceptable fiber wax levels could be used to improve tolerant genotypes in conventional breeding [139].

In quantitative genetic studies about heritability and gene action for different selection criteria of heat stress, multigenic inheritance and both additive and non-additive gene action in controlling cellular membrane thermostability was found to be higher under heat stress conditions [140]. The results of the scaling test, which is an important biometrical analysis method, indicated the significant dominance, additive x dominance, and dominance x dominance referred to as non-additive gene action for relative cell injury under heat stress [141]. Similarly, non-additive gene actions were estimated for fiber quality characters and ginning out-turn in heat stress (~38–39°C) at peak flowering time under field conditions by arranged sowing time [142]. In contrast, high heritability associated with high genetic advance for hydrogen peroxide content, catalase activity, total soluble proteins, carotenoids, and chlorophyll contents were found to have significant additive gene action under heat stress conditions [143]. In conventionally breeding of drought and heat stress tolerance, it was emphasized that instead of single plant selection in the F_2 generation, superior plants should be selected in further generations such as F_4 [144, 145].

PEG 6000 (polyethylene glycol) is defined as a rapid and effective method to observe the response of cotton genotypes for selection in drought tolerance breeding [146, 147]. Drought-tolerant species are defined with low maximum transpiration and photosynthetic rate, stomatal conductance, specific leaf area, small leaf size high leaf longevity, root mass ratio, and small leaf size [148]. The higher chlorophyll stability and relative water content exhibited drought tolerance due to photosynthate, which results in higher biomass [149] and fiber quality [150]. Furthermore, the presence of late embryogenesis abundant proteins is important for drought tolerance of cotton [151]. The physiological characteristics used for drought tolerance should be combined with yield, yield attributes, and fiber quality parameters in the breeding of cotton varieties with high adaptability [152]. Drought responsive genes were classified as induced (stress-related, metabolism, transcription factor, proline, and cellular transport) and repressed (mainly comprising metabolism, cellular transport, and stress-related) [153]. Drought tolerance genes such as RD2 (rice drought-responsive), HAT22 (homeobox from *Arabidopsis thaliana*), PIP2, PIP2C (plasma membrane intrinsic proteins), and GaTOP6B (encoding DNA topoisomerase from *G. arboreum*) were associated with drought in cotton [104]. Similarly, QTL analysis revealed that genes responsive to drought tolerance are spread over nine chromosomes while one QTL hotspot is concentrated on the eighth chromosome [154]. Anwar et al. [155] evaluated the selected varieties by molecular under drought stress, and MNH-886 cotton cultivar with high boll retention percentage, photosynthesis rate, and stomatal conductance was recommended against drought stress. The ratio of general combining ability and specific combining ability variance in line x tester analysis showed significant non-additive gene effects for proline content, total chlorophyll, canopy temperature, and cell membrane stability in drought stress conditions [156].

NaCl treatment at different doses is the most used method for screening cotton genotypes in order to determine suitable genotypes. Many researchers revealed significant variations in morphological, physiological, and biochemical characters under salinity stress [157–159]. The performance of genotypes, depending on genetic factors, compared with environmental factors and genotype x environment interactions, and high genetic gain by suitable selection increased the breeding success because of high additive effects under saline [160] and drought conditions [161]. On the other hand, non-additive gene actions were found significantly higher for chlorophyll content, K^+/Na^+ ratio and within boll yield components under saline conditions [144, 162]. At the same time, selection for K^+, Na^+, and K^+/Na^+ were recommended for salinity tolerance according to the results of factor analysis in cotton [163]. In addition, a reciprocal effect found in some studies indicated cytoplasmic and cytoplasmic x nuclear genes interaction for salinity tolerance breeding [160, 164]. Besides, the fact that genes with pleiotropic effects are effective for both salinity and drought indicated physiological characters are used in indirect selection can increase the success of multi-stress breeding [165, 166].

In many studies about engineering abiotic stress-tolerant crops, genes responsible for multiple stress factors found to confer tolerance in different plants are defined [167]. Co-overexpressing of SUMO E3 ligase (OsSIZ) and Vacuolar H^{+-} pyrophosphatase (AVP1) from Arabidopsis and vacuolar Na^{+}/H^{+} antiporter genes performed significantly in increasing tolerance to multiple stressors such as drought, salinity, and higher temperature [168]. Overexpression of mitogen-activated protein kinases (MAPK) from cotton GhMPK2 performed stress tolerance when induced by salt, ABA, and water scarcity [169]. Eight genes belonging to the GhHSP20 family are responsible for heat, drought, and salinity [170], and the expression of GHSP26 caused a significant increase in proteins under drought stress [171]. Similarly, it was verified that GhCIPK6a overexpressed cotton lines can reduce the negative effects of salinity, and the seeds of these lines exhibited higher water absorption capacity at the germination stage [172]. In another study, JAZ proteins were identified as inhibitors of JA, and overexpression of JAZ genes resulted in a higher performance to salt stress [173]. GaJAZ1 transgenic plants carrying genes transferred from *G. arboreum* to *G. hirsutum* were found to be different from wilt type in terms of salinity tolerance [174]. *GhPHD* genes (plant homeodomain genes from *G. hirsutum*) increased plant tolerance in adverse environmental conditions to alleviate abiotic and phytohormonal stresses [175]. Also, the overexpression of the 14–3-3 gene GF14λ from Arabidopsis resulted in a stay-green phenotype due to decreased wilting, delayed senescence, and higher photosynthetic capacity. It was suggested that this gene can be used successfully in drought resistance [176, 177].

7. Inferences (future considerations)

The nutrient relations of cotton plants were greatly affected by abiotic stress. The nitrogen, silicon, magnesium, and calcium uptake are limited under high temperatures, drought, and salinity. In addition, the architecture of cotton changed to capture the less mobile nutrients. Therefore, fertilization should be reconsidered under stress conditions. To reduce the negative effects of abiotic stress, the use of some osmoprotectant and hormones may be a solution. The sowing time and plant density should be rearranged especially for high temperatures and drought. Cluster-type cotton varieties suitable for High-Density Planting System (HDPS) should be developed. These varieties will also reduce the density of weeds and dry matter content that compete for available water under drought conditions.

The most effective way to overcome the adverse effects of multiple abiotic stress factors is to improve tolerant cultivars. The possible effects of drought, salinity, and high-temperature stress on physiological, morphological, and yield should be well resolved to mitigate stress and develop varieties. The crossing between standard cultivars and donor suitable parents is a basic stage for genetic variation in conventional breeding. Multiple stress factors are very complex and polygenic characters and are controlled by non-additive gene effects with low heritability. Therefore, bulk selection should be useful for traits with low heritability such as stress tolerance, and the selection of a single plant should be postponed to F_4 or F_5 generation. Drought and high temperature are effective stress factors from flowering and during the first boll formation period, whereas salinity is an important problem in the early development period. In areas where these conditions are created, indirect selection can be made in terms of physiological characteristics, whereas direct selection can be made for yield. The evaluation of genotypes to stress tolerance under the plastic tunnel at the reproductive stage and in saline and drought conditions is a more accurate approach compared to artificial conditions and delayed sowing date in cotton.

Genetic engineering studies are promising, but manipulation of a single gene does not seem to be sufficient as resistance to multiple stress conditions is controlled by multiple genes. However, future transgenic cultivar breeding may be a good approach for multiple stress tolerance. In both classical breeding and genetic engineering studies, tolerance to multiple stress factors should be pyramided with yield and fiber characteristics. The non-genetically CRISPR-Cas system could be evaluated for engineering multiple stress tolerance in future cotton cultivars for all cotton-growing regions.

It could be highlighted that temperature extremes and fluctuations in the cotton-growing season along with other stress appear to be major factors of future yield reductions.

Acknowledgements

Volkan Mehmet CINAR thanks the Higher Education Council of Turkey (YOK) for 100/2000 PhD scholarship.

Conflict of interest

The authors declare no conflict of interest.

Author statement

Volkan Mehmet CINAR: Conceptualization, Literature survey, Writing – Review & Editing. Serife BALCI: Writing – Review & Editing. Aydın UNAY: Conceptualization, Literature survey, Writing – Review, Supervision.

Author details

Volkan Mehmet Cinar[1], Serife Balci[2] and Aydın Unay[1]*

1 Faculty of Agriculture, Department of Field Crops, Aydın Adnan Menderes University, Aydın, Turkey

2 Department of Plant Breeding and Genetics, Cotton Research Institute, Nazilli-Aydın, Turkey

*Address all correspondence to: aunay@adu.edu.tr

References

[1] FAO. Global soil partnership. Soil salinization as a global major challenge. [Internet]. 2021. Available from: https://www.fao.org/global-soil-partnership/resources/highlights/detail/en/c/1412475/ [Accessed: April 11, 2021]

[2] Anonymous. Physical climate risk for global cotton production: Global analysis. [Internet]. 2021. Available from: https://www.acclimatise.uk.com/wp-content/uploads/2021/06/Cotton2040-GAReport-FullReport -highres.pdf [Accessed: October 11, 2021]

[3] Rahman HU. Number and weight of cotton lint fibres: Variation due to high temperatures in the field. Australian Journal of Agricultural Research. 2006;**57**(5):583-590. DOI: 10.1071/AR05135

[4] Zhou R, Yu X, Ottosen CO, Rosenqvist E, Zhao L, Wang Y, et al. Drought stress had a predominant effect over heat stress on three tomato cultivars subjected to combined stress. BMC Plant Biology. 2017;**17**(1):1-3. DOI: 10.1186/s12870-017-0974-x

[5] Jiao F, Shi XR, Han FP, Yuan ZY. Increasing aridity, temperature and soil pH induce soil CNP imbalance in grasslands. Scientific Reports. 2016;**6**(1):1-9. DOI: 10.1038/srep19601

[6] Rahman AK, Ahmed KM, Butler AP, Hoque MA. Influence of surface geology and micro-scale land use on the shallow subsurface salinity in deltaic coastal areas: A case from southwest Bangladesh. Environmental Earth Sciences. 2018;**77**(12):1-8. DOI: 10.1007/s12665-018-7594-0

[7] Gorham J, Läuchli A, Leidi EO. Plant responses to salinity. In: Stewart JM, Oosterhuis DM, Heitholt JJ, Mauney JR, editors. Physiology of Cotton. Dordrecht: Springer; 2010. pp. 129-141. DOI: 10.1007/978-90-481-3195-2_13

[8] Rehman A, Kamran M, Afzal I. Production and processing of quality cotton seed. In: Ahmad S, Hasanuzzaman M, editors. Cotton Production and Uses. Singapore: Springer; 2020. pp. 547-570. DOI: 10.1007/978-981-15-1472-2_27

[9] Oosterhuis DM, Loka D. Polyamines and cotton flowering. In: Oosterhuis DM, Cothren JT, editors. Flowering and Fruiting of Cotton. Memphis: The Cotton Foundation; 2012. pp. 109-132

[10] Basal H, Dagdelen N, Unay A, Yilmaz E. Effects of deficit drip irrigation ratios on cotton (*Gossypium hirsutum* L.) yield and fibre quality. Journal of Agronomy and Crop Science. 2009;**195**(1):19-29. DOI: 10.1111/j.1439-037X.2008.00340.x

[11] Loka DA, Oosterhuis DM. Water stress and reproductive development in cotton. In: Oosterhuis DM, Cothren JT, editors. Flowering and Fruiting in Cotton. Candova: The Cotton Foundaiton; 2012. pp. 51-58

[12] Lokhande S, Reddy KR. Reproductive and fiber quality responses of upland cotton to moisture deficiency. Agronomy Journal. 2014;**106**(3):1060-1069. DOI: 10.2134/agronj13.0537

[13] Reddy KR, Bheemanahalli R, Saha S, Singh K, Lokhande SB, Gajanayake B, et al. High-temperature and drought-resilience traits among interspecific chromosome substitution lines for genetic improvement of upland cotton. Plants. 2020;**9**(12):1747. DOI: 10.3390/plants9121747

[14] Reddy KR, Davidonis GH, Johnson AS, Vinyard BT. Temperature regime and carbon dioxide enrichment alter cotton boll development and fiber properties. Agronomy Journal. 1999;**91**(5):851-858. DOI: 10.2134/agronj1999.915851x

[15] Zhao D, Reddy KR, Kakani VG, Koti S, Gao W. Physiological causes of cotton fruit abscission under conditions of high temperature and enhanced ultraviolet-B radiation. Physiologia Plantarum. 2005;**124**(2):189-199. DOI: 10.1111/j.1399-3054.2005.00491.x

[16] Reddy KR, Brand D, Wijewardana C, Gao W. Temperature effects on cotton seedling emergence, growth, and development. Agronomy Journal. 2017;**109**(4):1379-1387. DOI: 10.2134/agronj2016.07.0439

[17] Patterson JH, Newbigin ED, Tester M, Bacic A, Roessner U. Metabolic responses to salt stress of barley (*Hordeum vulgare* L.) cultivars, Sahara and Clipper, which differ in salinity tolerance. Journal of Experimental Botany. 2009;**60**(14):4089-4103. DOI: 10.1093/jxb/erp243

[18] Krasensky J, Jonak C. Drought, salt, and temperature stress-induced metabolic rearrangements and regulatory networks. Journal of Experimental Botany. 2012;**63**(4):1593-1608. DOI: 10.1093/jxb/err460

[19] Dobra J, Motyka V, Dobrev P, Malbeck J, Prasil IT, Haisel D, et al. Comparison of hormonal responses to heat, drought and combined stress in tobacco plants with elevated proline content. Journal of Plant Physiology. 2010;**167**(16):1360-1370. DOI: 10.1016/j.jplph.2010.05.013

[20] De Ronde JA, Van Der Mescht A, Steyn HS. Proline accumulation in response to drought and heat stress in cotton. African Crop Science Journal. 2000;**8**(1):85-92

[21] Ahmed AH, Darwish E, Alobaidy MG. Impact of putrescine and 24-epibrassinolide on growth, yield and chemical constituents of cotton (*Gossypium barbadense* L.) plant grown under drought stress conditions. Asian Journal of Plant Sciences. 2017;**16**(1):9-23. DOI: 10.3923/ajps.2017.9.23

[22] Zahid Z, Khan MK, Hameed A, Akhtar M, Ditta A, Hassan HM, et al. Dissection of drought tolerance in upland cotton through morpho-physiological and biochemical traits at seedling stage. Frontiers in Plant Science. 2021;**12**:627107. DOI: 10.3389/fpls.2021.627107

[23] Rahman MU, Majeed A, Zulfiqar S, Ishfaq S, Mohsan M, Ahmad N. In: Rahman M, Zafar Y, Zhang T, editors. Cotton Precision Breeding. Cham: Springer; 2021. pp. 137-156. DOI: 10.1007/978-3-030-64504-5_6

[24] Najeeb U, Bange MP, Tan DK, Atwell BJ. Consequences of waterlogging in cotton and opportunities for mitigation of yield losses. AoB Plants. 2015;**1**:1-17. DOI: 10.1093/aobpla/plv080

[25] Wani SH, Kumar V. Plant stress tolerance: Engineering ABA: A potent phytohormone. Transcriptomics. 2015;**3**(2):1000113. DOI: 10.4172/2329-8936.1000113

[26] Fujita Y, Fujita M, Shinozaki K, Yamaguchi-Shinozaki K. ABA-mediated transcriptional regulation in response to osmotic stress in plants. Journal of Plant Research. 2011;**124**(4):509-525. DOI: 10.1007/s10265-011-0412-3

[27] Renault H, Roussel V, El Amrani A, Arzel M, Renault D, Bouchereau A, et al. The Arabidopsis pop2-1 mutant

reveals the involvement of GABA transaminase in salt stress tolerance. BMC Plant Biology. 2010;**10**(1):1-6. DOI: 10.1186/1471-2229-10-20

[28] Guo R, Shi L, Yang C, Yan C, Zhong X, Liu Q, et al. Comparison of ionomic and metabolites response under alkali stress in old and young leaves of cotton (*Gossypium hirsutum* L.) seedlings. Frontiers in Plant Science. 2016;**25**(7):1785. DOI: 10.3389/fpls.2016.01785

[29] Loka DA, Oosterhuis DM, Pilon C. Effect of water-deficit stress on polyamine metabolism of cotton flower and their subtending leaf. In: Oosterhuis D, editor. Summaries of Arkansas Cotton Research. Arkansas: Arkansas Agricultural Experiment Station; 2011. pp. 70-75

[30] Shallan MA, Hassan HM, Namich AA, Ibrahim AA. Effect of sodium nitroprusside, putrescine and glycine betaine on alleviation of drought stress in cotton plant. American-Eurasian Journal of Agricultural and Environmental Science. 2012;**12**(9):1252-1265. DOI: 10.5829/idosi.aejaes.2012.12.09.1902

[31] Ahmed AH, Darwish ES, Hamoda SA, Alobaidy MG. Effect of putrescine and humic acid on growth, yield and chemical composition of cotton plants grown under saline soil conditions. American-Eurasian Journal of Agricultural and Environmental Science. 2013;**13**(4):479-497. DOI: 10.5829/idosi.aejaes.2013.13.04.1965

[32] Sakamoto A, Murata N. The role of glycine betaine in the protection of plants from stress: Clues from transgenic plants. Plant, Cell & Environment. 2002;**25**(2):163-171. DOI: 10.1046/j.0016-8025.2001.00790.x

[33] Naidu BP, Williams RL. Seed Treatment and Foliar Application of Osmoprotectants to Increase Crop Establishment and Cold Tolerance at Flowering in Rice: A Report for the Rural Industries Research and Development Corporation. Brisbane, Australia: CSIRO Tropical Agriculture; 2004

[34] Chen TH, Murata N. Glycine betaine protects plants against abiotic stress: Mechanisms and biotechnological applications. Plant, Cell & Environment. 2011;**34**(1):1-20. DOI: 10.1111/j.1365-3040.2010.02232.x

[35] Cheng C, Pei LM, Yin TT, Zhang KW. Seed treatment with glycine betaine enhances tolerance of cotton to chilling stress. The Journal of Agricultural Science. 2018;**156**(3):323-332. DOI: 10.1017/S0021859618000278

[36] Bharwana SA, Ali S, Farooq MA, Iqbal N, Hameed A, Abbas F, et al. Glycine betaine-induced lead toxicity tolerance related to elevated photosynthesis, antioxidant enzymes suppressed lead uptake and oxidative stress in cotton. Turkish Journal of Botany. 2014;**38**(2):281-292. DOI: 10.3906/bot-1304-65

[37] Naidu BP, Cameron DF, Konduri SV. Improving drought tolerance of cotton by glycinebetaine application and selection. In: Michalk DL, Pratley JE, editors. Agronomy, Growing a Greener Future? Proceedings of the 9th Australian Agronomy Conference; Charles Sturt University, Wagga Wagga, Australia. Serpentine, Australia: The Australian Society of Agronomy Inc., Gosford, NSW, Australia: Published online by The Regional Institute Ltd.; 20-23 July 1998;1-5. Available from: http://www.regional.org.au/au/asa/1998/4/221naidu.htm

[38] Hamani AK, Wang G, Soothar MK, Shen X, Gao Y, Qiu R, et al. Responses

of leaf gas exchange attributes, photosynthetic pigments and antioxidant enzymes in NaCl-stressed cotton (*Gossypium hirsutum* L.) seedlings to exogenous glycine betaine and salicylic acid. BMC Plant Biology. 2020;**20**(1):1-4. DOI: 10.1186/s12870-020-02624-9

[39] Hamani AK, Li S, Chen J, Amin AS, Wang G, Xiaojun S, et al. Linking exogenous foliar application of glycine betaine and stomatal characteristics with salinity stress tolerance in cotton (*Gossypium hirsutum* L.) seedlings. BMC Plant Biology. 2021;**21**(1):1-2. DOI: 10.1186/s12870-021-02892-z

[40] Ahmad N, Michoux F, Nixon PJ. Investigating the production of foreign membrane proteins in tobacco chloroplasts: Expression of an algal plastid terminal oxidase. PLoS One. 2012;7(7):e41722. DOI: 10.1371/journal.pone.0041722

[41] Potters G, Pasternak TP, Guisez Y, Jansen MA. Different stresses, similar morphogenic responses: Integrating a plethora of pathways. Plant, Cell & Environment. 2009;**32**(2):158-169. DOI: 10.1111/j.1365-3040.2008.01908.x

[42] Rasheed R. Salinity and Extreme Temperature Effects on Sprouting Buds of Sugarcane (*Saccharum officinarum* L.): Some Histological and Biochemical Studies. Faisalabad: University of Agriculture; 2006

[43] Majeed S, Rana IA, Mubarik MS, Atif RM, Yang SH, Chung G, et al. Heat stress in cotton: A review on predicted and unpredicted growth-yield anomalies and mitigating breeding strategies. Agronomy. 2021;**11**(9):1825. DOI: 10.3390/agronomy11091825

[44] Burke JJ, Hatfield JL, Klein RR, Mullet JE. Accumulation of heat shock proteins in field-grown cotton. Plant Physiology. 1985;**78**(2):394-398. DOI: 10.1104/pp.78.2.394

[45] Sable A, Rai KM, Choudhary A, Yadav VK, Agarwal SK, Sawant SV. Inhibition of heat shock proteins HSP90 and HSP70 induce oxidative stress, suppressing cotton fiber development. Scientific Reports. 2018;**8**(1):1-7. DOI: 10.1038/s41598-018-21866-0

[46] Sarwar M, Saleem MF, Ullah N, Rizwan M, Ali S, Shahid MR, et al. Exogenously applied growth regulators protect the cotton crop from heat-induced injury by modulating plant defense mechanism. Scientific Reports. 2018;**8**(1):1-5. DOI: 10.1038/s41598-018-35420-5

[47] Law RD, Crafts-Brandner SJ. High temperature stress increases the expression of wheat leaf ribulose-1, 5-bisphosphate carboxylase/oxygenase activase protein. Archives of Biochemistry and Biophysics. 2001;**386**(2):261-267. DOI: 10.1006/abbi.2000.2225

[48] Law DR, Crafts-Brandner SJ, Salvucci ME. Heat stress induces the synthesis of a new form of ribulose-1, 5-bisphosphate carboxylase/oxygenase activase in cotton leaves. Planta. 2001;**214**(1):117-125. DOI: 10.1007/s004250100592

[49] Rokka A, Zhang L, Aro EM. Rubisco activase: An enzyme with a temperature-dependent dual function? The Plant Journal. 2001;**25**(4):463-471. DOI: 10.1046/j.1365-313x.2001.00981.x

[50] DeRidder BP, Salvucci ME. Modulation of Rubisco activase gene expression during heat stress in cotton (*Gossypium hirsutum* L.) involves post-transcriptional mechanisms. Plant Science. 2007;**172**(2):246-254. DOI: 10.1016/j.plantsci.2006.08.014

[51] Rai KK, Pandey N, Rai SP. Salicylic acid and nitric oxide signaling in plant heat stress. Physiologia Plantarum. 2020;**168**(2):241-255. DOI: 10.1111/ppl.12958

[52] Corrales I, Poschenrieder C, Barceló J. Boron-induced amelioration of aluminium toxicity in a monocot and a dicot species. Journal of Plant Physiology. 2008;**165**(5):504-513. DOI: 10.1016/j.jplph.2007.03.014

[53] Zahoor R, Zhao W, Dong H, Snider JL, Abid M, Iqbal B, et al. Potassium improves photosynthetic tolerance to and recovery from episodic drought stress in functional leaves of cotton (*Gossypium hirsutum* L.). Plant Physiology and Biochemistry. 2017a;**119**:21-32. DOI: 10.1016/j.plaphy.2017.08.011

[54] Sarwar M, Saleem MF, Ullah N, Ali S, Rizwan M, Shahid MR, et al. Role of mineral nutrition in alleviation of heat stress in cotton plants grown in glasshouse and field conditions. Scientific Reports. 2019;**9**(1):1-7. DOI: 10.1038/s41598-019-49404-6

[55] Brown PW. Cotton Heat Stress [Internet]. 1999. Available from: https://cals.arizona.edu/AZMET/az1448.pdf [Accessed: November 13, 2021]

[56] Hedhly A, Hormaza JI, Herrero M. Global warming and sexual plant reproduction. Trends in Plant Science. 2009;**14**(1):30-36. DOI: 10.1016/j.tplants.2008.11.001

[57] Zinn KE, Tunc-Ozdemir M, Harper JF. Temperature stress and plant sexual reproduction: Uncovering the weakest links. Journal of Experimental Botany. 2010;**61**(7):1959-1968. DOI: 10.1093/jxb/erq053

[58] Reddy KR, Hodges HF, Reddy VR. Temperature effects on cotton fruit retention. Agronomy journal. 1992;**84**(1):26-30. DOI: 10.2134/agronj1992.00021962008400010006x

[59] Snider JL, Oosterhuis DM, Kawakami EM. Diurnal pollen tube growth rate is slowed by high temperature in field-grown *Gossypium hirsutum* pistils. Journal of Plant Physiology. 2011;**168**(5):441-448. DOI: 10.1016/j.jplph.2010.08.003

[60] Echer FR, Oosterhuis DM, Loka DA, Rosolem CA. High night temperatures during the floral bud stage increase the abscission of reproductive structures in cotton. Journal of Agronomy and Crop Science. 2014;**200**(3):191-198. DOI: 10.1111/jac.12056

[61] Oosterhuis DM, Snider JL. High temperature stress on floral development and yield of cotton. In: Oosterhuis DM, editor. Stress Physiology in Cotton. Cordova: The Cotton Foundation; 2011. pp. 1-24

[62] Sanders PM, Bui AQ, Weterings K, McIntire KN, Hsu YC, Lee PY, et al. Anther developmental defects in Arabidopsis thaliana male-sterile mutants. Sexual Plant Reproduction. 1999;**11**(6):297-322. DOI: 10.1007/s004970050158

[63] Li X, Gao X, Wei Y, Deng L, Ouyang Y, Chen G, et al. Rice APOPTOSIS INHIBITOR5 coupled with two DEAD-box adenosine 5′-triphosphate-dependent RNA helicases regulates tapetum degeneration. The Plant Cell. 2011;**23**(4):1416-1434. DOI: 10.1105/tpc.110.082636

[64] Min L, Zhu L, Tu L, Deng F, Yuan D, Zhang X. Cotton Gh CKI disrupts normal male reproduction by delaying tapetum programmed cell death via inactivating starch synthase. The Plant Journal. 2013;**75**(5):823-835. DOI: 10.1111/tpj.12245

[65] Crafts-Brandner SJ, Salvucci ME. Rubisco activase constrains the photosynthetic potential of leaves at high temperature and CO_2. Proceedings of the National Academy of Sciences. 2000;**97**(24):13430-13435. DOI: 10.1073/pnas.230451497

[66] Tian JS, Hu YY, Gan XX, Zhang YL, Hu XB, Ling GO, et al. Effects of increased night temperature on cellulose synthesis and the activity of sucrose metabolism enzymes in cotton fiber. Journal of Integrative Agriculture. 2013;**12**(6):979-988. DOI: 10.1016/S2095-3119(13)60475-X

[67] Peng J, Liu J, Zhang L, Luo J, Dong H, Ma Y, et al. Effects of soil salinity on sucrose metabolism in cotton leaves. PLoS One. 2016;**11**(5):e0156241. DOI: 10.1371/journal.pone.0156241

[68] Ruan YL, Jin Y, Yang YJ, Li GJ, Boyer JS. Sugar input, metabolism, and signaling mediated by invertase: Roles in development, yield potential, and response to drought and heat. Molecular Plant. 2010;**3**(6):942-955. DOI: 10.1093/mp/ssq044

[69] Andersen MN, Asch F, Wu Y, Jensen CR, Næsted H, Mogensen VO, et al. Soluble invertase expression is an early target of drought stress during the critical, abortion-sensitive phase of young ovary development in maize. Plant Physiology. 2002;**130**(2):591-604. DOI: 10.1104/pp.005637

[70] Xiao YH, Li DM, Yin MH, Li XB, Zhang M, Wang YJ, et al. Gibberellin 20-oxidase promotes initiation and elongation of cotton fibers by regulating gibberellin synthesis. Journal of Plant Physiology. 2010;**167**(10):829-837. DOI: 10.1016/j.jplph.2010.01.003

[71] Leyser O. Auxin signaling. Plant Physiology. 2018;**176**(1):465-479. DOI: 10.1104/pp.17.00765

[72] Ding Y, Ma Y, Liu N, Xu J, Hu Q, Li Y, et al. micro RNA s involved in auxin signalling modulate male sterility under high-temperature stress in cotton (*Gossypium hirsutum*). The Plant Journal. 2017;**91**(6):977-994. DOI: 10.1111/tpj.13620

[73] Stokes ME, Chattopadhyay A, Wilkins O, Nambara E, Campbell MM. Interplay between sucrose and folate modulates auxin signaling in Arabidopsis. Plant Physiology. 2013;**162**(3):1552-1565. DOI: 10.1104/pp.113.215095

[74] Yuan Y, Mei L, Wu M, Wei W, Shan W, Gong Z, et al. SlARF10, an auxin response factor, is involved in chlorophyll and sugar accumulation during tomato fruit development. Journal of Experimental Botany. 2018;**69**(22):5507-5518. DOI: 10.1093/jxb/ery328

[75] Min L, Li Y, Hu Q, Zhu L, Gao W, Wu Y, et al. Sugar and auxin signaling pathways respond to high-temperature stress during anther development as revealed by transcript profiling analysis in cotton. Plant Physiology. 2014;**164**(3):1293-1308. DOI: 10.1104/pp.113.232314

[76] Tariq M, Yasmeen A, Ahmad S, Hussain N, Afzal MN, Hasanuzzaman M. Shedding of fruiting structures in cotton: Factors, compensation and prevention. Tropical and Subtropical Agroecosystems. 2017;**20**(2):251-262

[77] Guinn G, Brummett DL. Changes in abscisic acid and indoleacetic acid before and after anthesis relative to changes in abscission rates of cotton fruiting forms. Plant Physiology. 1988;**87**(3):629-631. DOI: 10.1104/pp.87.3.629

[78] Zhang L, Zhang G, Wang Y, Zhou Z, Meng Y, Chen B. Effect of soil salinity on physiological characteristics of functional leaves of cotton plants. Journal

of Plant Research. 2013;**126**(2):293-304. DOI: 10.1007/s10265-012-0533-3

[79] Naderiarefi A, Ahmadi A, Tavakoli AR, Vafaietabar MAR, Sabikdast M. The effect of issue stress on physiological characteristics of leaf and drought resistance of different cotton genotypes. To Agricultural Crop Journal. 2016;**18**(4):987-999

[80] Anagholi A, Rousta MJ, Azari A. Selection of salt tolerance varitieis of cotton by using of tolerant indices. Arid Bio Scientific and Research Journal. 2016;**2**:1-9

[81] Afrasiab P, Delbari M, Asadi R, Mohammadi E. Effect of soil section and water salinity on yield and components yield of cotton. Journal of Plant Production Research. 2015;**22**(3):295-311

[82] Razaji A, Paknejad F, Moarefi M, Mahdavi Damghani A, Ilkaee M. Meta-analysis of the effects of salinity stress on cotton (*Gossypium* spp.) growth and yield in Iran. Journal of. Agricultural Sciences. 2020;**26**(1):94-103

[83] Wang Z, Wang J, Bao Y, Wu Y, Zhang H. Quantitative trait loci controlling rice seed germination under salt stress. Euphytica. 2011;**178**(3):297-307. DOI: 10.1007/s10681-010-0287-8

[84] Zhang F, Zhu G, Du L, Shang X, Cheng C, Yang B, et al. Genetic regulation of salt stress tolerance revealed by RNA-Seq in cotton diploid wild species, *Gossypium davidsonii*. Scientific Reports. 2016;**6**(1):1-5. DOI: 10.1038/srep20582

[85] Khan AN, Qureshi RH, Ahmad N, Rashid A. Response of cotton cultivars to salinity at various growth development stages. Sarhad Journal of Agriculture. 1995;**11**(6):729-731

[86] Niu X, Bressan RA, Hasegawa PM, Pardo JM. Ion homeostasis in NaCl stress environments. Plant Physiology. 1995;**109**(3):735. DOI: 10.1104/pp.109.3.735

[87] Roy SJ, Negrão S, Tester M. Salt resistant crop plants. Current Opinion in Biotechnology. 2014;**26**:115-124. DOI: 10.1016/j.copbio.2013.12.004

[88] Deinlein U, Stephan AB, Horie T, Luo W, Xu G, Schroeder JI. Plant salt-tolerance mechanisms. Trends in Plant Science. 2014;**19**(6):371-379. DOI: 10.1016/j.tplants.2014.02.001

[89] Chen X, Lu X, Shu N, Wang D, Wang S, Wang J, et al. GhSOS1, a plasma membrane Na^+/H^+ antiporter gene from upland cotton, enhances salt tolerance in transgenic *Arabidopsis thaliana*. PLoS One. 2017;**12**(7):e0181450. DOI: 10.1371/journal.pone.0181450

[90] Faiyue B, Al-Azzawi MJ, Flowers TJ. A new screening technique for salinity resistance in rice (*Oryza sativa* L.) seedlings using bypass flow. Plant, Cell & Environment. 2012;**35**(6):1099-1108. DOI: 10.1111/j.1365-3040.2011.02475.x

[91] Thorne SJ, Hartley SE, Maathuis FJ. Is silicon a panacea for alleviating drought and salt stress in crops? Frontiers in plant science. 2020;**11**:1221. DOI: 10.3389/fpls.2020.01221

[92] Liu S, Guo X, Feng G, Maimaitiaili B, Fan J, He X. Indigenous arbuscular mycorrhizal fungi can alleviate salt stress and promote growth of cotton and maize in saline fields. Plant and Soil. 2016;**398**(1-2):195-206. DOI: 10.1007/s11104-015-2656-5

[93] Khaitov B, Teshaev S. The effect of arbuscular mycorrhiza fungi on cotton growth and yield under salinated soil condition. Cotton Genomics and

Genetics. 2015;**17**:6. DOI: 10.5376/cgg.2015.06.0003

[94] Evelin H, Kapoor R, Giri B. Arbuscular mycorrhizal fungi in alleviation of salt stress: A review. Annals of Botany. 2009;**104**(7):1263-1280. DOI: 10.1093/aob/mcp251

[95] Egamberdieva D, Jabborova D, Hashem A. Pseudomonas induces salinity tolerance in cotton (*Gossypium hirsutum*) and resistance to Fusarium root rot through the modulation of indole-3-acetic acid. Saudi Journal of Biological Sciences. 2015;**22**(6):773-779. DOI: 10.1016/j.sjbs.2015.04.019

[96] Kang A, Zhang N, Xun W, Dong X, Xiao M, Liu Z, et al. Nitrogen fertilization modulates beneficial rhizosphere interactions through signaling effect of nitric oxide. Plant Physiology. 2021;**1**:kiab555

[97] Jiang D, Lu B, Liu L, Duan W, Meng Y, Li J, et al. Exogenous melatonin improves the salt tolerance of cotton by removing active oxygen and protecting photosynthetic organs. BMC Plant Biology. 2021;**21**(1):1-9. DOI: 10.1186/s12870-021-03082-7

[98] Venturas MD, Sperry JS, Hacke UG. Plant xylem hydraulics: What we understand, current research, and future challenges. Journal of Integrative Plant Biology. 2017;**59**(6):356-389. DOI: 10.1111/jipb.12534

[99] Li X, Smith R, Choat B, Tissue DT. Drought resistance of cotton (*Gossypium hirsutum*) is promoted by early stomatal closure and leaf shedding. Functional Plant Biology. 2019;**47**(2):91-98. DOI: 10.1071/FP19093

[100] Choat B, Brodribb TJ, Brodersen CR, Duursma RA, López R, Medlyn BE. Triggers of tree mortality under drought. Nature. 2018;**558**(7711):531-544. DOI: 10.1038/s41586-018-0240-x

[101] Azhar MT, Rehman A. Overview on Effects of Water Stress on Cotton Plants and Productivity. In: Biochemical, Physiological and Molecular Avenues for Combating Abiotic Stress Tolerance in Plants. Cambridge: Academic Press; 2018. pp. 297-316. DOI: 10.1016/B978-0-12-813066-7.00016-4

[102] Hameed A, Goher M, Iqbal N. Drought induced programmed cell death and associated changes in antioxidants, proteases, and lipid peroxidation in wheat leaves. Biologia Plantarum. 2013;**57**(2):370-374. DOI: 10.1007/s10535-012-0286-9

[103] Fradin EF, Thomma BP. Physiology and molecular aspects of Verticillium wilt diseases caused by *V. dahliae* and *V. albo-atrum*. Molecular Plant Pathology. 2006;**7**(2):71-86. DOI: 10.1111/j.1364-3703.2006.00323.x

[104] Hou S, Zhu G, Li Y, Li W, Fu J, Niu E, et al. Genome-wide association studies reveal genetic variation and candidate genes of drought stress related traits in cotton (*Gossypium hirsutum* L.). Frontiers. Plant Science. 2018;**9**:1276. DOI: 10.3389/fpls.2018.01276

[105] Zhang JB, Wang XP, Wang YC, Chen YH, Luo JW, Li DD, et al. Genome-wide identification and functional characterization of cotton (*Gossypium hirsutum*) MAPKKK gene family in response to drought stress. BMC Plant Biology. 2020;**20**(1):1-4. DOI: 10.1186/s12870-020-02431-2

[106] Wang C, Lu W, He X, Wang F, Zhou Y, Guo X, et al. The cotton mitogen-activated protein kinase 3 functions in drought tolerance by regulating stomatal responses and root growth. Plant and

Cell Physiology. 2016;**57**(8):1629-1642. DOI: 10.1093/pcp/pcw090

[107] Li J, Meng B, Chai H, Yang X, Song W, Li S, et al. Arbuscular mycorrhizal fungi alleviate drought stress in C3 (*Leymus chinensis*) and C4 (*Hemarthria altissima*) grasses via altering antioxidant enzyme activities and photosynthesis. Frontiers in Plant Science. 2019;**10**:499. DOI: 10.3389/fpls.2019.00499

[108] Majid M, Ali M, Shahzad K, Ahmad F, Ikram RM, Ishtiaq M, et al. Mitigation of osmotic stress in cotton for the ımprovement in growth and yield through ınoculation of rhizobacteria and phosphate solubilizing bacteria coated diammonium phosphate. Sustainability. 2020;**12**(24):10456. DOI: 10.3390/su122410456

[109] Iqbal A, Dong Q, Wang X, Gui H, Zhang H, Zhang X, et al. High nitrogen enhance drought tolerance in cotton through antioxidant enzymatic activities, nitrogen metabolism and osmotic adjustment. Plants. 2020;**9**(2):178. DOI: 10.3390/plants9020178

[110] Singh M, Singh VP, Prasad SM. Responses of photosynthesis, nitrogen and proline metabolism to salinity stress in *Solanum lycopersicum* under different levels of nitrogen supplementation. Plant Physiology and Biochemistry. 2016;**109**:72-83. DOI: 10.1016/j.plaphy.2016.08.021

[111] Zahoor R, Zhao W, Abid M, Dong H, Zhou Z. Potassium application regulates nitrogen metabolism and osmotic adjustment in cotton (*Gossypium hirsutum* L.) functional leaf under drought stress. Journal of Plant Physiology. 2017b;**215**:30-38. DOI: 10.1016/j.jplph.2017.05.001

[112] Wu S, Hu C, Tan Q, Li L, Shi K, Zheng Y, et al. Drought stress tolerance mediated by zinc-induced antioxidative defense and osmotic adjustment in cotton (*Gossypium hirsutum*). Acta Physiologiae Plantarum. 2015;**37**(8):1-9. DOI: 10.1007/s1173 8-015-1919-3

[113] Zahoor R, Dong H, Abid M, Zhao W, Wang Y, Zhou Z. Potassium fertilizer improves drought stress alleviation potential in cotton by enhancing photosynthesis and carbohydrate metabolism. Environmental and Experimental Botany. 2017c;**137**:73-83. DOI: 10.1016/j.envexpbot.2017.02.002

[114] Ratnayaka HH, Molin WT, Sterling TM. Physiological and antioxidant responses of cotton and spurred anoda under interference and mild drought. Journal of Experimental Botany. 2003;**54**(391):2293-2305. DOI: 10.1093/jxb/erg251

[115] Shallan MA, Hassan HM, Namich AA, Ibrahim AA. Biochemical and physiological effects of TiO2 and SiO2 nanoparticles on cotton plant under drought stress. Research Journal of Pharmaceutical Biological and Chemical Sciences. 2016;**7**(4):1540-1551

[116] Dabbert TA, Gore MA. Challenges and perspectives on improving heat and drought stress resilience in cotton. Journal of Cotton Science. 2014;**18**(3):393-409

[117] Dever JK. Cotton breeding and agro-technology. In: Handbook of Natural Fibres. Sawston: Woodhead Publishing; 2012. pp. 469-507. DOI: 10.1533/9780857095503.2.469

[118] Ashraf M, Harris P. Breeding for abiotic stress tolerance in cotton. In: Ashraf M, Harris PJC, editors. Abiotic stresses: Plant Resistance through Breeding and Molecular Approaches. Binghamton: Food Products Press; 2005. pp. 595-613

[119] Wei Y, Xu Y, Lu P, Wang X, Li Z, Cai X, et al. Salt stress responsiveness of a wild cotton species (*Gossypium klotzschianum*) based on transcriptomic analysis. PLoS One. 2017;**12**(5):e0178313. DOI: 10.1371/journal.pone.0178313

[120] Chen Y, Feng S, Zhao T, Zhou B. Overcoming obstacles to interspecific hybridization between *Gossypium hirsutum* and *G. turneri*. Euphytica. 2018;**214**(2):1-5. DOI: 10.1007/s10681-018-2118-2

[121] Hu G, Grover CE, Yuan D, Dong Y, Miller E, Conover JL, et al. Evolution and diversity of the cotton genome. In: Rahman M, Zafar Y, Zhang T, editors. Cotton Precision Breeding. Cham: Springer; 2021. pp. 25-78. DOI: 10.1007/978-3-030-64504-5_2

[122] Fan X, Guo Q, Xu P, Gong Y, Shu H, Yang Y, et al. Transcriptome-wide identification of salt-responsive members of the WRKY gene family in *Gossypium aridum*. PLoS One. 2015;**10**(5):e0126148. DOI: 10.1371/journal.pone.0126148

[123] Oluoch G, Zheng J, Wang X, Khan MK, Zhou Z, Cai X, et al. QTL mapping for salt tolerance at seedling stage in the interspecific cross of *Gossypium tomentosum* with *Gossypium hirsutum*. Euphytica. 2016;**209**(1):223-235. DOI: 10.1007/s10681-016-1674-6

[124] Mammadov J, Buyyarapu R, Guttikonda SK, Parliament K, Abdurakhmonov IY, Kumpatla SP. Wild relatives of maize, rice, cotton, and soybean: Treasure troves for tolerance to biotic and abiotic stresses. Frontiers in Plant Science. 2018;**9**:886. DOI: 10.3389/fpls.2018.00886

[125] CICR. Abiotic Stresses in Cotton: A Physiological Approach [Internet]. 2000. CICR Tech Bull. 2:1-13. Available from: http://www.cicr.org.in/pdf/abiotic_stress.pdf [Accessed: November 16, 2021]

[126] Ahmad F, Perveen A, Mohammad N, Ali MA, Akhtar MN, Shahzad K, et al. Heat stress in cotton: Responses and adaptive mechanisms. In: Ahmad S, Hasanuzzaman M, editors. Cotton Production and Uses. Singapore: Springer; 2020. pp. 393-428. DOI: 10.1007/978-981-15-1472-2_20

[127] Moreno AA, Orellana A. The physiological role of the unfolded protein response in plants. Biological Research. 2011;**44**(1):75-80. DOI: 10.4067/S0716-97602011000100010

[128] Zafar MM, Manan A, Razzaq A, Zulfqar M, Saeed A, Kashif M, et al. Exploiting agronomic and biochemical traits to develop heat resilient cotton cultivars under climate change scenarios. Agronomy. 2021 Sep;**11**(9):1885. DOI: 10.3390/agronomy11091885

[129] Hassan IS, El-Shaarawy SA, Abou-Tour HB. Response of some Egyptian genotypes to varied climatic measurements over varied environments. In: Proceedings Beltwide Cotton Conferences; 2000; San Antonio, USA, 553-560.

[130] Abro S, Rajput MT, Khan MA, Sial MA, Tahir SS. Screening of cotton (*Gossypium hirsutum* L.) genotypes for heat tolerance. Pakistan Journal of Botany. 2015;**47**(6):2085-2091

[131] Ekinci R, Başbağ S, Karademir E, Karademir C. The effects of high temperature stress on some agronomic characters in cotton. Pakistan Journal of Botany. 2017;**49**(2):503-508

[132] Goren H. Comparison of Heat Stress Response in Cotton (*Gossypium* spp.) Genotypes in Diyarbakir Conditions.

Aydin-Turkey: Aydin Adnan Menderes University; 2017

[133] McDowell AJ, Bange MP, Tan DK. Cold temperature exposure at 10 C for 10 and 20 nights does not reduce tissue viability in vegetative and early flowering cotton plants. Australian Journal of Experimental Agriculture. 2007;**47**(2):198-207. DOI: 10.1071/ea05371

[134] Cottee NS, Tan DK, Bange MP, Cothren JT, Campbell LC. Multi-level determination of heat tolerance in cotton (*Gossypium hirsutum* L.) under field conditions. Crop Science. 2010;**50**(6):2553-2564. DOI: 10.2135/cropsci2010.03.0182

[135] Jaconis SY, Thompson AJ, Smith SL, Trimarchi C, Cottee NS, Bange MP, et al. A standardised approach for determining heat tolerance in cotton using triphenyl tetrazolium chloride. Scientific Reports. 2021;**11**(1):1-4. DOI: 10.1038/s41598-021-84798-2

[136] Saha S, Jenkins JN, Wu J, McCarty JC, Gutiérrez OA, Percy RG, et al. Effects of chromosome-specific introgression in upland cotton on fiber and agronomic traits. Genetics. 2006;**172**(3):1927-1938. DOI: 10.1534/genetics.105.053371

[137] Zhai H, Gong W, Tan Y, Liu A, Song W, Li J, et al. Identification of chromosome segment substitution lines of *Gossypium barbadense* introgressed in *G. hirsutum* and quantitative trait locus mapping for fiber quality and yield traits. PLoS One. 2016;**11**(9):e0159101

[138] Kakani VG, Reddy KR, Koti S, Wallace TP, Prasad PV, Reddy VR, et al. Differences in in vitro pollen germination and pollen tube growth of cotton cultivars in response to high temperature. Annals of Botany. 2005;**96**(1):59-67. DOI: 10.1093/aob/mci149

[139] Birrer KF, Conaty WC, Cottee NS, Sargent D, Francis ME, Cahill DM, et al. Can heat stress and water deficit affect cotton fiber wax content in field-grown plants? Industrial Crops and Products. 2021;**168**:113559. DOI: 10.1016/j.indcrop.2021.113559

[140] Ur-Rahman H. Environmental interaction, additive and non-additive genetic variability is involved in the expression of tissue and whole-plant heat tolerance in upland cotton (*Gossypium hirsutum* L). Genetics and Molecular Biology. 2006;**29**(3):525-532. DOI: 10.1590/S1415-47572006000300022

[141] Salman M, Zia ZU, Rana IA, Maqsood RH, Ahmad S, Bakhsh A, et al. Genetic effects conferring heat tolerance in upland cotton (*Gossypium hirsutum* L.). Journal of Cotton Research. 2019;**2**(1):1-8. DOI: 10.1186/s42397-019-0025-2

[142] Amjad F, Amir S, Saghir A, Khan TM, Afzal AI. Evaluation of breeding potential of cotton germplasm of Pakistan origin for fiber quality traits under heat stress. International Journal of Agriculture and Biology. 2020;**23**(2):311-318. DOI: 10.17957/IJAB/15.1290

[143] Manan A, Zafar MM, Ren M, Khurshid M, Sahar A, Rehman A, et al. Genetic analysis of biochemical, fiber yield and quality traits of upland cotton under high-temperature. Plant Production Science. 2021;**9**:1-5. DOI: 10.1080/1343943X.2021.1972013

[144] Shakeel A, Naeem M, Imtiaz AL, Allah SU, Afzal I, Saeed A, et al. Genetic mechanism controlling selected within boll yield components and physiological traits of *Gossypium hirsutum* L. under salinity stress. Turkish Journal of Field Crops. 2017;**22**(1):89-95

[145] Blum A. Breeding methods for drought resistance. In: Jones HG, Flowers TJ, Jones MB, editors. Plant under Stress. Cambridge: Cambridge University Press; 1989. pp. 197-216

[146] Xue-yan ZH, Chuan-liang LI, Jun-juan WA, Fu-Guang L, Wu-wei YE. Drought-tolerance evaluation of cotton with PEG water-stress method. 棉花学报. 2008;**20**(1):56-61

[147] Jaafar KS, Mohammed MA, Mohammed SM. Screening for drought tolerance in cotton (*Gossypium hirsutum* L.) using in vitro technique. Journal of Dryland Agriculture. 2021;7(4):52-59. DOI: 10.5897/JODA2021.0067

[148] Lambers H, Chapin FS, Pons TL. Photosynthesis. In: Plant Physiological Ecology. New York: Springer; 2008. pp. 11-99. DOI: 10.1007/978-0-387-78341-3_2

[149] Mohan MM, Narayanan SL, Ibrahim SM. Chlorophyll stability index (CSI): Its impact on salt tolerance in rice. International Rice Research Notes. 2000;**25**(2):38-39

[150] Baytar AA, Peynircioğlu C, Sezener V, Basal H, Frary A, Frary A, et al. Identification of stable QTLs for fiber quality and plant structure in Upland cotton (*G. hirsutum* L.) under drought stress. Industrial Crops and Products. 2018;**124**:776-786. DOI: 10.1016/j.indcrop.2018.08.054

[151] Magwanga RO, Lu P, Kirungu JN, Lu H, Wang X, Cai X, et al. Characterization of the late embryogenesis abundant (LEA) proteins family and their role in drought stress tolerance in upland cotton. BMC Genetics. 2018;**19**(1):1-31. DOI: 10.1186/s12863-017-0596-1

[152] Basal H, Unay A. Water stress in cotton (*Gossypium hirsutum* L.). Journal of Agricultural Faculty of Ege University. 2006;**43**(3):101-111

[153] Rodriguez-Uribe L, Abdelraheem A, Tiwari R, Sengupta-Gopalan C, Hughs SE, Zhang J. Identification of drought-responsive genes in a drought-tolerant cotton (*Gossypium hirsutum* L.) cultivar under reduced irrigation field conditions and development of candidate gene markers for drought tolerance. Molecular Breeding. 2014;**34**(4):1777-1796. DOI: 10.1007/s11032-014-0138-8

[154] Shukla RP, Tiwari GJ, Joshi B, Song-Beng K, Tamta S, Boopathi NM, et al. GBS-SNP and SSR based genetic mapping and QTL analysis for drought tolerance in upland cotton. Physiology and Molecular Biology of Plants. 2021;**27**(8):1731-1745. DOI: 10.1007/s12298-021-01041-y

[155] Anwar M, Saleem MA, Dan M, Malik W, Ul-Allah S, Ahmad MQ, et al. Morphological, physiological and molecular assessment of cotton for drought tolerance under field conditions. Saudi Journal of Biological Sciences. 2022;**29**(1):444-452. DOI: 10.1016/j.sjbs.2021.09.009

[156] Mahmood T, Ahmar S, Abdullah M, Iqbal MS, Yasir M, Khalid S, et al. Genetic potential and inheritance pattern of phenological growth and drought tolerance in cotton (*Gossypium hirsutum* L.). Frontiers in Plant Science. 2021;**12**:705392. DOI: 10.3389/fpls.2021.705392

[157] Gouia H, Ghorbal MH, Touraine B. Effects of NaCl on flows of N and mineral ions and on NO3-reduction rate within whole plants of salt-sensitive bean and salt-tolerant cotton. Plant Physiology. 1994;**105**(4):1409-1418. DOI: 10.1104/pp.105.4.1409

[158] Higbie SM, Wang F, Stewart JM, Sterling TM, Lindemann WC, Hughs E, et al. Physiological response to salt (NaCl) stress in selected cultivated tetraploid cottons. International Journal of Agronomy. 2010;**4**:643475. DOI: 10.1155/2010/643475

[159] Munawar W, Hameed A, Khan MK. Differential morphophysiological and biochemical responses of cotton genotypes under various salinity stress levels during early growth stage. Frontiers in Plant Science. 2021;**12**:622309. DOI: 10.3389/fpls.2021.622309

[160] Ashraf M, Ahmad S. Genetic effects for yield components and fibre characteristics in upland cotton (*Gossypium hirsutum* L.) cultivated under salinized (NaCl) conditions. Agronomie. 2000;**20**(8):917-926

[161] Abdelraheem A, Kuraparthy V, Hinze L, Stelly D, Wedegaertner T, Zhang J. Genome-wide association study for tolerance to drought and salt tolerance and resistance to thrips at the seedling growth stage in US Upland cotton. Industrial Crops and Products. 2021;**169**:113645. DOI: 10.1016/j.indcrop.2021.113645

[162] Zafar MM, Razzaq A, Farooq MA, Rehman A, Firdous H, Shakeel A, et al. Genetic variation studies of ıonic and within boll yield components in cotton (*Gossypium hirsutum* L.) under salt stress. Journal of Natural Fibers. 2020;**25**:1-20. DOI: 10.1080/15440478.2020.1838996

[163] Hosseini G, Thengane RJ. Salinity tolerance in cotton (*Gossypium hirsutum* L.) genotypes. International Journal of Botany. 2007;**3**(1):48-55

[164] Dehdari A, Rezai A, Maibody MM. Nuclear and cytoplasmic inheritance of salt tolerance in bread wheat plants based on ion contents and biological yield. Iran Agricultural Research. 2007;**24**(1):15-26

[165] Fan Y, Shabala S, Ma Y, Xu R, Zhou M. Using QTL mapping to investigate the relationships between abiotic stress tolerance (drought and salinity) and agronomic and physiological traits. BMC Genomics. 2015;**16**(1):43. DOI: 10.1186/s12864-015-1243-8

[166] Abdelraheem A, Fang DD, Zhang J. Quantitative trait locus mapping of drought and salt tolerance in an introgressed recombinant inbred line population of upland cotton under the greenhouse and field conditions. Euphytica. 2018;**214**(1):1-20. DOI: 10.1007/s10681-017-2095-x

[167] Turan S, Cornish K, Kumar S. Salinity tolerance in plants: Breeding and genetic engineering. Australian Journal of Crop Science. 2012;**6**(9):1337-1348

[168] Esmaeili N, Cai Y, Tang F, Zhu X, Smith J, Mishra N, et al. Towards doubling fibre yield for cotton in the semiarid agricultural area by increasing tolerance to drought, heat and salinity simultaneously. Plant Biotechnology Journal. 2021;**19**(3):462-476. DOI: 10.1111/pbi.13476

[169] Zhang L, Xi D, Li S, Gao Z, Zhao S, Shi J, et al. A cotton group C MAP kinase gene, GhMPK2, positively regulates salt and drought tolerance in tobacco. Plant Molecular Biology. 2011;**77**(1):17-31. DOI: 10.1007/s11103-011-9788-7

[170] Ma W, Zhao T, Li J, Liu B, Fang L, Hu Y, et al. Identification and characterization of the GhHsp20 gene family in *Gossypium hirsutum*. Scientific Reports. 2016;**6**(1):1-3. DOI: 10.1038/srep32517

[171] Maqbool A, Zahur M, Irfan M, Qaiser U, Rashid B, Husnain T, et al.

Identification, characterization and expression of drought related alpha-crystalline heat shock protein gene (GHSP26) from Desi Cotton. Crop Science. 2007;**47**(6):2437-2444. DOI: 10.2135/cropsci2007.03.0120

[172] Su Y, Guo A, Huang Y, Wang Y, Hua J. GhCIPK6a increases salt tolerance in transgenic upland cotton by involving in ROS scavenging and MAPK signaling pathways. BMC Plant Biology. 2020;**20**(1):1-9. DOI: 10.1186/s12870-020-02548-4

[173] Yao D, Zhang X, Zhao X, Liu C, Wang C, Zhang Z, et al. Transcriptome analysis reveals salt-stress-regulated biological processes and key pathways in roots of cotton (*Gossypium hirsutum* L.). Genomics. 2011;**98**(1):47-55. DOI: 10.1016/j.ygeno.2011.04.007

[174] Delgado C, Mora-Poblete F, Ahmar S, Chen JT, Figueroa CR. Jasmonates and plant salt stress: Molecular players, physiological effects, and improving tolerance by using genome-associated tools. International Journal of Molecular Sciences. 2021;**22**(6):3082. DOI: 10.3390/ijms22063082

[175] Wu H, Zheng L, Qanmber G, Guo M, Wang Z, Yang Z. Response of phytohormone mediated plant homeodomain (PHD) family to abiotic stress in upland cotton (*Gossypium hirsutum* spp.). BMC Plant Biology. 2021;**21**(1):1-20. DOI: 10.1186/s12870-020-02787-5

[176] Yan J, He C, Wang J, Mao Z, Holaday SA, Allen RD, et al. Overexpression of the Arabidopsis 14-3-3 protein GF14λ in cotton leads to a "stay-green" phenotype and improves stress tolerance under moderate drought conditions. Plant and Cell Physiology. 2004;**45**(8):1007-1014. DOI: 10.1093/pcp/pch115

[177] Abdelrahman M, El-Sayed M, Jogaiah S, Burritt DJ, Tran LS. The "STAY-GREEN" trait and phytohormone signaling networks in plants under heat stress. Plant Cell Reports. 2017;**36**(7):1009-1025. DOI: 10.1007/s00299-017-2119-y

Chapter 15

Influence of Soil Moisture Stress on Vegetative Growth and Root Yield of Some Cassava Genotypes for Better Selection Strategy in Screen House Conditions and Different Agro-Ecologies in Nigeria

Najimu Adetoro and Sikirou Mouritala

Abstract

Cassava is a vital staple crop for many African populations particularly in Nigeria. This study was conducted to determine the effect of soil moisture on the performance of selected 12 cassava genotypes that were evaluated for yield and related traits under three percentages of field capacity (75% – control, 50%, and 25%) in the screen house and field conditions in three agro-ecologies (Ibadan-Derived Savanna, Mokwa-Southern Guinea Savanna, and Zaria-Northern Guinea Savanna) and randomized complete block design was used. Data were collected on plant height, stem girth, number of nodes and leaves, shoot weight, stomata conductant, stay-green, fresh root weight, and dry matter percentage and were analyzed using descriptive statistics and ANOVA. Genotypes differed significantly across and within locations. The higher stress level (25% field capacity – F.C.) resulted in a more significant reduction in vegetative growth than the moderate stress level of 50% F.C.; moisture levels were uniform over time for plant height and stem girth. The response to moisture levels varied widely among genotypes, indicating that they experienced a higher stress condition. Genotypes IITA-TMS-IBA980581, IITA-TMS-IBA010040, and IITA-TMS-IBA010034 were identified with good drought tolerance. Integrating physiological research with breeding efforts will help in the selection of suitable varieties for release.

Keywords: soil moisture, cassava, agro-ecologies, root yield, selection

1. Introduction

Cassava is a major staple food and widely grown across Nigeria owing to its wide adaptability, economic importance, and acceptance both in rural and urban regions being a common staple consumed by Nigerians. It is also increasingly becoming raw materials for food, feed, and industrial applications. In 2018, worldwide production of cassava stood at about 278 million tons (t). In the same period, Nigeria produced about 60 million t [1], and Africa's total production was about 170 million t (about 56% of the world production) [1]. It is a source of calcium, vitamins B and C, and other essential minerals [2]. However, several biotic and abiotic constraints, such as drought, pests, diseases, low soil fertility, shortage of planting material, postharvest physiological deterioration, and access to markets, limit cassava production [3, 4].

A major impact of climate change is drought or water deficit, which imposes limited water environment on plants [5]. Global monitoring and analysis of climatic variables have provided evidence that the countries where cassava is cultivated are experiencing impacts of climate change [6]. Under drought conditions, water available for plant uptake for metabolic reactions falls below requirement, thus adversely impacting growth and physiological processes. The effects of water deficit on cassava plants are many and vary depending on length and intensity of drought and stage of growth of the plant [7].

Drought or water deficit remains the major impact of climate change, which imposes limited water environment on plants and seriously affected tuber yield [5]. Crops are dependent on rainfall, and so water scarcity is the primary productivity constraint in arid and semiarid tropical areas [8]. As a meteorological event, drought is a period in which the potential evaporation exceeds the rainfall. Agricultural drought is the result of water flow imbalance between the environmental demands of evapotranspiration and water transport in the soil-root system [9]. Water stress increases abscisic acid (ABA) concentration in plant, which in turn increases root resistance by affecting membrane permeability and root tuberization (**Figure 1**) [12, 13].

The morphophysiological responses to drought stress increases abscisic acid (ABA) concentration, ion transport, and the induction of the associated signaling pathway genes in plant, which in turn increases root resistance by affecting membrane permeability and root tuberization [14]. Under water deficit, cassava leaves rapidly accumulate large amounts of ABA and young leaves halt leaf expansion growth and transpiration rate decreases. Young leaves accumulate more ABA than mature leaves, but the high ABA levels under water deficit are completely reversed to control levels after one day of re-watering, corresponding with a rapid recovery of leaf area growth rate. The rapid reduction in leaf area growth and stomatal closure might be due to cassava's ability to rapidly synthesize and accumulate ABA at an early phase of a water deficit episode [15]. Plants have developed defense mechanisms, which enable them to adapt and survive under drought condition in their life cycle [16].

1.1 Biochemical and molecular mechanisms known to contribute to water-deficit stress tolerance in cassava plants

The defense strategies against drought environment also vary from different cassava cultivars. During a prolonged drought stress condition, reactive oxygen species (ROS) generate excessively and cause oxidative damage [17]. ROS can damage multiple cellular components such as proteins and lipids, and unlimited disruption will finally lead to cell death [18].

1.2 Use of crop diversity in plant breeding for drought-tolerance traits

Valuable genes from natural inter- and intraspecific diversity can be used to take advantage of several mechanisms of survival and coadaptation in plants produced by

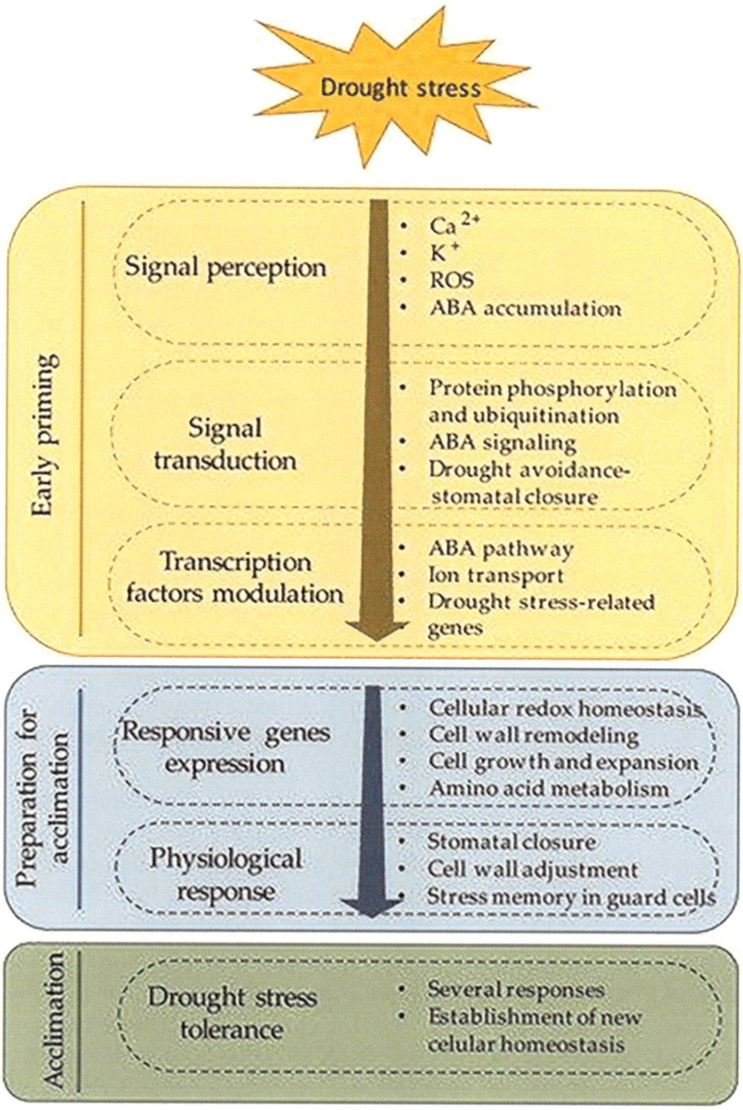

Figure 1.
General description of physiological responses of plants to drought stress conditions. ROS: Reactive oxygen species and ABA: abscisic acid [10, 11].

natural selection [19]. Some of these genes are conserved by farmers (in landraces) or are present in crop wild relatives and the narrow genetic base of modern cultivars; therefore, crop wild relatives have been extremely valuable in adapting crop varieties to changing climatic conditions [20].

Cassava grows and produces well in the Nigerian environment but shows different growth behavior and yields in different years due to differences in annual weather conditions. The water regime of an environment is an essential factor that affects the growth of crops. Differential soil water and nutrient regimes have been reported to affect yield stability in cassava [21]. Although it is incredibly tolerant to water stress, a long dry period has been reported to decrease yields [22]. Similarly, prolonged moisture deficiency leads to a reduction in growth, development, and root yields [23].

In the past, decade the International Institute of Tropical Agriculture (IITA) had developed improved varieties, which were selected from diverse source crosses; that are resistant to the crops, major pests, and diseases. As these new genotypes are developed, there is need to evaluate their performance under different moisture regimes to identify those that are stable across varying moisture environments. This study evaluates the vegetative growth parameters and root yield of selected genotypes under different moisture conditions in the screen house and three different agro-ecologies. The objectives of the study are to determine the influence of soil moisture stress on vegetative growth and root yield of selected cassava genotypes.

2. Materials and methods

Twelve genotypes (IITA-TMS-IBA010040, IITA-TMS-IBA011086, IITA-TMS-IBA011663, IITATMSIBA020131, IITA-TMS-IBA30572, IITA-TMS-IBA91934, IITA-TMS-IBA920067, IITA-TMS-IBA920326, IITA-TMS-IBA950166, IITA-TMS-IBA980510, IITA-TMS-IBA980581, and TMEB 1) were selected based on their field performance for root yield, root numbers, and dry matter and were evaluated in the screen house at IITA, Ibadan. The plants were raised in large polythene bags of 36 cm length and 156 cm circumference for 6 months under three moisture conditions: 75% (control), 50%, and 25% F.C. using the procedure of Anderson and Ingram [24].

The F.C. moisture levels at 75% (well-watered), 50% (moderate), and 25% (severe) were used to simulate stress conditions in the field. The polythene bags were filled with 86 kg of topsoil (obtained from Ibadan) to a height of 36 cm and made firm by being doubled. The soil used was classified as Ferric Luvisol with sandy-loam texture (USDA); pH (water) was 5.4, organic carbon (C) 1.26%, and total nitrogen (N) 0.12%. Available phosphorus (P) was 34.4 mg/kg, calcium (Ca) 5.8 cmol/kg, and magnesium (Mg) 0.7 cmol/kg. The experiment was laid out in a 3 by 12 factorial arrangement (moisture level × clones) in a completely randomized design (CRD) and replicated three times. Each of the replicates had a total of 36 bags in three rows of 12 bags per row. Healthy stakes of uniform length (25 cm) were planted vertically in the central portion of the bags. Holes at the bottom of the bags allowed easy draining. Following the procedure of Anderson and Ingram [24], plants were watered to field capacity for the first 4 weeks to ensure good plant establishment, after which moisture treatments were imposed by irrigation once a week with 5.58 liters for 75% F.C., 3.72 liters for 50% F.C., and 1.86 liters for 25% F.C. Measurements on vegetative traits were taken at 4, 8, 12, 16, and 20 weeks after planting (WAP) for plant height (cm), and stem girth (cm). Yield parameters were taken at 24 WAP.

These 12 genotypes were also evaluated on the field using healthy stakes of 25 cm length planted in a slanting position on ridges 30 cm high with two-thirds of the length buried in the soil. Mokwa [(Southern Guinea Savanna (SGS), Lat.9°29^{1}N and Long. 5°04^{1}E and 152 masl] and Zaria [Northern Guinea Savanna (NGS), Lat.11°11^{1}N and Long.11°78^{1}E and 610 masl]. The three locations represent different agro-ecologies with varying climatic and soil characteristics. Planting was done in each location when soil moisture was sufficient to sustain establishment. Weeding was manual at 1 month after planting (MAP), and herbicides were applied at 3, 6, and 9 MAP. Harvesting was done at 12 MAP.

2.1 Data collection

Data collected in the screen house were **Plant height** at 4, 8, 12, 16, and 20 WAP; **Stem girth** at a uniform stem length of 50 cm from ground level; **Fresh shoot weight** was obtained in kg as the fresh weight of shoots per plot; **Fresh root weight** per plot and estimated in t/ha at 24 WAP; **Number of stems per plant; Number of leaves** per plant was counted per plot and Screening genotypes for resistant to Cassava Mosaic Disease (CMD), Cassava Bacteria Blight (CBB), and Cassava anthracnose (CAD). The incidence and severity of genotypes to the African cassava mosaic disease (CMD) were evaluated at 1, 3, and 6 months after planting (MAP) at 3 and 6 MAP for cassava bacterial blight (CBB) and CAD at 6 and 9 MAP; since the symptom would not have expressed at 1 month after planting. Disease incidence was taken as proportion of plants units that are visibly diseased relative to total number of plants, while disease severity = volume of plant parts affected compared with the whole plant unit.

Disease incidence = Number of infected plants/plot/Total number of plants/plot.
Severity of genotypes to CMD was recorded based on a scale of 1–5 [25].

Leaf Chlorophyll: The leaf chlorophyll contents of four selected leaves in each plot were measured at 4, 8, 12, 16, and 20 WAP using Chlorophyll Meter Model SPAD-502 (Minolta Co. Ltd. Japan).

Leaf stomata conductance: A steady-state porometer (Licor Instrument Corporation, Model Li-1600) was used to measure diffusive resistance and transpiration rate on the abaxial surface of the uppermost fully expanded leaves of four plants per plot. The sensor head with a narrow leaf aperture (LI 1600–01) with an area of 1cm^2 was used. Measurement was taken when sun was not too low or high in the morning (900–1100 h) and afternoon (1330–1530 h) on a clear sunny day at 3, 6, and 9 months after planting (MAP) in all locations [26].

The dried sample was weighed, and root dry matter percentage was calculated as follows.

$$\text{Percent root cortex DM} = \text{dry weight/Fresh weight} \times 100$$

***Gari* production**: Cassava roots were converted to fermented roasted granules called *Gari*. Ten kilograms of roots were taken from each genotype harvested, washed with water, and grated with a grating machine. The pulp was put in a jute bag, and pressure was exerted on it to remove water from the pulp. Dewatering took up to 3 days and the pulp was also undergoing fermentation while being dewatered. The pulp was sieved to remove chaff and toasted in a pot until gelatinized grains were formed.

2.2 Statistical analysis

All data obtained were analyzed using the Statistical Analytical System (SAS) (9.2 version). The model used was the factorial arrangement in a CRD. Least Significant Difference (LSD) at 5% probability was used to separate mean squares.

Sources of variation	DF	Plant height	Stem girth	Fresh Root weight	Leaf number	Fresh Shoot weight	Stem number	Leaf chlorophyll
Rep(R)	2	84.15^{ns}	0.02^{ns}	14.81^{ns}	127.86^{ns}	399.98^{ns}	0.18^{ns}	7.95^{ns}
Trt (T)	2	16492.35***	0.44***	16336.57***	3412.78***	19628.18***	3.13**	505.89***
Rep×Trt	4	722.37^{ns}	0.01^{ns}	302.34^{ns}	39.16^{ns}	3367.64*	0.65^{ns}	5.03^{ns}
genotype	12	2758.90***	0.08**	642.83^{ns}	245.47**	1593.05^{ns}	2.60***	27.10***
Trt × genotype	22	311.85^{ns}	0.05^{ns}	449.85^{ns}	92.22^{ns}	1610.94^{ns}	0.63^{ns}	12.77*
Error	65	408.34	0.04	530.91	88.86	1227.67	0.43	6.57
R^2		0.74	0.54	0.59	0.68	0.58	0.66	0.80

****, **, * means significant at $p \geq (0.001, 0.01, 0.05)$ and ns means not significant.*

Table 1.
Pooled analysis of variance for yield and yield related traits of 12 cassava genotypes evaluated in the screen house Ibadan Nigeria.

3. Results

3.1 Performance of selected genotypes in different moisture conditions in the screen house and field environments

Mean squares (MS) from the combined analysis of variance (ANOVA) for root and shoot characters of the twelve cassava genotypes in the screen house at Ibadan are presented in **Table 1**. The result shows highly significant ($p < 0.001$) mean squares (MS) for all sources of variations. However, MS for all traits studied were significant for more than one source of variations. The mean squares (MS) for treatment were highly significant so also mean square for except for fresh root weight and fresh shoot weight. Whereas the interaction between replicate and the treatments was not significant except for fresh shoot weight as well as the interaction between treatment and genotype except for leaf chlorophyll. But the mean square for replicate was not significant for all the traits studied. Mean values at 75% and 25% FC differ significantly for all parameters at 24 WAP, except for leaf number, leaf chlorophyll content, and fresh root weight.

When mean values at 25% FC were compared with mean values at 75% FC, it was observed that mean plant height decreased by 29.33%, stem girth by 17.32%, root weight by 61.76%, and shoot weight by 29.27% at harvest (**Table 2**). There was no significant difference in mean values at 75% and 50% FC for plant height at harvest

	Plant	% diff.	Stem	% diff.	Root	% diff.	Shoot	% diff.
	height		girth		weight		weight	
TRT	(cm)		(cm)2		(Kg)		(kg)	
75%FC (T3)	144.98		1.27		0.068		0.164	
50%FC (T2)	127.93	11.7	1.2	5.5	0.044	35.3	0.142	13.41
25% FC (T1)	102.46	29.33	1.05	17.32	0.026	61.76	0.116	29.27

*% diff = (T3-T1/T3) *100.*

Table 2.
Mean and percentage differences of 12 cassava genotypes in the greenhouse at different moisture levels for different traits at 24WAP.

24WAP. Similar growth pattern was observed for plant height and stem girth over time at 75% and 50% FC (**Table 3**).

IITA-TMS-IBA010040 had the highest fresh root weight of 70.2 g, followed by IITA-TMS-IBA920326(55.4 g) and IITA-TMS-IBA980581 had fresh root weight of 50.9 g; while IITA-TMS-IBA30572 had the least fresh root weight of 38.4 g (**Table 4**).

Treatment	pltht	stmgrth	stmno	lfno	lfchlorph	rtfrhwt	shtfrhwt
High water treatment (75%)	145.0^{a}	1.3^{a}	1.7^{a}	43.3^{b}	33.4^{b}	68.7^{a}	116.5^{b}
Low water treatment (25%)	102.5^{b}	1.1^{b}	1.7^{b}	30.8^{c}	39.3^{a}	26.3^{c}	142.1^{ab}
Medium water treatment (50%)	127.9^{a}	1.2^{a}	2.2^{a}	50.1^{a}	40.4^{a}	44.1^{b}	164.6^{a}

Same numbers are not significant from each other while numbers together mean they are over lapse and a separate number is significantly different.
pltht: plant hight, stmgrth: stem girth, lfchorph: leaf chlorophyll, rtfrhwt: root fresh weight, shtfrhwt: shoot fresh weight, stmno: stem number.

Table 3.
Mean differences of 12 cassava genotypes in the screen house at different moisture levels for different traits.

clone	mpltht	mstmgrth	mlfno	mlfchlorph	mrtfrhwt	mshtfrhwt	mstmno	Genotype mean
010040	109.3	1.2	37.9	37.3	70.2	140.4	1.5	56.8
011086	108.6	1.1	39.0	39.9	41.3	107.1	1.4	48.3
011663	149.4	1.2	34.8	38.4	40.9	142.9	1.2	58.4
020131	133.2	1.2	43.2	34.8	41.9	146.8	1.7	57.5
30572	144.0	1.2	46.9	39.7	38.4	145.8	1.8	59.7
91934	102.8	1.0	52.0	38.8	45.2	151.6	2.6	56.3
920067	140.4	1.1	36.9	38.3	41.0	131.0	1.7	55.8
920326	106.1	1.1	48.4	38.0	55.4	150.4	3.2	57.5
010034	102.9	1.2	42.3	37.1	42.3	141.9	2.0	52.8
980510	121.8	1.4	42.0	38.9	44.1	126.2	1.6	53.7
980581	141.4	1.2	36.9	34.1	50.9	152.2	1.9	59.8
TME 1	141.6	1.3	36.8	37.3	44.8	156.3	1.8	60.0
Mean	125.1	1.2	41.4	37.7	46.4	141	1.9	56.4
SE	5.3	0	1.6	0.5	2.6	4	0.2	2.0
Min	102.8	1	34.8	34.1	38.4	107.1	1.2	45.6
Max	149.4	1.4	52	39.9	70.2	156.3	3.2	67.5
CV(%)	16.2	17	22.8	6.8	49.7	24.8	35.1	24.6
Pr. > F	***	*	**	***	ns	ns	***	
LSD(0.)	24.25	0.24	11.31	3.08	27.65	42.05	0.79	7.9

****, **, * means significant at* $p \geq (0.001, 0.01, 0.05)$ *and ns means not significant.*
Mpltht: mean plant hight, mstmgrth: mean stem girth, mlfchorph: mean leaf chlorophyll, mrtfrhwt: mean root fresh weight, mshtfrhwt: mean shoot fresh weight, mstmno: mean stem number.

Table 4.
Overall means of yield and yield related traits of 12 cassava genotypes evaluated in the screen house at 24 WAP.

The mean reduction in genotypes performance at moisture stress levels of 25% and 50% FC ranged from 15.9% (30572) to 44.3% (010040) for plant height, between 7.7% (020131) and 41.6% (920326) for stem girth, between 40.8% (011086) and 85.7% (30572) for root weight, and between 12.8% (TMEB 1) and 63.9% (011086) for shoot weight. The mean plant height was 125.1 cm with 011663 having the tallest while 980510 and 91934 being the shortest 102.8 cm. Stem girth ranged from 1 cm (91934) to 1.4 cm (980510); leaf number ranged from 34.8(011663) to 52 cm (91934). The coefficient of variation for the traits was quite low (**Table 5**).

3.2 Overall fields disease means scores, yield, and yield-related traits of 12 cassava genotypes evaluated for 2 years at three locations in Nigeria

There was significant ($P \leq 0.05$) difference in genotypes performance with respect to stay green, mean diseases score, fresh root yield, harvest index, root yield and size, shoot weight, and gari yield except for dry matter content. 80% of tested genotypes performed better than the checks in terms of stay green, mean diseases score, fresh root yield, and shoot weight while 70% and 30% of the genotypes outperformed checks with respect to root size and gari weight. The genotypes with outstanding field performance across the three locations were IBA980581, IBA010034, and IBA010040 (**Table 6**).

3.3 Morphological and physiological field performance of 12 cassava genotypes evaluated at three locations in Nigeria for 2 years

There is significant difference ($p \geq 0.05$) in genotypes morphological and physiological traits except for stem girth. More than half of tested genotypes perform better than checks in all traits except for stomata, level of branching, and number of leaves. The genotypes that recorded highest and least score for plant height were (IBA920067, IBA91934), number of nodes (IBA980510, IBA91934), chlorophyll content (IBA011663, IBA980581), and Leaf Area (IBA011663, TME 1). Genotypes IBA011663, IBA980510, IBA010040, IBA010034, IBA30572 were identified for being stable across three locations for physiological and morphological traits (**Table 7**).

3.4 Location means recorded by 12 genotypes evaluated for 2 years at three locations in Nigeria

Ibadan location recorded the highest score for level of branching, chlorophyll content, plant height, gari weight, number of leaf, fresh root yield, harvest index with least score for cassava bacteria blight severity (CBBS), Mokwa location had highest mean score for shoot weight, number of nodes, stem girth, stomata, and least mean score for cassava anthracnose disease severity (CADS) while Zaria location recorded highest mean score for stay green, dry matter content, and least mean score for cassava mosaic disease severity score (CMDS). Zaria and Mokwa location recorded the same mean score for Harvest Index while Ibadan and Mokwa location also recorded the same score for root size (**Table 8**).

3.5 The correlation coefficients for traits measured for 2 years at three locations in Nigeria

Most of the traits show significant correlation *inter se*. Notably, Fresh root yield was positively correlated garri, harvest index, and cassava mosaic disease, Fresh root

Genotypes	Plant Height (cm)				Stem girth (cm^2)				Root weight (Kg)				Shoot weight (Kg)			
	T1	T2	T3	% diff.	T1	T2	T3	% diff.	T1	T2	T3	% diff.	T1	T2	T3	% diff.
010040	76.5	114.1	137.3	44.3	0.9	1.3	1.3	30.8	0.055	0.049	0.105	47.6	0.114	0.14	0.166	31.3
011086	85.5	122.7	117.6	27.3	0.9	1.3	1.3	30.8	0.04	0.041	0.042	40.8	0.056	0.109	0.155	63.9
011663	127.7	142.6	177.9	28.2	1	1.3	1.2	16.7	0.02	0.044	0.058	65.5	0.103	0.152	0.172	40.1
020131	107.5	142.4	149.8	28.2	1.2	1.2	1.3	7.7	0.02	0.042	0.062	67.7	0.121	0.152	0.166	27.1
30572	129.5	148.6	154	15.9	1.1	1.2	1.2	8.3	0.01	0.034	0.07	85.7	0.145	0.111	0.181	19.9
91934	83.6	100.8	124.1	32.6	1	0.9	1.1	9.1	0.025	0.035	0.074	66.2	0.099	0.186	0.169	41.4
920067	110.8	155.5	154.7	28.4	1	1	1.3	23.1	0.022	0.034	0.066	66.7	0.091	0.138	0.162	43.8
920326	87.1	106.4	124.7	30.2	0.7	1.4	1.2	41.6	0.024	0.07	0.071	66.2	0.139	0.142	0.169	17.8
010034	88.8	92.8	127.2	30.2	1	1.1	1.4	28.6	0.023	0.031	0.072	68.1	0.116	0.142	0.166	30.1
980510	104.3	121.1	139.8	25.4	1.3	1.3	1.5	13.3	0.027	0.033	0.071	70	0.08	0.129	0.168	52.4
980581	107.5	160.1	156.6	31.4	1	1.3	1.2	16.7	0.019	0.072	0.06	68.3	0.135	0.155	0.165	18.2
TMEB1	120.7	127.9	176	31.4	1.2	1.2	1.4	14.3	0.027	0.038	0.068	62.3	0.143	0.16	0.164	12.8
Means	102.46	127.93	144.98		1.05	1.2	1.27		0.03	0.04	0.07		0.12	0.14	0.16	
SE	5.19	6.37	5.76		0.04	0.04	0.04		0.003	0.003	0.004		0.009	0.005	0.003	
LSD	25.88	35.3	39.29		0.31	0.64	0.25		33.06	48.12	33.01		64.76	34.65	71.4	

*T1 = 25% FC, T2 = 50% FC, T3 = 75% FC, % diff = (T3-T1/T3) *100.*

Table 5.
Overall mean performance and percentage difference among 12 cassava genotypes under different moisture conditions in the greenhouse for four traits.

Genotypes	S.G.	mean diseases score	FYLD(t/ha)	HI	DM (%)	nrt	rtsz	htwt (kg)	*Gari* wt (kg)	Rank_Total	Final rank
IITA-TMS-IBA980581	2.5(8)	1.97(7)	22.32(1)	0.5	32(1)	86.7(4)	5.9(2)	32.5(1)	1.7(2)	21	1
IITA-TMS-IBA010034	3(3)	2(8)	21(3)	0.5	32(1)	88(3)	5.8(4)	30.8(2)	1.9(1)	31	2
IITA-TMS-IBA010040	2.2(12)	1.87(5)	21.6(2)	0.5	29(8)	76.7(6)	5.9(2)	29(6)	1.4(4)	34	3
IITA-TMS-IBA920067	2.3(11)	1.67(1)	17.4(10)	0.5	31(3)	76.1(7)	5.6(5)	24.4(11)	1.2(9)	48	4
IITA-TMS-IBA020131	2.7(5)	1.9(6)	18.07(7)	0.4	31(3)	68.4(10)	5.4(9)	30.7(3)	1.5(3)	48	5
IITA-TMS-IBA011086	2.7(5)	1.73(2)	18.68(5)	0.5	27(10)	95.7(1)	5(12)	30.2(4)	1.2(9)	50	6
IITA-TMS-IBA980510	2.5(8)	1.80(3)	18.05(8)	0.5	27(10)	82.5(5)	5.3(10)	27.5(8)	1.1(11)	55	7
IITA-TMS-IBA91934	3(3)	2.4(12)	18.13(6)	0.5	30(5)	75.4(8)	6.2(1)	25.4(10)	1.4(4)	55	8
IITA-TMS-IBA920326	2.6(7)	2.07(9)	19.6(4)	0.5	29(8)	73.3(9)	5.6(6)	28.5(7)	1.3(8)	57	9
IITA-TMS-IBA30572	3.1(2)	2.23(11)	17.65(9)	0.5	30(5)	88.6(2)	5.1(11)	25.8(9)	1.4(4)	62	10
IITA-TMS-IBA011663	2.5(8)	1.83(4)	15.02(11)	0.4	24(12)	61(11)	5.6(7)	29.6(5)	1.1(11)	64	11
TMEB1	4.7(1)	2.2(10)	13.48(12)	0.5	30(5)	53.8(12)	5.5(8)	20(12)	1.4(4)	75	12
Mean	2.8	1.9	18.3	0.5	29	74.3	5.6	27.8	1.4		
Min	2.2	1.67	13.48	0.4	24	53	5	20	1.1		
Max	4.7	2.4	22.32	0.5	32	95.7	6.3	32.5	1.7		
SE	0	0.06	0.67	0	3	3	0.1	0.98	0.05		
CV	14	16.5	28.8	59	13	26.8	17	27.3	27.6		
Pr. > F	*	***	***	**	ns	***	***	**			

***, **, * *means significant at* $p \geq (0.001, 0.01, 0.05)$ *and* ^{ns} *means not significant, Number in parenthesis represents the rank/position of each genotype, S.G.: Stay-green, FYLD: Fresh root yield, nrt: number of roots harvested, rtsz: root size, shtwt: shoot weight, HI: harvest index,*

Table 6.
Overall disease mean scores, yield, and yield-related traits of 12 cassava genotypes evaluated for 2 years at three locations in Nigeria.

Genotypes	Pltht (cm)	Stem girth (cm)	Nnode	Stomata	Chlorophyll content	Lbrch	Nleaf	LA (m^2)	Rank_ Total	Final Rank
IITA-TMS-IBA920067	100 (4)	8.7(1)	40.6 (11)	125(6)	36.7(4)	1.2 (10)	34 (12)	0.01 (6)	54	7
TMEB1	103 (8)	7.7(4)	41.7 (10)	138.7(1)	34.6(9)	1.1 (12)	36.7 (11)	0.01 (6)	61	12
I IITA-TMS-IBA920326	94.2 (10)	7.5(7)	45(6)	127.8(4)	35.5(5)	1.2 (10)	38.4 (10)	0.01 (6)	58	9
IITA-TMS-IBA980581	105 (6)	7.6(6)	45(6)	132.6(2)	31.1(12)	1.4(8)	40.6 (9)	0.01 (6)	55	10
IITA-TMS-IBA91934	87.3 (12)	6.3(12)	38.2 (12)	127.5(5)	37.3(2)	1.8(3)	41(8)	0.01 (6)	60	11
IITA-TMS-BA980510	109 (3)	7.7(4)	50.5 (1)	123.5(7)	37.3(2)	1.8(3)	49(2)	0.01 (6)	28	1
IITA-TMS-IBA30572	92.1 (11)	8.6(2)	44.4 (9)	120.5 (10)	35(7)	2(1)	49.3 (1)	0.01 (6)	47	2
IITA-TMS-IBA010034	110 (2)	7.4(9)	46.8 (4)	120.2 (11)	35(7)	1.7(6)	48.3 (4)	0.012 (4)	47	2
IITA-TMS-IBA010040	104 (7)	7.4(9)	48.8 (2)	122.1(9)	33.4(10)	1.8(3)	48.8 (3)	0.012 (4)	47	2
IITA-TMS-IBA020131	109 (3)	7.5(7)	45.1 (5)	114.4 (12)	33.2(11)	1.7(6)	43.5 (7)	0.013 (3)	54	7
IITA-TMS-IBA011086	132 (1)	6.9(11)	44.5 (8)	123.2(8)	35.4(6)	1.9(2)	48.1 (5)	0.014 (2)	42	6
IITA-TMS-IBA011663	109 (3)	7.8(3)	47.7 (3)	132.4(3)	38(1)	1.2(9)	45.1 (6)	0.015 (1)	28	1
Grand Mean	105	7.59	44.9	125.7	35.21	1.6	43.6	0.01		
Min	87.3	6.3	38.2	114.4	31.1	1.1	34	0.01		
Max	132	8.7	50.5	138.7	38	2	49.3	0.02		
SE	2.97	0.15	0.97	2.42	0.55	0.1	0.85	0		
CV	18.2	18.97	12.1	14.4	6.85	20.5	17.67	51.9		
Pr. > F	*	ns	**	*	***	*	***	*		

****, **, * means significant at $p \geq (0.001, 0.01, 0.05)$ and ^{ns} means not significant, Number in parenthesis represents the rank/position of each genotype Nnode: number of nodes, Lbrch: level of branching, Nleaf: number of leaves, LA: leaf area.*

Table 7.
Overall means morphological and physiological traits of 12 cassava genotypes evaluated for 2 years in three locations in Nigeria.

yield was however negatively correlated to cassava bacterial blight (−0.28) and cassava anthracnose disease (−0.29).

Chlorophyll content was positively correlated to fresh root yield (0.25), but negatively correlated to dry matter (−0.14).

Stomata had significantly negative correlation with harvest index (−0.11) and number of root (−0.15).

Shoot weight correlated with chlorophyll content (0.19), and fresh root yield (0.79) but had negative correlation with dry matter (−0.17) (**Table 9**).

Traits	Location means			Across mean
	Ibadan	Mokwa	Zaria	
Level of branching	2 (1)	1.6 (2)	1.4 (3)	1.60
Chlorophyll	36.5 (1)	36.1 (2)	33.5 (3)	35.40
Stomata	120.7 (3)	133.7 (1)	129.1 (2)	127.80
Stem girth	4.4(2)	14.3 (1)	3.4 (3)	7.30
Plant height (cm)	134.1 (1)	115.1 (2)	68.5 (3)	105.90
Number of nodes	49.7 (2)	54.6 (1)	29.9(3)	44.70
Gari weight (kg)	2.1 (1)	0.9 (3)	1.5 (2)	1.50
Number of leaves	86 (1)	29.7 (2)	24.5 (3)	46.70
CMDS	1.6 (3)	1.4 (2)	1.1 (1)	1.40
CBBS	1.8 (1)	2.1 (2)	2.6 (3)	2.20
CADS	2.2 (2)	0 (1)	2.3 (3)	1.50
Fresh root yield (t/ha)	28.9 (1)	20.3 (2)	8 (3)	19.10
Harvest index	0.6 (1)	0.5 (2)	0.5 (2)	0.50
Shoot weight (kg)	34 (2)	35.7 (1)	12.2 (3)	27.30
Stay-green	2.7 (2)	2.7 (2)	2.9 (1)	2.80
Root size	6 (1)	6 (1)	5 (3)	5.70
Dry matter content (%)	29.7 (2)	26.7 (3)	33.9 (1)	30.10
Number of roots	96.8 (2)	97.7 (1)	42.4 (3)	79.00

Table 8.
Overall mean location values for morphological, physiological, and yield traits at three locations for 2 years in Nigeria.

4. Discussions

As an important environmental limitation, drought has become a rising concern due to its harm to the development and productivity of crop plants [27]. Cassava is a major staple food to resource-limited people in marginal areas because of its ability to survive and produce in such poor land with infrequent rainfall and low fertility [28]. The present study by using 12 cassava genotypes can provide a fundamental basis for the identification of drought-tolerant germplasm resources.

Plants are known to respond to water deficit with some adjustment at morphological, physiological, cellular, and metabolic levels. These responses are, however, dependent upon the duration and severity of stress, the type of genotype/the stage of development, and the organ and cell in question [29]. Reduction in photosynthesis results in the inability of the genotypes to produce tuberous roots compared with conditions with relatively higher moisture levels. Any factor in the plant's environment that is not the optimum, being either deficient or in excess, will limit plant growth. Moisture stress at 25% F.C. in the screen house led to a reduction in root weight by over 61%, and this is indicative that severe moisture stress that occurs within a period of 8–24 WAP can lead to a very high level of yield loss [30]. Porto [30] also reported that water stress from 4 to 20 WAP led to a reduction of storage root

	mcbbs	Mcmds	Mcads	chlrphy	Stomata	stgth	pltht	Nnode	Nleaf	*Gari* wt	FYLD	HI	DM	shtwt	Rtno	Rtsz
mcbbs	1	-0.34^{***}	-0.14^{**}	-0.14^{***}	-0.01^{***}	-0.02^{ns}	-0.31^{***}	-0.15^{***}	-0.57^{***}	-0.39^{***}	-0.28^{***}	-0.23^{***}	0.27^{***}	-0.19^{***}	-0.06^{ns}	1
mcmds	-0.32^{***}	1	0.19^{***}	0.03^{ns}	0.01^{ns}	-0.05^{ns}	0.18^{***}	0.02^{ns}	0.52^{***}	0.29^{***}	0.13^{***}	0.12^{**}	-0.24^{***}	0.08^{ns}	-0.02^{ns}	-0.32^{***}
mcads	0.19^{***}	-0.06^{ns}	1	-0.049^{ns}	-0.455^{***}	0.19^{***}	-0.03^{ns}	-0.54^{***}	0.29^{***}	0.01^{ns}	-0.29^{***}	-0.01^{ns}	-0.05^{ns}	-0.21^{***}	-0.28^{***}	0.19^{***}
chlrphy	0.09^{ns}	-0.02^{ns}	-0.19^{***}	1	-0.15^{***}	0.19^{***}	0.23^{***}	0.11^{**}	0.08^{ns}	-0.18^{***}	0.25^{***}	0.05^{ns}	-0.14^{***}	0.19^{***}	0.19^{***}	0.09^{ns}
stomata	0.11^{***}	-0.28^{***}	0.08^{ns}	0.27^{***}	1	-0.11^{***}	0.01^{ns}	0.55^{***}	-0.32^{***}	0.26^{***}	-0.04^{ns}	-0.11^{**}	-0.04^{ns}	0.029^{ns}	-0.15^{***}	0.11^{***}
stgth	0.12^{***}	0.19^{***}	-0.21^{***}	-0.24^{***}	-0.23^{***}	1	0.24^{***}	0.25^{***}	-0.1^{*}	-0.63^{***}	0.24^{***}	-0.12^{**}	-0.16^{***}	0.55^{***}	0.49^{***}	0.12^{***}
pltht	0.18^{***}	-0.15^{***}	0.19^{***}	-0.05^{ns}	0.12^{**}	-0.28^{***}	1	0.41^{***}	0.68^{***}	0.18^{**}	0.49^{***}	0.09^{*}	-0.25^{***}	0.54^{***}	0.37^{***}	0.18^{***}
Nnode	0.24^{***}	0.49^{***}	0.029^{ns}	-0.14^{***}	-0.01^{ns}	0.13^{***}	-0.39^{***}	1	-0.17^{***}	0.07^{ns}	0.44^{***}	-0.08^{*}	-0.19^{***}	0.56^{***}	0.33^{***}	0.24^{***}
LA	0.22^{***}	0.37^{***}	0.55^{***}	-0.04^{ns}	0.05^{ns}	-0.29^{***}	0.29^{***}	-0.57^{***}	-0.22^{***}	0.18^{**}	0.09^{*}	-0.01^{ns}	0.01^{ns}	0.31^{***}	0.06^{ns}	0.22^{***}
Nleaf	-0.01^{ns}	0.33^{***}	0.54^{***}	-0.16^{***}	-0.11^{**}	0.25^{***}	0.01^{ns}	0.52^{***}	1	0.58^{***}	0.46^{***}	0.31^{***}	-0.35^{***}	0.26^{***}	0.12^{**}	-0.01^{ns}
Gari wt	0.40^{***}	0.06^{ns}	0.56^{***}	-0.25^{***}	-0.12^{**}	-0.04^{ns}	-0.18^{***}	0.29^{***}	-0.15^{***}	1	0.23^{***}	0.35^{***}	0.40^{***}	-0.09^{ns}	-0.12^{*}	0.40^{***}
Fyld	0.17^{**}	0.12^{**}	0.31^{***}	-0.19^{***}	0.09^{*}	0.24^{***}	0.26^{***}	0.08^{ns}	0.02^{ns}	-0.31^{***}	1	0.43^{***}	-0.09^{***}	0.79^{***}	0.77^{***}	0.17^{**}
HI	0.43^{***}	-0.12^{*}	0.26^{***}	0.01^{ns}	-0.08^{*}	0.49^{***}	-0.63^{***}	-0.32^{***}	-0.54^{***}	0.18^{***}	-0.02^{ns}	1	0.05^{ns}	-0.06^{ns}	0.24^{***}	0.43^{***}
DM	0.26^{***}	0.77^{***}	-0.09^{ns}	-0.35^{***}	-0.01^{ns}	0.44^{***}	0.18^{**}	-0.1^{*}	0.11^{**}	-0.03^{ns}	-0.05^{ns}	-0.01^{***}	1	-0.17^{***}	-0.02^{ns}	0.26^{***}
shtwt	-0.27^{ns}	0.24^{***}	0.79^{***}	0.40^{***}	0.31^{***}	0.09^{*}	0.07^{ns}	0.68^{***}	0.55^{***}	0.23^{***}	0.19^{***}	0.01^{ns}	-0.14^{***}	1	0.78^{***}	-0.27^{ns}
rtno	0.39^{***}	-0.02^{ns}	-0.06^{ns}	-0.09^{***}	0.35^{***}	0.46^{***}	0.18^{**}	-0.17^{***}	0.25^{***}	0.01^{ns}	0.19^{***}	-0.455^{***}	0.03^{ns}	-0.14^{**}	1	0.39^{***}
rtsz	1	0.78^{***}	-0.17^{***}	0.05^{ns}	0.43^{***}	0.23^{***}	0.58^{***}	-0.22^{***}	0.41^{***}	0.24^{***}	-0.11^{***}	-0.15^{***}	-0.049^{ns}	0.19^{***}	-0.34^{***}	1

***,**,*: significant at ($p \le 0.001$, 0.01 and 0.05) while ns represents not significant at ($p \le 0.05$), cbbs: cassava bacteria blight mean severity, cassava mosaic disease severity, cads: cassava anthracnose disease severity, chlrphy: chlorophyll, stmata: stomata, stmgrt: stem-girth, ptlht: plant height, LA: leaf area, fyld: fresh yield, Nleaf: Number of leaves, HI: harvest index, DM: Dry Matter, Nroot: Number of roots, rtsz: root size, shtwt: shoot weight, Gari *wt:* Gari *weight.*

Table 9.
Correlation coefficient for morphological, physiological, yield, and related traits of 12 cassava genotypes evaluated for 2 years at three locations in Nigeria.

yield by 32–60%. Shoot weight at 25% F.C. was also reduced by 22.4%. Reduction in plant height was 29.3%, leaf chlorophyll content was reduced by 17.7%, leaf number had a reduction of 28.9%, whereas stem girth had a reduction of 15.4%. The implication is that moisture stress could hinder the manifestation of the genetic potentials of cassava. Therefore, a variety that is the best for a trait offers the opportunity (genetic base) for improving such trait through selection or hybridization and further selection. Even with significant differences in the traits that should contribute to root yield, fresh root weight was still not significantly different among genotypes. This also goes along with the result of the screen house experiment by [7]. The presence of the tonoplast sugar transporter (for roots and tubers) to transport the assimilates from the source to the sink (roots) enhances root formation and development. If not, the yield would not be significantly different as shown in this result. Ludewig and Flügge [31] and Cho *et al.* [32] also report similar observations. There should be continued work on several of the implied and not yet confirmed transport steps within plants, which frequently turn out to be the rate-limiting step to production of valuable compounds in storage sinks.

Plants respond to drought conditions either by increasing or decreasing their root growth.

Relatively higher stomata conductance observed in most genotypes in this study implies that the genotypes would have high photosynthetic potential although not translated to the highest levels of dry matter and root yield. The essence is that this attribute alone might not be a strong contributor to high yield under water-limited conditions. Measurements of leaf diffusive resistance reflected bulk resistance to water loss, combining activities of both the stomata and genotypes. In the dry savannas, genotypes with the ability to optimize water use are desirable due to limited availability of cultivable soil during the dry season [33]. A major component parameter associated with high Water Use Efficiency (WUE) is rapid stomata closure. Generally, stomata resistance of all the genotypes was low in the morning but increased in the afternoon and showed an increase in response to light and vapor pressure deficit (VPD) [34]. The morphophysiological responses to drought stress increases abscisic acid (ABA) concentration, ion transport, and the induction of the associated signaling pathway genes in plant, which in turn increases root resistance by affecting membrane permeability and root tuberization [14]. The rapid reduction in leaf area growth and stomatal closure might be due to cassava's ability to rapidly synthesize and accumulate ABA at an early phase of a water deficit episode [15].

Plants obtained energy and virtually all their structural materials by photosynthesis and the leaves are the main photosynthetic organ. Jarvis and Morison [35] and Akparobi *et al.* [36] had similar results. This may indicate that the higher the number of photosynthetic organs, the higher the photosynthetic rate. In this study, correlation between leaf number and root yield was positive and significant. Also, chlorophyll, harvest index, and plant height correlated positively with leaf number.

However, genotypes IITA-TMS-IBA30572 and IITA-TMS-IBA980510 with higher leaf numbers yielded less than IITA-TMS-IBA980581, which had fewer leaves. High leaf number in these genotypes is largely due to their branching habit. Shoot apices and storage roots compete for available carbohydrate in cassava [37, 38]. The high number of branches (level 2) may have caused higher competition for available carbohydrate in IBA30572 than in the other improved genotypes, which probably reduced the root yield in the former genotype despite its high number of leaves. Experiments have shown that production (formation) and growth (elongation) of cassava's adventitious and lateral roots are suppressed by deficit in soil moisture [7, 39].

The higher leaf area observed in genotype IITA-TMS-IBA011663 did not translate to higher root and dry matter yield. This did not go along with the results of Enyi [37] and [40], who observed that high yielding cultivars had high leaf area but also validated the findings of (Ludewig and Flügge [31] and Cho *et al.* [32] that a plant with a good number of stems, better plant height, enough leaf chlorophyll content without the presence of the tonoplast sugar transporter (for roots and tubers) that transports the assimilates from the source to the sink (roots) will not translate to high yields as shown in this result. This might suggest that other parameters of canopy function such as leaf, carbon fixation, and assimilate use and partition could interact significantly in yield formation. Among the improved genotypes, IITA-TMS-IBA011663 possibly partitioned the highest amount of dry matter into leaves, petioles, stems, and fibrous roots and was not the highest in root yield; thus, partitioning of dry matter partially explained the lower yield of IITA-TMS-IBA011663 than in other improved genotypes even though its L.A. was higher.

A better value was observed for stay-green in the improved genotypes IITA-TMS-IBA010040 and IITA-TMS-IBA980581 compared with the value in the local variety TME 1. This implies that the improved genotypes would be expected to be more efficient in supporting photosynthesis, stomata conductance, and carbon fixation. The better stay-green values for IITA-TMS-IBA010040 and IITA-TMS-IBA980581 may be a contributor to their root yield being the best compared with TMEB1, which had the overall worst root yield and the worst stay-green value. However, this finding does not go along with the finding by Oluwafemi *et al.* [41], which reported that the local variety with high stay-green yielded better than the improved ones.

The lower values that were observed in CMD for new improved genotypes IITA-TMS-IBA980581, IITA-TMS-IBA010040, and IITA-TMS-IBA010034 compared with the checks (local variety TMEB1 and the old-improved varieties IITA-TMS-IBA30572) implied that the new improved genotypes would have healthy clean leaves for better photosynthesis compared with the local variety and the old-improved genotypes. This might be a contributing factor for these genotypes exhibiting better root yields than the old but improved varieties while the local variety had the worst root yield. Cassava mosaic disease was positively correlated with plant height, *Gari* weight, harvest index, root size, and fresh root yield, whereas CBB was negatively correlated with yield and yield-related traits such as harvest index, plant height, chlorophyll, stomata, root size, and root number.

The significant positive correlations that existed between storage root yields and harvest index indicated that these traits were important in improving root yield. Makame [42] also reported a strong association between storage root yield and root size, while the works of (Radhakrishnan and Gopakumar [43] and Rubaihayo *et al.* [44] also revealed a strong association between storage root yield and harvest index.

Dry matter in this study was not correlated with storage root yield, thus signifying that dry matter was not an important indicator of storage root yield. Similar observations were also made by Ntawuruhunga [45], Varma and Mathura [46] and Makame [42]. Stomata conductance also showed negative association with storage root yield, indicating this was not an important index of yield. Selections in favor of any of these traits would result in simultaneous decrease in root yield. Negative association between root yield and other traits has also been reported by Makame [42] and Rubaihayo *et al.* [44].

The high variation that was observed in this study for storage root yield across the different environments indicated that these environments differed greatly as reflected in each location being categorized as a unique agro-ecological zone. The expression of yield, therefore, depends on the genetic factors and on the environmental factors. The genetic factors, which promoted yield, will only be able to express themselves to full capacity when the environmental conditions needed for its expression are found and where this is limiting, yield will be reduced.

Furthermore, effect of storage roots on storage root yield appeared to be positively influenced by root size, root number, and harvest index. [47] made similar assertions and reported that yield increase was mainly due to increase in both numbers of storage roots and individual root weight. Mahungu [48], however, observed that number of tuberous roots contributed more to the final yield than root size. Radhakrishnan and Gopakumar [43] also observed that a high value for harvest index indicated a correspondingly higher yield for storage roots. These three characters can thus be regarded as the most reliable components of yield for selection of high yielding genotypes. A similar assertion was made by [47], who reported that storage root number was more closely correlated with root dry weight than with the individual weight of storage roots and consequently yield increase was mainly due to increase in both numbers of storage roots and individual root weight. Storage root yield, number of storage roots, and root sizes had the greatest contribution to total observed variability, thus suggesting that these characters were the most important of all that were evaluated and improvement on them will lead to improvement in yield and hence, bring progress in crop improvement. The plant breeders and agronomists are, however, interested not only in high yields but also on the ability of genotypes to produce such high yields across diverse environments over several years. Therefore, breeders are interested in identifying genotypes that are stable across environments as well as those that are well adapted and suitable for a specific agro-ecology [49, 50]. Results from the field experiment were able to identify IITA-TMS-IBA980581, IITA-TMS-IBA010034, and IITA-TMS-IBA010040 as having good drought tolerance.

This experiment explained the use of eco-physiological research to improve the genetic base and develop clones that are more adaptable to the drought environments of the tropics. Interdisciplinary research that combines physiology and breeding approach will help in selection of good varieties for release since the performance of a genotype is a function of its adaptability and the availability of conducive environment along with better genetic composition.

Further research is needed and investigation on nutrient transportation, the effect of source and sinks, and their relationship as these affect yield.

Acknowledgements

The authors thank the technical staff of International Institute of Tropical Agriculture (IITA) cassava breeding program that assisted with the fieldwork, screen house evaluation, and laboratory analyses. We also extend our appreciation to the Bill & Melinda Gates Foundation and the United Kingdom's Foreign, Commonwealth & Development Office (FCDO) for funding this research through the "Next Generation Cassava Breeding Project."

Contributions

ANA, conceived, designed the study, implemented field trials, performed analyses, and wrote the manuscript, ANA and SM, performed analysis. SM, edited the manuscript, generated, and curated data.

Funding

This research work was generously funded by the Bill & Melinda Gates Foundation and the United Kingdom's Foreign, Commonwealth & Development Office (FCDO) through the "Next Generation Cassava Breeding project" (award number OPP1048542) managed by Cornell University.

Declaration of interest

The authors have not declared any conflict of interest.

Data availability statement

All relevant data supporting findings in this study are available on request from the corresponding author.

Author details

Najimu Adetoro[1,2]* and Sikirou Mouritala[2,3]

1 International Institute of Tropical Agriculture (IITA), Ibadan, Oyo State, Nigeria

2 International Institute of Tropical Agriculture (IITA), Bukavu, Democratic Republic of the Congo

3 International Institute of Tropical Agriculture, Kinshasa, Democratic Republic of the Congo

*Address all correspondence to: n.adetoro@cgiar.org

References

[1] FAO. The State of Food Security and Nutrition in the World: Safeguarding against Economic Slowdowns and Downturns. Italy: FAO; 2019

[2] Montagnac JA, Davis CR, Tanumihardjo SA. Nutritional value of cassava for use as a staple food and recent advances for improvement. Comprehensive Reviews in Food Science and Food Safety. 2009;**8**:181-194

[3] Asante-Pok A. Analysis of Incentives and Disincentives for Cassava in Nigeria. Rome: FAO; 2013

[4] FAO. Save and Grow Cassava: A Guide to Sustainable Production Intensification. Rome: Food and Agriculture Organization of the United Nations Publishers; 2013

[5] Muiruri SK, Ntui VO, Tripathi L, Tripathi JN. Mechanisms and approaches towards enhanced drought tolerance in cassava (*Manihot esculenta*). Current Plant Biology. 2021;**28**:100227

[6] IPCC. The Physical Science Basis. Cambridge, UK: Cambridge University Press; 2007

[7] Aina OO, Dixon AGO, Akinrinde EA. Effect of soil moisture stress on growth and yield of cassava in Nigeria. Pakistan Journal of Biological Sciences. 2007;**10**: 3085-3090

[8] Simelton ES, Fraser EDG, Termansen M, Benton TG, Gosling SN. The socioeconomics of food crop production and climate change vulnerability: A global scale quantitative analysis of how grain crops are sensitive to drought. Food Security. 2012;**4**:163-179

[9] Lipiec J, Doussan C, Nosalewicz A, Kondracka K. Effect of drought and heat stresses on plant growth and yield: A review. International Agrophysics. 2013; **27**:463-477

[10] Harb A, Krishnan A, Ambavaram MM, Pereira A. Molecular and physiological analysis of drought stress in Arabidopsis reveals early responses leading to acclimation in plant growth. Plant Physiology. 2010;**154**:1254

[11] Shabala S, Bose J, Fuglsang AT, Pottosin I. On a quest for stress tolerance genes: Membrane transporters in sensing and adapting to hostile soils. Journal of Experimental Biology. 2015;**67**:1015-1031

[12] Cai S, Chen G, Wang Y, Huang Y, Marchant DB, Wang Y, et al. Evolutionary conservation of ABA signaling for stomatal closure. Plant Physiology. 2017;**17**:732-747

[13] Wilkinson S, Davies WJ. Drought, ozone, ABA and ethylene: New insights from cell to plant to community. Plant, Cell & Environment. 2020;**33**:510-525

[14] Osakabe Y, Osakabe K, Shinozaki K, Tran L-SP. Response of plants to water stress. Frontiers in Plant Science. 2014;**5**:86

[15] Ding S, Zhang B, Qin F. Arabidopsis RZFP34/CHYR1, a ubiquitin E3 ligase, regulates stomatal movement and drought tolerance via SnRK2.6-mediated phosphorylation. Plant Cell. 2015;**27**: 3228-3244

[16] Carvalho MHCD. Drought stress and reactive oxygen species. Plant Signaling & Behavior. 2008;**3**:156-165

[17] Smirnoff N. The role of active oxygen in the response of plants to water

deficit and desiccation. New Phytologist. 1993;**125**:27-58

[18] Mittler R. Oxidative stress, antioxidants and stress tolerance. Trends in Plant Science. 2002;7:405-410

[19] Ford-Lloyd BV, Schmidt M, Armstrong SJ, Barazani O, Engels J, Hadas R, et al. Crop Wild Relatives—Undervalued, underutilized and under threat? Bioscience. 2011;**61**: 559-565

[20] Dempewolf H, Baute G, Anderson J, Kilian B, Smith C, Guarino L. Past and future use of wild relatives in crop breeding. Crop Science. 2017;**57**: 1070-1082

[21] Cock JH. Cassava: New Potential for a Neglected Crop. Colorado, USA: Westview Press, IADS Publication; 1985. p. 192

[22] Connor DJ, Cock JH, Parra GE. Response of cassava to water shortage I. Growth and yield. Field Crops Research. 1981;**4**:181-200

[23] El-Sharkawy MA, Hernández ADP, Hershey C. Yield stability of cassava during prolonged mid-season water stress. Experimental Agriculture. 1992; **28**:165-174

[24] Anderson JM, Ingram JSI. A Handbook of Methods. Vol. 221. Wallingford Oxfords: CAB International; 1993. pp. 62-65

[25] International Institute of Tropical Agriculture (IITA). Cassava in Tropical Africa: A Reference Manual. Ibadan, Nigeria: IITA; 1990 176 p

[26] Ekanayeke IJ. Screening for Abiotic stress resistance in root and tuber crops. IITA Research Guide 68; elongation of stem cuttings of cassava. Field Crop Research. 1988;**2**:224-252

[27] Putpeerawit P, Sojikul P, Thitamade S, Narangajavana J. Genome-wide analysis of aquaporin gene family and their responses to water-deficit stress conditions in cassava. Plant Physiology and Biochemistry. 2017;**121**:118-127

[28] Zhao PJ et al. Analysis of different strategies adapted by two cassava cultivars in response to drought stress: Ensuring survival or continuing growth. Journal of Experimental Botany. 2015; **66**:1477-1488

[29] Bray EA. Alteration in gene expression in response to water deficit. In: Basra AS, editor. Stress-Induced Gene Expression in Plants. Chur, Switzerland: Hardwood Academic Publishers; 1994. pp. 1-23

[30] Porto MCM. Leaf conductance in cassava cultivars [PhD dissertation]. Rev Bras Fisiol Veg Braz. Tucson: University of Arizona; 1989

[31] Ludewig F, Flügge U-I. Role of metabolite transporters in source-sink carbon allocation. Frontiers in Plant Science. 2013;**4**:231

[32] Cho J-H, Wang K, Galas DJ. An integrative approach to inferring biologically meaningful gene modules. BMC Systems Biology. 2011;**5**:117

[33] Ekanayake I, Osiru D, Porto M. Morphology of Cassava: IITA Research Guide, No. 61. 1997

[34] Ekanayake I, Porto M, Dixon A. Response of cassava to dry weather: Potential and genetic variability. In: African Crop Science Conference Proceedings. Vol. 1. 1994. pp. 115-119

[35] Jarvis P, Morison J. The control of transpiration and photosynthesis by the stomata. In: Stomatal Physiology.

Cambridge: Cambridge University Press; 1981. pp. 247-279

[36] Akparobi SO, Togun AO, Ekanayake IJ. Temperature effects on leaf growth of cassava (*Manihot esculenta* Crantz) in controlled environments. African Journal of Root and Crops. 2000; **4**:1-4

[37] Enyi BAC. Effect of shoot number and time of planting on growth, development and yield of Cassaya (*Manihot Esculenta* Crantz). Journal of Horticultural Sciences. 1972;**47**:457-466

[38] Lian T, Cock JH. Cassava plant forms and their associated morphophysiological characters. MARDI Research Bulletin. 1979;**7**:55-69

[39] Bergantin RV, Yamauchi A, Pardales JR Jr, Bolatete-Al DM. Screening cassava genotypes for resistance to water deficit during crop establishment. Philippines Journal of Crop Science. 2004;**29**:29-39

[40] Lahai M, George J, Ekanayake I. Cassava (Manihot esculenta Crantz) growth indices, root yield and its components in upland and inland valley ecologies of Sierra Leone. Journal of Agronomy and Crop Science. 1999;**182**: 239-248

[41] Oluwafemi Z, Omonona B, Adepoju A, Sowunmi F. Cassava productivity growth in Nigeria. Asian Journal of Research and Agriculture. 2019:1-9

[42] Makame M. Genetic variation, stability of performance of cassava clones and their responses to intercropping with sweet potato in Tanzania [Ph.D. Thesis]. Ibadan, Nigeria: University of Ibadan; 1995; 248 p

[43] Radhakrishnan V, Gopakumar K. Correlation between yield and its components in tapioca. Indian Journal of Agricultural Sciences. 1984;**54**:975-978

[44] Rubaihayo P, Whyte J, Dixon A, Osiru D. Inter-relationships among traits and path analysis for yield components of cassava: A search for storage root yield indicators. African Crop Science Journal. 2001;**9**:599-606

[45] Ntawuruhunga P. Assessment of dry matter determination and its accumulation in cassava (*Manihot esculenta* Crantz) [Ph.D. thesis]. University of Ibadan; 1992

[46] Varma S, Mathura R. Genetic variability and inter-relation in cassava (Manihot esculenta Crantz) under rainfed conditions of Tripura. Journal of Root Crops. 1993;**19**:77-80

[47] Kasele I, Ekanayake IJ, Dixon AGO. Preliminary investigations on photosynthetic capacity and nutrient uptake of colloid cassava (*Manihot esculenta Crantz)*. In: Akoroda MO, Ekanayake IJ, editors. Root Crop and Poverty Alleviation Proceedings of Sixth Symposium of ISTRC-AB. Lilongwe, Malawi; 1998

[48] Mahungu N. Relationship among selected agronomic characters and their effects on tuberous root yield of cassava (*Manihot esculenta* Cratz) [Ph.D. Thesis]. Nigeria: University of Ibadan; 1983; p. 193

[49] Acuña TB, Lafitte H, Wade LJ. Genotype× environment interactions for grain yield of upland rice backcross lines in diverse hydrological environments. Field Crops Research. 2008;**108**:117-125

[50] Nassir AL, Adewusi KM. Genotype x environment analysis of root traits of upland rice (*Oryza sativa* L.) in a drought prone tropical rainfed ecology. Tropical Agriculture. 2015;**92**:16-26

Chapter 16

Tolerance of Plant Cell Wall to Environment

Olena Nedukha

Abstract

Drought and flooding of soil are negatively factors for growth and development of plants. Exogenous factors, including moisture of soil, intensity of sun light, temperature, salinization, the content and diffusion rate of CO_2 and O_2 is main that influence terrestrial and flood plants. Cell walls actively participate in the mechanisms of plant adaptation to drought and flooding. It has been established that the resistance of plants to unfavorable environmental conditions is due to the plasticity of the structural, biochemical and functional characteristics of plant cell walls, that manifests itself in a change of ultrastructure cell walls, density of stomata and wax in leaf epidermis, compacting or loosening of cell walls, presence of cuticle pores, change of content of crystalline and amorphous cellulose, hemicellulose, callose and lignin and change in a ratio of syringyl/quajacyl monolignols and also expression of the specific genes.

Keywords: cell wall structure, lignin, cellulose, drought, flooding, stress, genes

1. Introduction

It is known that external environmental conditions provoke to the phenotypic and genetic plasticity of plant during vegetative and generative growth and lead to change of duration of ontogenesis of both individual species and populations [1]. Given that exposed to a specific exogenous factor, some plants experience stress, and for other plant species this factor is the optimal condition for life, the definition of stress for the plant is quite complex and problematic. According to many definitions, stress is a harmful adverse force or condition that inhibits the normal functioning of a biological system, such as a plant. According to [2], stress for a plant is its response to the action of adverse or even detrimental to growth and development of plant. For the plant, stress is measured by both signs of survival and signs of adaptation, yield, growth parameters and assimilation. External signals of the environment, such as light, temperature, water status of the soil–these are the most important signals that affect the growth of the plant. The perception of these signals and the plant's response to them affects a whole cascade of events that require knowledge of the signal and its transduction into a physiological response [3]. In the perception of signals of adverse abiotic stresses, primarily involved protein receptors of the cell wall, that send this signal to the transport system into the cytoplasm. Such receptors of ell wall appear to be arabinogalactan protein molecules that bind the cell wall to the plasmalemma, cytoskeleton elements, and apoplast components. In addition to these proteins, stress receptors can be mitogen-activated protein, numerous kinases, and several transcription factors [4]. Stress is first perceived by cell wall receptors, which send a signal to the receptors of the cytoplasmic membrane, then the signal is reformed and reduced, and the result of this transformation is the participation of secondary mediators [3–6].

Determination of the plant state in a changing environment in conditions of increased anthropogenic pressure and global climate changes is becoming one of the main problems of plant biology and ecology. In natural conditions plants can be influenced by a complex of unfavorable environmental factors. Despite the long list of abiotic and biotic stresses, including: cold, high temperature, salinity, drought, floods, radiation, air and soil pollutants, pathogens and others, we will consider the most significant adverse environmental factors: drought and flooding, which negatively affect plants' growth, up to their death. The search for universal biomarkers that would make it possible to determine the state of plants regardless of nature and number of stress factors is urgent. The cell wall of plants can be such a marker, since it is the growth and differentiation of the cell wall during primary and secondary growth that undergoes significant changes under conditions of changes in the water balance of the plant. The basis of this section is the idea that the stability of ontogenesis under conditions of unfavorable climatic and anthropogenic changes in the environment is due to the plasticity of the structural and functional organization of plant cell walls. We put forward a hypothesis about the existence of a coordinated response of the structural and functional systems of the cell wall and the cytoplasm of plant cells, which is involved in the adaptation of the plant to the action of extreme natural factors—drought and flooding.

2. Drought

Drought is a deficit of water in the soil, which affects the growth and development of the plant. Drought stress is seen as a condition in which water potential and turgor of a cell are reduced, although the plant can function normally. Water stress is considered as the loss of water by the plant, which leads to the closure of the stomata and restriction of gas exchange by the plant. Wilting of plants is characterized by an intensive loss of water, which leads to next changes, including of plant metabolism and cell structure, to change of activation of catalytic enzymatic reactions, to inhibiting the process of photosynthesis and destructed metabolism, which can lead to cell death [7, 8]. Drought can be chronic or temporary. The latter is observed when the weather changes rapidly and unpredictably. Moderate drought is a phenomenon in which the plant begins to feel the effects of drought. Under such conditions, plants have developed specific mechanisms of acclimatization and adaptation in response to the short-term or long-term action of the factor [9, 10].

In this respect, the reaction of plants to drought is well studied in psammophytes growing on sand dunes has been better studied. Psammophytes develop mechanisms and specific features that ensure not only a normal state of life, but also functioning under stressful conditions. These mechanisms are reflected in the morpho-anatomical changes in the vegetative organs of plants [7, 11] that help psammophytes to adapt to environmental conditions, and manifested in a decrease in the size of leaf blades, the formation of water-retaining parenchyma, a change in the size of the leaf conducting system, twisting of leaf blades, a change in the cell wall structure, change of density of stomata, an optimization of transpiration, enhanced synthesis of wax and lignin, the formation of trichomes and silicon inclusions in cell walls and formation a thick cuticle [7, 12].

2.1 Growth processes during drought

It is showed that even with a slight drought, the growth rate of plant organs decreases: roots and aboveground organs react very strongly to such stress, their growth reduce [13, 14] that connected with structural-functional changes of cell walls [15]. Drought cell growth decreases have been described for leaves for psammophytes, including *Zygophyllum album* and *Nitraria retusa* [16], *Tragopogon borysthenicus* and *T. orientalis* [11], *Helichrysum arenarium* and *H. corymbiforme* [17]. The mechanisms of this phenomenon in psammophytes have not been studied. However, in mesophytes and halophytes, cell and organ sizes have been shown to be mediated by the rate of cell division and stretching [7]. That is, the decrease in cell growth is mediated by changes in the synthesis of polysaccharides of cell walls. This phenomenon has been studied in *Sium latifolium* leaves [18]. When studying the development of the ephemeral desert plant *Gymnarrhena micranatha* for 50 days, it was found that its inflorescence is located in the depth of the soil. The apical bud is drawn from the soil surface into the soil to a depth of 10 mm. It is shown that the direct retraction of the main apex of the stem occurs from the soil surface into its depth, while the stem rotates, changes its direction of growth by 180°, continues to grow and develop. What processes take place in cells' walls? There was compression of cells in the endoderm, pericycle and primary phloem of the root, which decreased in length (by ½ initial lengths). This was due to the formation of radial cell walls. The shrunken and compressed cell walls had a wavy appearance. Partial disintegration of polysaccharides was observed in cell walls, including lysis of cellulose wavy-like micro fibrils [19].

Under drought conditions in the roots there revealed a decrease in the size of the parenchyma [20]; in the endoderm, cell walls thicken, and additional layers of cells were with strongly suberinized cell walls are formed around the stele [21]. In the periderm, cell walls were also impregnated with suberin, which reduces the penetration of water through the cells of the cortex. Special lacunae for water storage were formed in root [22]. Whereas in leaves the effect of drought is manifested in the reduction of sugars in the fraction of cell walls, which should certainly be reflected in the composition of polysaccharides in the walls. Studies of the effects of drought on crops have shown that the cell walls of aboveground photosynthetic organs are also sensitive to this factor. Studies of polysaccharides of cell wall matrix in reduced coleoptiles of wheat seedlings under drought from 6 to 15 weeks shown that during the first week of drought exposure, drought-sensitive varieties showed a decrease sugar in the fractions of wall matrix: rhamnose, mannose, galactose, arabinose, xylose, and glucose and uronic acids [23]. In addition, in the hemicellulose fraction of drought-resistant variety was shown decrease in arabinose, mannose, galactose and increase in rhamnose, xylose, glucose, uronic acids in comparison with drought-sensitive variety. These changes were accompanied by an increase in the activity of glucoside hydrolysing enzymes: α-galactosidase, α-L-arabinofuranosidase and 1.3–1.4-β-glucanase in drought-resistant varieties. The observed changes in the matrix of cell wall of coleoptiles of two varieties of wheat under the action of drought reflect changes in cell metabolism, which directly affected the growth rate [23]. Similar changes in the content of sugars (glucose, fructose and sucrose) and the activity of

1.3–1.4-β-glucanase have been previously noted by other researchers in studying the effects of water and salt stress on wheat stalks [24].

2.2 The role of wax and cuticle

Wax and cutin are involved in the regulation of water and lipids transport through the cell wall [25]. Plant' wax is a mixture of aliphatic and cyclic hydrocarbons and their derivatives. The composition of waxes varies depending on the species and organs' plant. Cutin is involved in the regulation of the diffusion of gases and moisture in the main cells of the epidermis and the stomata. It is known that the cuticular membrane can be both hydrophobic and hydrophilic. If the cuticular membrane is hydrophobic, the functions of the cuticle are to reduce water loss by the organs; and if the cuticle is hydrophilic, then the function is to transport water, aqueous solutions, and lipids (waxes) [26]. It is known that the aboveground organs of plants that grow in dry climates synthesize a significant amount of wax and cuticle, which are a barrier to transpiration [27]. Wax and cuticle are the main barriers against "uncontrolled" water loss by leaves. Therefore, in the adaptive responses of above-ground bodies, to action of a drought, the strengthened synthesis of these two components of cellular components of epidermis plays a certain role.

Wax can be located the inside cutin layer, or be situated on top of the cuticle. A two-year study of the long-term effects of drought on pine needles (*Pinus pinaster*) showed that pine needles activated the synthesis of cuticular wax, which was accompanied by the expression of transcription factor (*SHINE*), which is involved in cuticle synthesis [28]. It was found that the content of newly deposited wax under such conditions depends on the duration of exposure and the plant state [29]. Thus, a study of the effects of drought on tobacco plants (*Nicotiana glauca* L.) showed an increase of one and a half to two times the wax content in the leaves during 3 days of drought compared with unstressed plants [30]. Under conditions of enhanced drought, *Arabidopsis thaliana* plants increase not only the wax and cutin content, but it was accompanied by an increase of almost 50% in the thickness of cell walls. Wax deposition and extension of the cuticle layer in the epidermis increased plant tolerance to prolonged drought. A similar phenomenon has been noted by other researchers in studying the effects of drought on the leaves [31]. Even in temperate climates (under conditions of natural water deficit in the soil) there is an acceleration of wax synthesis on the surface of leaf blades. Carrying out a comparative study of the ultrastructure of the epidermis of leaves *Alisma plantago-aquatuca*, which grew on the river bank and in conditions of moderate soil drought, the researchers found an increase in the density of wax on the upper epidermis 2.5 times, on the lower epidermis—eight times [32]. In roots, drought or salt stress causes an increase in the content of suberin in the cell walls of the exo- and endoderm, which was shown in the species *Pistacia integerrima* and a hybrid, *P. atlantica x Plagiochila integerrima* [33].

For plants that grow in drought conditions is characterized by the participation of cell walls of the epidermis of the leaves in the water intake. It is known that the above ground organs of desert plants can absorb water from the leaf surface, intercept precipitation and absorb fog, using an atmosphere saturated with water [34, 35]. To do this, plants use trichomes [36], the specialized glands [37], and also form a hydrophilic surface in specialized epidermal cells that contain water pores [38]. It is shown that the leaves of *Reaumuria soongorica*, a super-xerophytic desert plant, are characterized by the presence of water-absorbing cells in the epidermis, which are closed by scales. Such scales cover water-absorbing cells during the day and open water-absorbing pore channels at night, rising above the surface of epidermis. During the day, the valves of the water-absorbing structure of the scales are compressed, leaving a small central hole. At night, when the humidity rises, the basal cells raise the upper cells of the lid; after which the cells of the porous channels (capillaries) expand, forming a hole through which they begin to absorb atmospheric water [39]. The depth of the stomata is also an important structural feature of plants that have adapted to drought. Slightly sunken, submerged stomata are a typical feature of numerous psammophytes growing on the Mediterranean coast, including the coasts of Crete, Lampione, Tavolari and Malta, where extreme values of annual temperature have been recorded [40]. The results indicate that modifications in leaf architecture, including deepening the stomata into the epidermis, are important anatomical and physiological strategies that help psammophytes reduce water consumption.

Another feature of leaf structure to optimize water balance is twisting and/or folding of leaves. Leaf twisting is designed to maintain the optimal water balance of plants growing in inadequate water supply conditions [40, 41]. The twisting of the leaves of many psammophyte grasses is due to the specialized structure of the epidermis of the leaf blades and the presence of bulliform (motor) cells, the cell walls of which function to enter and exit water, reducing leaf area affected by drought [41, 42]. The cell walls of the bulliform cells of the epidermis synthesize guajacyl monolignol and callose, which helps to quickly change the entry or exit of water from these cells [43]. Twisting preserves optimal heat transfer and optimal water-vapor density in leaf tissues [41].

The presence of trichomes and increased cuticle density in cell walls are typical features of the leaf blades of psammophyte plants growing on coastal dunes [44]. Psammophytes have two types of trichomes: glandular and non-glandular. Glandular trichomes were found in the leaves and stems of psammophyte *Silene thymifolia*, which grew in Romania [45], in the leaves of *Tragopogon borysthenicus* and *T. orienthalis*, which grew in the Black Sea Reserve (Ukraine) and sand terraces of the Desna River, Ukraine [11].

2.3 Role of lignin, pectin and cellulose

Lignin is a branched biopolymer that, together with hemicellulose and pectin, acts as an adhesive matrix for cellulose microfibrils. Lignin provides mechanical strength of tissues and organs, impermeability of water and aqueous solutions through the cell walls. Lignin is a complex of phenylpropanoids (monolignols) [46]. Early work (Barnett, 1976) on the effect of drought on wood lignification showed that the tracheid rings stuck together because the secondary walls of young trees did not contain lignin. The formation of false rings in drought-stricken trees is a well-known phenomenon [47]. According to Lloyd Donald [48], who studied the anatomy of wood and the characteristics of cell walls in *Pinus radiate* under conditions of water stress (drought), false growth rings with bundles of weakly lignified tracheids, were found. It has been shown that wood exfoliation is due to poor adhesion between the tracheids due to a decrease in lignin content in the middle plate. The author explains this phenomenon in such a way that, apparently, there is an abnormal lignification due to dehydration of the outer cell walls. Since the formation of lignin occurs with the movement of water, the emergence of water or its insufficiency in the cell walls should interfere with or prevent both the transport and inclusion of lignin precursors and the process of lignification of the wall.

It has been shown that even a slight drought (up to 12 days) caused an increase in lignin precursors (coumaric and caffeic acids) in xylem maize juice, and this was due to a decrease in anionic peroxidase activity, indicating the effect of drought on lignin biosynthesis [49]. Different areas of the corn root respond differently to drought: in the basal part of the roots, growth is inhibited compared to the apical part of the roots, which is associated with the expression of two genes involved in lignin biosynthesis: shinamyl-CoA reductase-1 and -2. Such decrease in growth is due to an increase in lignin deposits, which increase the stiffness of the cell wall and reduce the growth rate, which may also be due to changes in factors such as water, minerals and sugars.

It was shown that after 28 days of drought, *Trifolium repens* reduced growth and increased lignin synthesis in the leaves, accompanied by increased activity of guajacol peroxidase, syringaldazine peroxidase and coniferol alcohol peroxidase [50]. It has been found that the activity of enzymes associated with lignin synthesis changes in plant leaves during drought. Thus, in the study of prolonged (35 and 47 days) drought on the leaves of *Ctenanthe setosa*, it was found that in parallel with the increase in lignin activity of enzymes involved in its synthesis also increased, in particular: phenylalanine ammonium lyase, soluble covalently bound peroxidase and polyphenol oxidase [51].

It is established that the impact of drought depends on the duration of its action, the species of plants and the growth stage. It has been shown that even a slight drought (up to 12 days) caused an increase in lignin precursors (coumaric and caffeic acids) in xylem maize juice, and this was due to a decrease in anionic peroxidase activity, indicating the effect of drought on lignin biosynthesis [49]. Roig-Oliver et al. showed for the first time that during long-term water deficiency, changes in the content of lignin, cellulose and hemicellulose in the cell walls of *Helianthus annuus* were accompanied by a significant decrease in phenols associated with the wall (coumaric, ferulic and caffeic acids) and with a negative correlation with photosynthesis (conductivity of the mesophyll to CO_2), and with a positive correlation with palisade mesophyll thickness [52, 53].

Abiotic stress, including drought, cause a change in the mechanical strength of the cell wall due to the synthesis of lignin and activation of several the types of reactive oxidative species (ROS). Cell walls become stiffer and the overall mechanical stability of tissues and cells increases provided of an increase of wall peroxidases activity, increase in H_2O_2 concentration and/or an excess of peroxidase substrates [54]. The resulting increase in mechanical strength of the cell wall is occurred the change of cell's turgor that enable plant cells to endure the osmotic stress caused by drought [55].

Cell walls not only change their structure in response to drought, to reduce water evaporation by cells, but also act as structures that, accumulate water for the needs of the cell. In particular, plants increase the content of pectins as a wet absorbing structure. This has been shown in the laboratory in the study of roots and stems of wheat seedlings (*Triticum durum*) of two varieties (drought-resistant and drought-resistant) [14]. Comparison between the two genotypes showed some differences in the content of polysaccharides of the wall matrix and the content of α-cellulose. It was found that the residues of xylose, glucose and arabinose in the matrix are more than 90 mole%; the level of xyloglucans—was 23–39 mole%, arabinoxylans—38–48 mol%, while the content of pectins and 1–3,1–4 β-D-glucans—was very low. It has been shown that in drought-stable plants the content of rhamno-galacturonans I and II significantly increased under conditions of water stress, while in the second genotype such an increase was not observed [14]. The obtained results indicate that in drought-resistant wheat varieties adaptation to drought occurs due to the increase of pectin chains, which leads to an increase in pectin gel that is the wet absorbing structure of the wall.

The study of the effect of water deficiency on the content of pectins in sunflower leaves showed that this polysaccharide is the most sensitive to water stress, it is the first to react to stress, reducing its content after a short exposure to stress (5 hours), while hemicellulose and lignin changed its contents only after 24 hours of stressful influence [56]. Early was established that pectins are crucial to determine wall characteristics. Changes in pectin physicochemical properties during stress induce the rearrangement of cell wall compounds, thus, modifying wall architecture and influencing on photosynthetic characterization of leaves of *A. thaliana atpme17.2* (SALK_059908) and *atpae11.1* (GK 505H02) mutants from wild-type Columbia (Col-0) [57]. This study provides insights on how different cell wall architecture could influence the photosynthetic efficiency in *A. thaliana atpme17.2* and *atpae11.1* mutants in comparison to wild-type Col-0. Thus, it was established that cell wall composition modification could lead to reduced photosynthetic traits in *atpme17.2* and *atpae11.1* mutants maybe because of alterations in However, there was a strong reduction in the cell wall pectin fraction, expressed by the amounts of galacturonic acid in both *atpae11.1* and *atpme17.2* mutants.

The use of a model object, in particular *Penium margaritaceum*, which is a unicellular zygnematophyte to study the effect of a hyperosmotic environment on the formation of cell walls, showed that rhamnogalacturan-1 is one of the first components of the wall, which take part in acclimatization to hyper osmotic stress. This phenomenon was established using the method the labeled antibody [33, 58]. The study of cell walls composition in six Antarctic mosses (*Brachytecium austrosalebrosum, Warnstorfia sarmentosa, Bryum pseudotriquetrum*, *Polytrichum juniperinum*, *P. alpinum*, and *Sanionia uncinata)* showed a clear positive correlation between cell wall thickness and pectin content, which apparently determines the porosity of cell walls in arctic species, and contributes to adaptation and optimal photosynthesis in such conditions. The investigators found that the less pectin in the leaves, the higher the values of CO_2 assimilation [59].

2.4 Role of silicon

It is established that the resistance of plants to drought is due not only to changes in the structure of cell walls of epidermal tissue, but also the deposition of silica in cell walls in the form of amorphous or crystalline inclusions [60]. According to Wang [61] silicon inclusions in epidermal cells reduce the influence of thermal effect on the leaves by reflecting the heat flow in the far infrared region of the sun light flux. This provides a passive mechanism for cooling the leaves in high sunlight. Although the mechanism of this action is not yet known, these issues need further to study. Silicon can deposit in leaf epidermis trichomes giving these structures are hardness and rigidity, making the leaves inedible to animals [62]. As a rule, most silicon is contained in cell wall protopectin, a water-soluble pectin fraction [63].

It has been established that silicon decrease the cuticle transpiration of aboveground organs. This chemical element, which accumulates in the cells of the epidermis of leaves and stems, forms a thickened cuticle-silicon wall, which protects the plant from excessive moisture consumption by reducing the cuticle transpiration. In addition, the plant's walls can form hydrophilic silicate-galactose complexes that bind free water, thereby increasing the water retention capacity as in specific cells, as and in different tissues and in the organs of plant [64, 65].

Because of the density of cell walls and their ability to retain moisture, silicon compounds can significantly increase plant resistance to drought and protect plants from being lodged (fallen) [66]. Silicon reduces of water evaporation on the leaf surface, as has been shown, for example, on rice seedlings [67], on other crops, in particular in drought-resistant wheat [68] and sorghum [69]. Silicon can also influence water transport by regulating the osmotic potential of cells by increasing synthesis and accumulation of osmotic active substances (e.g., proline, sugars and inorganic ions) [70, 71].

2.5 Involvement of genes in adaptation process to drought

Over the years, significant progress has been made in discovering the cell wall-specific genes related to drought tolerance [72, 73]. These researches were carried out at rice in vegetative and reproductive stages [72]. In the reviews [72, 73] shown the major candidate genes underlying the function of quantitative trait loci directly or indirectly associated with the cell wall plasticization-mediated under drought tolerance or salinity stress of plants. On rice plant during of drought stress was identifying series genes, which take part in tolerance of this species to both drought or salinity stress: 1) drought inducible AP2/ERF family TF gene *OsERF48,* including cell wall related genes such as *OsXTH9*, *OsAGP24, OsEXPA4, OsEXPA8, OsEXPB2, OsEXPB3, OsEXPB6* and *OsAGP3,* which associated with cell expansion and cell wall plasticization-mediated root growth under abiotic stress; 2) a lignin biosynthesis gene *OsCCR10* (*Oryza sativa CINNAMOYL-COA REDUCTASE 10*) is also highly induced by drought in the roots of rice/ and 3) genes are associated with cell wall loosening (*OsEXP1, OsEXP2, EGase*, and two *XETs*), with lignin biosynthesis (*PAL, C3H, 4CL, CCoAOMT, CAD*, and *peroxidase*), and with the metabolism of cell wall proteins (*GRP* and *UDP-GlcNAc pyrophosphorylase*) and polysaccharides (*OsCslF2, GMPase, xylose isomerase,* and *beta-1,3-glucanase*), and 4) genes, including drought responsive, ABA-responsive, superoxide dismutase and cell wall-related genes (*LOC_Os01g64860; LOC_Os01g72510; LOC_Os05g35320; LOC_Os12g36810*, etc. [72].

Molecular methods have shown that during drought, increased wax and cutin synthesis is accompanied by activation of genes (*Ltps* and *WAX9*) that express proteins involved in the synthesis and deposition of wax and cutin, as well as the synthesis of transport proteins [74]. In a study of prolonged drought on the model plant *A. thaliana* [75, 76] was identified the presence of 11 genes, which were divided into groups according to the functions of the proteins they encode:

1. Transcription factor that regulates the biosynthesis of the cuticle (*SHINE1*

2. Genes (*CER6 / KCS6, KCS4, KCR1, ECR / CER10*), which regulate protein synthesis in VLCFA (very long-chain fatty acids);

3. Genes involved in the reduction and decarboxylation pathways (*CER1, CER2, CER3*);

4. Gene involved in the biosynthesis of wax and cuticle (*LACS3*).

An early response of the *pDr* (*mDr*- drought) genes to stress drought in *A. thaliana* seedlings was detected and it was shown that drought reduces the expression of four expansin genes on the first day of stress. The fifth gene, *EXLB1*, has been deducting since the second day of stress [76]. It was shown that these genes take part in an early response to drought. Other researchers have found the expression of expansin genes (*Exp1, Exp5* and *ExpB8*) by reducing water potential in the apex of roots, corn leaves, and in the apex of tomato stems [77–79].

The physical properties of the cell wall are also play an important role in water deficiency [80]. Analysis of the *pDr* transcript showed repression of numerous genes involved in wall elongation stretching [81], whereas the action of mean osmotic stress induced elongating genes [82]. Microray analysis (qRT-PCR) revealed the regulation of cell stretching genes during *mDr* treatment. It has been established that most drought genes reduce their regulation during drought. The repression of four *EXLB* genes begins on the first day, while the fifth *EXLB1* gene peaks on the first day and decreases on the second day; it is on the first day that acclimatization to stress may begin, depending on the organ, species and tissue [29, 83]. These studies were performed on *A. thaliana* [76].

The study of physical properties, stiffness in particular, cell wall from the root elongation zone using atomic force microscopy in *A. thaliana* Columbia-0 (Col-0) wild-type and mutant plants (with *TETRATRICOPEPTIDE THIOREDOXIN-LIKE 1, ttl1* gene, cause root swelling and root growth arrest under NaCl and osmotic stress) revealed that root of mutant (with *ttl1* gene) increase of the stiffness of the cell wall in root elongation zone [84].

A study of the effects of drought on Arabidopsis mutant plants (with cellulose synthase genes—*AtCesA8 / IRX1*), which were resistant to drought, NaCl, mannitol and other osmotic stresses, showed that cellulose synthesis under drought and osmotic stress is due to the expression of cellulose-synthase [85]. These researchers showed the effect of drought, by the absence of watering for 2 weeks, on the genetic traits of the cellulose-synthesizing complex in the leaves of *A. thaliana*. It was found the next following physiological and molecular changes in the leaves of two mutants Arabidopsis (with genes withering, leaf wilting 2–1 and leaf wilting 2–2; genes: lew2–1 and lew 2–2), which were resistant to drought, salt salinity, mannitol and other osmotic stresses: the lew2 mutant was shown to accumulate more ABA, proline, and soluble sugars compared to the control (wild type). New alleles of the AtCesA8 / IRX1 gene encoding subunits of the cellulose-synthesizing complex have been discovered in this mutant. The data obtained suggest that cellulose synthesis is quite important for the response to osmotic stress and drought [85].

Rui and Finneny [86] proposed a model for regulating the cell wall response to stress; according to this model, certain aspects of the wall itself can act as growth-regulating signals. The molecular components of the signaling pathways that determine and maintain cell wall integrity are shown, including sensors that detect changes on the cell surface and downstream signal transduction modules. There are several cell wall receptors that sense stress, including drought or salinity. Such receptors, according to the authors, may be the receptor-like kinase THESEUS1 (THE1) and FERONIA (FER) localized on the plasma membrane or Ca^{2+}.ATPase. Kinase THE1 has been identified by suppressor screening in a cellulose-deficient mutant background; and FER is widely expressed and serves as a signaling node that functions in a wide range of processes, including plant growth, vacuole morphology, mechanosensing, hormonal signaling, and others. In contrast, the FER protein exhibits defects in growth recovery under salt stress as a result of failure to reverse salt-induced softening of the wall and increased frequency of cell rupture.

Summarizing the above material of numerous experimental works, we can propose the following scheme of response of cell walls of plants growing in drought

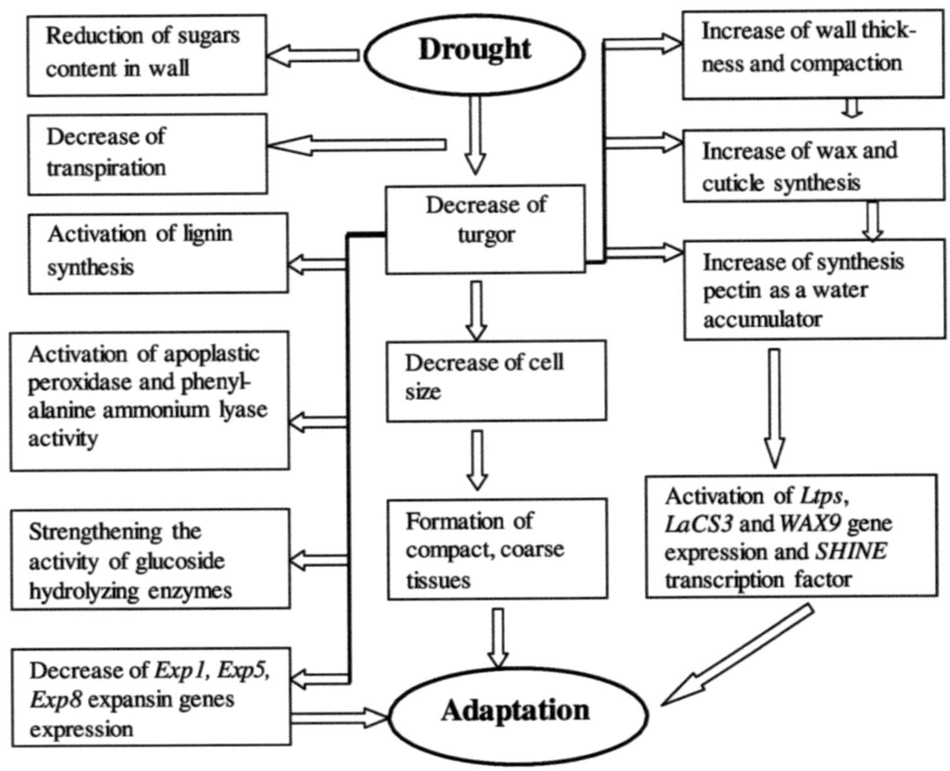

Figure 1.
Schematic representation of the main functional changes of plant cell wall during adaptation to drought.

or deserts: perception of drought signal (high air temperature and low soil moisture) leaves and roots → stopping or inhibiting growth of root and leaves → reduction of cell size → closure of stomata in leaves → reduction of stomatal conductivity for CO_2 (or cessation of stomata and shedding of leaves) → in the roots of the formation of water lacunae; in stems of succulents (during leaf shedding) water storage in specialized lacunae of the parenchyma → thickening of cell walls, their lignification and suberinization, intensified synthesis of wax, expression of genes associated with the synthesis of extensins, dehydrins and cellulose, activation of enzymes for synthesis of lignin, suberin due to changes in the expression of the corresponding genes (**Figure 1**).

3. Flooding

Flooding is a potentially detrimental stress for many terrestrial plants; flooding occur when water covers the area, caused by both natural (river floods, heavy rainfall, tides) and artificial causes (construction of reservoirs, ponds); it can be short-term, intermittent (during river floods) or long-term, in which many species may die. Peculiarity of flooding as a stress factor is a combination of significant changes in water availability of plant and oxygen respiration in the root system, and as a result there is inhibition of aerobic processes, impaired absorption of ions and nutrients, changes in metabolism and growth processes [87]. The next factors are affected on the flood plant: a decrease in illumination and change in the light spectrum, a lack of acidity and CO_2. It is known that water absorb flow of light and disperse of light [88, 89]. In flood conditions, the diffusion of gases $_{is}$ much slower than in air, and this is what limits normal photosynthesis and aerobic respiration [87]. Some plants that are resistant to flooding use the acceleration of stem growth to get out of the water and such a stem rises above the flooded part of the plant. The part of the plant that emerges from the water begins to come into contact with the air environment, renovating aerobic metabolism and photosynthesis [90]. Hydrophytes and wetland plants, which have adapted to both the lack of oxygen in the soil and the constant aquatic environment, have for millennia developed certain mechanisms of adaptation at different levels of the organization. The main signs of rearrangement are a decrease in the thickness of the leaf blade, rearrangement of the mesophyll, the presence of chloroplasts in the epidermis and changes in the structure of cell walls [91]. Cell wall of the epidermis of flooding plants is the first to react to the water environment, changing their structural- functional characteristics to optimize the water balance of plants. Therefore, the analysis of comparative structural and functional studies of flooded and above-water leaves is important for understanding the role of cell wall in the adaptation of plants to the aquatic environment.

3.1 Structural and functional changes of cell walls

The greatest stress for under-water plants is the weakening of gas exchange, which causes a decrease in oxygen in the stem and root, and also [92, 93] can induce enhanced growth by elongation, which promotes the release of leaves from the water to the surface and accelerates their contact with air [94]. Modification of cell walls for underwater growth and elongation requires energy, but, as a rule, such plants are characterized by limited aerobic metabolism. It is studied the structural changes in *Rumex palustris* stems and leaves that were induced by flooding [95]; authors have shown the decrease of size leaf and cuticle thickness in comparison with that in above-water leaves.

In cell walls of flooding leaves is occurred in protein synthesis. Under-water growth of rice is characterized by more elastic cell walls, which are usually characteristic of walls with increased synthesis of expansin [96–98]. In the cell wall noted protein modification, including expansins, which are activated at acidic pH [99, 100]. Rapid regulation of apoplastic pH provides a rapid way to regulate and modify apoplast expansin activity. The association between decreased cell wall elongations has been attributed to decreased tissue sensitivity to expansins [101]. Changes in the composition and nature of cross-links between cell wall polysaccharides may be limited by the mobility of expansins or their availability to the substrate polymer. The study showed a change in the ability of expansions to bind to cellulose depending on the properties of the hemicelluloses that cover the microfibrils [102].

In low-growing rice, flood resistance is explained by the activation of two genes: *SUB1A-1* gene and gene *ERF* (ethylene response factor) [103]. The *SUB1A-1* gene exhibits prolonged resistance to flooding associated with anaerobic metabolism and suppression of genes associated with high-energy processes, such as growth involving expansin [104]. The expression of *SUB1A-1e* is induced by ethylene, which is synthesized in a plant growing under water. Restriction of expansin transcripts by induction of *SUB1A-1* occurs as a result of suppression of stem tension and dormancy of the plant. Modification of the growth rate is mediated by the modification of the cell wall in the plant organs that is not only under water but also above water. In particular, when the internode of rice came out of the water into the air, the rate of growth decreased.

The composition of the wall can also determine the effectiveness of expansins to elongation of a wall under conditions of flooding the plant. The decrease in the elongation of cell walls in the segments of underwater rice stalks at the exit from the water to the air correlates with the changes in the composition of walls: an increase in xylose and pectic acids, such as ferule acid [101], which has the ability to form cross-links between polysaccharides of a cell wall [105]. Deposits of xylose-enriched polysaccharides can change the composition of the cell wall by limiting the action of expansin. It was found that the composition of polysaccharides of flooded plants differs from that of surface organs, as shown by Little [106] in stems of *Ludwigia repens*: along with a significant decrease in cellulose content, the content of hemicelluloses and lignin in underwater stems increased.

The outer cell walls of the epidermis of submerging and the above-water leaves are the first barrier, the first transport route of CO_2 and water, as well as the point of contact of plant organs with the environment. Cell walls of flooding leaves became thinner and their structure is characterized by loosening. Regarding the loosening of the cell wall, there are many models of this process. The first hypothesis about the acid-induced loosening mechanism was proposed by Cleland [107, 108]. It was later shown that the hydrolysis of polysaccharides during loosening is a complex process in which the enzymatic hydrolysis of polysaccharides of the wall matrix occurs with the participation of endoglucanases and expansins. The latter shown that hydrolyse polysaccharides induce cell expansion and increase the plasticity of the wall depending on the pH of the apoplast [109, 110]. The mechanism of formation of thin cell walls in various plant tissues is explained by changes in cell turgor and a decrease in the activity of enzymes involved in the synthesis of wall polysaccharides [109].

In submerged plants in the epidermis is also synthesized and deposited cuticle in the periclinal walls, and cuticle structure change [111]. In the cuticle of leaves and stems, which grow rapidly by elongation under water, there is an accelerated hydrolysis of cutin polymers [112]. In aquatic plants (hydrophytes) the cuticle of the epidermis of leaves and stems causes the presence of super hydrophobicity; it is this property that prevents the formation of an aqueous film on the surface of the organs submerged in water, which greatly reduces the gas exchange between the surface of the leaf and the gases dissolved in water. Despite the fact that CO_2 absorption for photosynthesis is reduced in flooded plants, the air layer or gas film on the surface of underwater leaves continues to exchange O_2 and CO_2 through the cuticle from the surrounding water layer, and therefore underwater photosynthesis and underwater respiration occur in epidermal cells [113].

Most underwater leaves of hydrophytes have no stomata. Transport functions mainly fall on the cell walls of the epidermis and pores in epidermis. Cuticular pores were revealed on the cross-sections of epidermal cells of underwater *Sagittaria sagittifolia, Trapa natans, Myriophyllum spicatum, Potamogeton pectinatus* and *Potamogeton perfoliatus* [113, 114]. It was shown that average high of the cuticular pores in the cells of the epidermis *of M. spica*tum was about 130 nm; pore density ~ 12–15 per 1 μm of cuticle length; in *P. pectinatus* cells—the pore length depending on the plane of the section is ranged from 80 to 100 nm, the average pore density ~ 20 per 1 μm of cuticle length; in *P. perfoliatus*, the cuticular pores had low contrast and were barely visible in the form of rounded electronically transparent structures with a diameter near ~4–5 nm. Besides, author shown the absence of stomata on both surfaces of flooding leaves, the decrease of high of cuticular ridges in anticlinal walls and absent of wax in cell walls. The mechanism of the absence of wax on the surface of the periclinal wall of the underwater leaves can be explained by the next, it is established that genes (*Ltp, LTPs* and *WAX9*) that are responsible for the transcription of lipid-transporting proteins have recently been identified in wax-enriched epidermal cells. In the case of expression of the corresponding genes in the cells of the epidermis, the effect of the accumulation of the corresponding mRNA was found [115, 116], and possible that absent of wax in the epidermis of underwater leaves may be a consequence of genetic plasticity, which lead to the inhibition of the synthesis of precursors of wax (C12-, C14- and C16-ω-hydroxy fatty acids) and wax synthase activity (fatty alcohol acyl-transferase) [117, 118].

3.2 The role of cellulose

Under-water leaves are characterized by the increase of amorphous cellulose and the decrease of its crystalline form. It is known that water is adsorbed by amorphous zones of cellulose, which are dominated by hydrogen bonds [119]. The crystalline component of cellulose micro fibrils is not involved in the transport or absorption of water molecules [120]. Given the above literature, we hypothesized that one of the adaptive features of the plant to flooding should be not only differences in cellulose content, but also advantages in the synthesis of its amorphous form. An optimal example of such adaptation to flooding can be the data of comparative structural and functional studies of cellulose in plants with underwater and above-water (surface) leaves, in particular in *Sium latifolium, T. natans* and *S. sagittifolia*, as well as leaves of hydrophytes growing only under water: *Potamogeton perfoliat, P. pectinatus* and *Myriiophyllum spicatum*. The above-water leaves of studied species, particular *S. latifolium, T. natans* and *S. sagittifolia* contained twice more of crystalline cellulose [18, 121, 122]. Considering the data on the identification of *PhEXP1* gene (in *Petunia hybrida* mutant) responsible for the synthesis of amorphous cellulose [123], we believe that increased content of amorphous cellulose in underwater leaves, due to genetic differences, is an adaptive change of flooded plants. Literature data on the cellulose content in flooded plants are different. Métraux and Kende [124] found no differences in cellulose content in flooded and above-water internodes of rice stems, whereas Little [106] showed a 1.5-fold decrease in cellulose content in flooded shoots of *L. repens* compared to above-water shoots. It is possible, that it depends on the species, tissue and stage of development of the plant. We hypothesize that the decrease in cellulose content in the underwater leaves of the studied species may be due to inhibition of thegene *CesA* (cellulose synthase catalytic subunit) encoding cellulose synthesis enzymes in both primary [125] and secondary cell walls [126].

3.3 The role callose, lignin and pectin

Callose—a polysaccharide of the cell walls, formed by glucose residues, connected at the base of β-1-3-glucoside bonds and in the lateral branches—1-6 connections. It is known that β—1,3—glucan plays a key role in intercellular water transport, cell growth and differentiation, osmotic stretching of cells, plant protection under biotic and abiotic stresses [127] and increases the elasticity and flexibility of leaves and stems [128–130]. It was established the effects of natural flooding on callose content in *Alisma plantago-aquatica* leaves with laser confocal microscopy. It was shown that the content of callose in the cell walls of submerged leaves was more in three times in epidermis walls and in 8 times—in mesophyll cell walls in comparison with above-water leaves [131]. Increase of callose content in cell walls of other species was noted also: in submerged leaves of *S. sagittifolia*—in six times, in floating leaves of *T. natans*-in 1.4 times in comparison with above-water leaves. It was established that increase of callose content in submerged leaves are accompanied by change of calcium ions content in walls [132].

Lignin is a polymer of aromatic alcohols, which is synthesized in the cell walls, is completed the growth by tension, and it is involved in the adaptation of plants to flooding and in the change of the structure of the matrix of cell wall, providing obstruction of water and aqueous solutions through the cell walls and also form the barrier for pathogens. Lignin is a complex of monolignols formed from p-hydroxy-phenyl, guajacyl, syringyl and H-phenylpropanoids components [133], which are involved in the polymerization of lignin, and they differ in the degree of methoxylation [46]. Flooding and siltation affect the lignification of cell walls. The study of mechanisms of adaptation of the root system of rice to flooding and siltation shown the main effect is the deficiency of oxygen, resulting in roots forming aerenchyma for storing of oxygen [134]. Lignin deposition, which counteracts the penetration of ions such as Fe^{2+}, Cu^{2+} and NaCl [135] has been observed during of flooding roots. It is considered that lignin and suberin can form a barrier to the penetration of oxygen and ions.

The effect of flooding on the lignification of rice stems was found by comparing the stems of three varieties of rice. It was found that the lignin content in rice stems and the activity of two enzymes of the lignification (coniferol alcohol dehydrogenase (CAD) and phenylalanine ammonium lyase (PAL) were reduced after flooding in the flood-sensitive variety and in control. Lignin and the activities of the studied enzymes were interrelated. According to researchers [136, 137], underwater plant organs are stressed due to the tension of the water column and the mechanical action of waves, which should cause stress in flooded organs. Lignin of dicotyledonous plants consists of guajacyl (G), syringyl (S) and phenylpropanoids (H) components, Lignin of most monocotyledons have G and S units, the content of which is almost the same, they may also contain H units [138].

The question of the distribution of lignin in various tissues of submerged plant organs, the role of monolignols and their ratio in cell walls in the process of natural adaptation of plants to flooding has remained open until recently. Recently it was established that in floating leaf walls underwater leaves of *T. natans* the level of lignin fluorescence intensity increased 1.52 times in the anticlinal walls of the epidermis and 1.2 times—in the periclinal walls, and decreased 1.6 times in the cell walls of photosynthetic parenchyma compared to the corresponding cell walls of floating leaf [18]. Cytochemical studies of monolignols, their localization and content in the leaves of the four studied species of hydrophytes (*Sagittaria sagitifolia, T. natans, M. spicatum; and P. perfoliatus*) showed both common and different features. Common features were: (1) the presence of syringyl and guajacyl in the studied species, regardless of the conditions of leaf growth; (2) almost identical (low) values of the S/G ratio in the cells of the vessels of the above-water leaves of *Sagittaria sagitifolia* and water nut; (3) the highest S/G values for *M. spicatum*; and (4) a certain polarity S/G, which is characteristic of each species. Distinctive features were: (1) the relative content of syringyl and guajacyl, as well as the ratio of S/G in the cell walls of underwater and above-water leaves of arrowroot and water nut; and (2) high S/G ratio in the periclinal walls of floating leaves compared to those in submerged leaves of water nut [132].

Similar to the increase in lignin in flooded stems *Ludviga repens* was established [106]. Little S. showed that when *L. repens* stems were flooded, the lignin content became 1.6 times higher than in stems that came out (from) of the water. Why is this happening? Lignin is known to be a highly branched polymer of phenylpropanoid components synthesized in a complex cycle [139], the passage of which depends on numerous endogenous (phytohormones) and exogenous factors, including exposure to light, temperature, different gases and biotic stresses [140]. It is believed that the functional value of lignin is the mechanical support of tissues, which allows the plant to stay upright relative to the Earth and not fall [141]. At the same time, underwater organs of flooded plants (stems and leaves) must withstand water pressure (its weight) and the action of waves [136, 137] which affects the growth and structural and functional parameters of tissues and cells as this is described for flooded leaves of *Veronica anagallis-aquatica* [141]. On the other hand, it is known that underwater leaves are devoid of trichomes, stomata and thick cuticle) [132], so the surface of submerged plant organs becomes more sensitive to pathogen invasion. It has been shown that leaves and stems synthesize lignin in response to attack by pathogens (bacteria and fungi) [142]. It is possible that the above exogenous parameters and some endogenous factors cause increased synthesis of lignin in the underwater leaves of the studied species of *T. natans* and *S. sagittifolia* [132].

At flooding of terrestrial plants leads to the formation of aerenchyma in roots, nodules, stem or submerged leaves. Aerenchyma helps the plant to survive in conditions of hypoxia by reducing the number of oxygen-consuming cells in vegetative organs [143, 144]. It was established that at lysigenous type of formation of an aerenchyma occurred the lysis not only of cytoplasmic organelles in tissues, but also lysis of their cell walls. The increases in aerenchyma air volume may enable prolonged functioning of aerobic metabolic processes in tissues exposed to low-oxygen conditions. Cellulose, hemicellulose and pectin lysis are occurs during aerenchyma formation. Probably, that modification of the pectin homogalacturonan backbone structure through de-methyl-esterification appears to be one mechanism by which cell walls and middle lamella of tissues is degradate of pectin and enable cavity formation of aerenchyma in roots [143]. Additionally, presence of fully and partially de-methyl-esterified homogalacturonan residues in cell walls of forming tylose-like cells suggests these pectin structures are essential to development of the cells that occlude aerenchyma of *P. sativum*, *P. coccineus* and *C. arietinum*. The investigators think that aerenchyma formation may depend on activity of cellulase, xylanase working together to achieve cell wall degradation. Specifically, xylanases and cellulases may degrade xylan and cellulose polysaccharides in advance of de-methyl-esterification of pectin by PME enzymes and subsequent degradation by pectinases [143].

That is, the constant aquatic environment is one of the main exogenous factors of increased synthesis of lignin in the studied hydrophytes. In addition, we see that the presence of syringyl and guajacyl monolignols, as well as their relationship in the cell walls of the epidermis, mesophyll and leaf vessels of hydrophytes is similar to that described for dicotyledonous angiosperms [138, 145]. We do not rule out that the cell walls of the underwater leaves of the studied plants contain a third monolignol—phenylpropanoid (*p*-hydroxyphenyl) which will need to be investigated by other methods. Summarizing the whole section on the impact of flooding on the structural and functional changes of cell walls, in particular on the synthesis of cellulose, callose and lignin, we can schematically present the course of major events occurring in the apoplast of most cells (**Figure 2**).

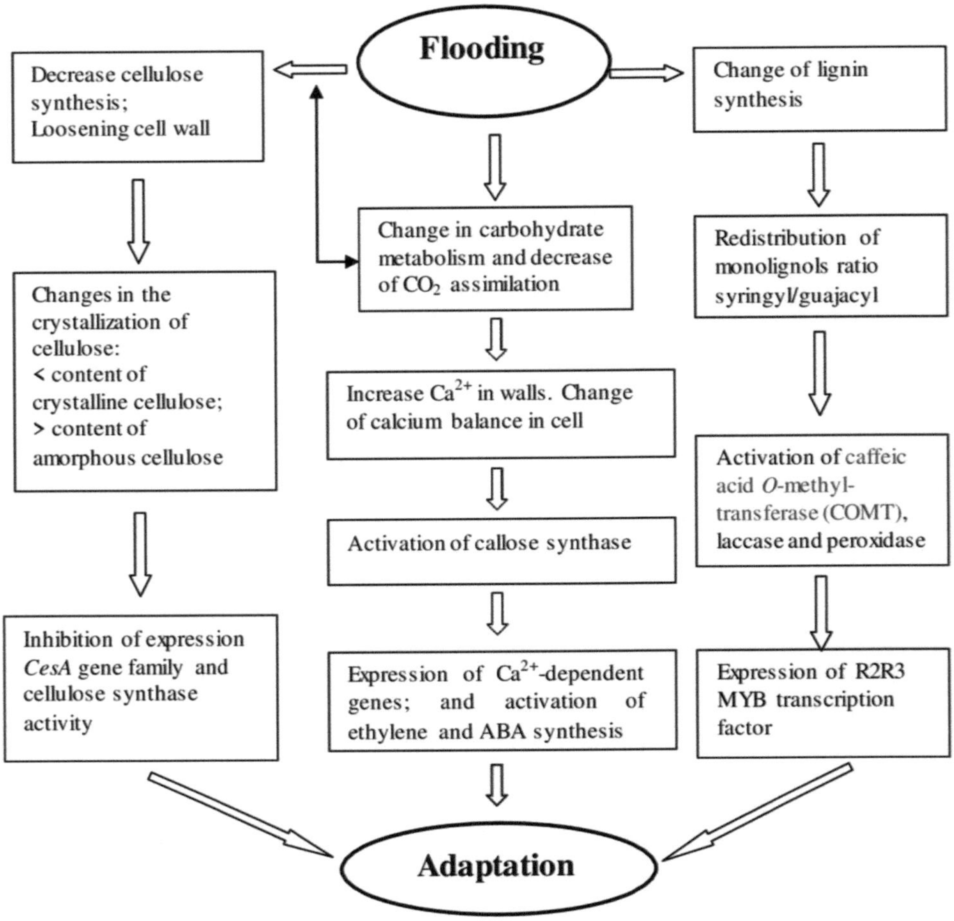

Figure 2.
Schematic representation of the main functional changes of plant cell wall during adaptation to flooding.

4. Concluding remarks

The results of researches concerning on the role of cell walls in plant response to natural unfavorable conditions influences show that cell wall is one of the compartments of a plant cell that responds to drought and flooding. In most wild species and in cultivated species, cell walls stand a marker of such influence. The inhibition of plant growth, the change of plant morphological and anatomical signs, change of cell wall ultrastructure, its composition is occurred under prolonged drought or flooding. Changes in the structural and functional characteristics of cell walls allow plants to survive. Plant adaptation to these factors is depended on species, stage of growth plant and influence duration. Numerous studies have shown that drought effects negatively on cell walls. The main mechanisms of plant adaptation to the effects of drought involve a decrease in the intensity of transpiration, an increase in the synthesis of wax, suberin, and lignin, as well as the compaction of the walls of the epidermis tissues for preservation of optimal water balance. Upon exposure to flooding, adaptation mechanisms are expressed in the next: decrease stomata density

DOI: http://dx.doi.org/10.5772/intechopen.105452

and wax in leaf epidermis; a loosening of cellulose micro fibrils in walls of epidermal tissue and a present of cuticle pores; the decrease of common cellulose content and crystalline form of cellulose; an increase of content of amorphous cellulose, hemicelluloses in a cell wall; an intensification of callose synthesis; the change of a ratio of monolignols (syringyl and quajacyl) in walls; the activation of peroxidase and expansin, an intensification of ethylene synthesis and a change of calcium balance in apoplast. However, the sequence of these processes has not been fully disclosed. The question of the launch of adaptative processes also remains open. These issues require further research. The question of the relationship between the water balance of the cell, photosynthesis and the values of energy of light photons on the surface of the leaves, which launch an adaptive response in the plant under adverse natural changes or under stress, also remains open.

Competing interests

The author declares that there is no conflict interest.

Author details

Olena Nedukha
Department of Cell Biology and Anatomy, M.G. Kholodny Institute of Botany, of National Academy of Sciences of Ukraine, Ukraine

*Address all correspondence to: o.nedukha@hotmail.com

References

[1] Kordyum EL. Phenotypic plasticity in plants: general characteristics of adaptive value, possible mechanisms, open questions. Ukrainian Botanical Journal. 2001;**58**(1):141-151

[2] De Olivera AB, Alencar NL, Gomes-Fiho E. Comparison between the Water and Salt Stress Effects on Plant Growth and Development. TECH; 2013. pp. 67-94

[3] Knight H, Trewavas AJ, Knight MR. Calcium signalling in arabidopsis thaliana responding to drought and salinity. Plant Journal. 1997;**12**(5):1067-1078

[4] Tuteja N, Sopory S. Chemical signaling under abiotic stress environment in plants. Plant Signaling and Behavior. 2008;**3**(8):525-536

[5] Tuteja R, Tuteja N. Serial analysis of gene expression (SAGE): unraveling the bioinformatics tools. BioEssays. 2004;**26**(8):916-922

[6] Yadov SK. Cold stress tolerance mechanisms in plants. A review. Agronnomy for Sustainable Developmant. 2010;**30**(3):515-527

[7] Grigore MN, Toma C. Histo-anatomical strate-gies of *Chenopodiaceae* halophytes; adaptive, eco-logical and evolutionary implications. WSEAS Transactions on Biology and Biomedicine. 2007;**12**(4):204-218

[8] Hong-Bo S et al. Water-deficit stress-induced anatomical changes in higher plants. Comptes Rendus Biologies. 2008;**331**(3):215-225

[9] Grene R, Vasquez-Robinet C, Bohnert HJ. Molecular biology and physiological genomics of dehydration stress. In: Lüttge U, Beck E, Bartels D, editors. Plant Desiccation Tolerance. Heidelberg: Springer; 2011. pp. 255-288

[10] Waraich EA et al. Alleviation of temperature stress by nutrient management in crop plants: a review. Journal of Soil Science and Plant Nutrition. 2012;**12**(2):221-244

[11] Futorna OA, Badanina VA, Zhygalova SL. Ecological-anatomical characteristics of some *Tragopogon* (Asteraceae) species of the flora of Ukraine. Biosystem Diversity. 2017;**25**(4):274-281

[12] Luo Y et al. Physiological acclimation of two psammo-phytes to repeated soil drought and rewatering. Acta Physiologiae Plantarum. 2011;**33**(1):79-91

[13] Aguirrezabal L et al. Plasticity to soil water deficit in *Arabidopsis thaliana*: dissection of leaf development into underlying growth dynamic and cellular variables reveals invisible phenotypes. Plant Cell and Environment. 2006;**29**(12):2216-2227

[14] Leucci M et al. Water stress and cell wall polysaccharides in the apical root zone of wheat cultivars varying in drought tolerance. Journal of Plant Physiology. 2008;**165**(11):1168-1180

[15] Neumann PM. The role of cell wall adjustment in plant resistance to water deficits. Crop Science. 1995;**35**(5):1258-1266

[16] Abd Elhalim Mona E et al. The anatomical features of the desert halophytes *Zygophyllum album* L.F. and *Nitraria retusa* (FORSSK.) ASCH

[2016]. Annals of Agricultural Sciences. 2018:**61**(8):97-104

[17] Futorna OA. Anatomical structure of the vegeta-tive organs *Helichrysum arenarium* and *H. corym-biforme* (Asteraceae Dumort.). Ukrainian Botanical Journal. 2001;**58**(1):111-120

[18] Nedukha OM, Kordyum EL. The participation of cell wall polysaccharides in cellular mechanisms of leaf tolerance to nature flooding of plants. In: Grzesiak MC, Rzepka A, Hura T, Grzesiak S, editors. Plant Functions under Environmental Stress. Cracow, Poland: The F. Gorski Institute of Plant Physiology, Polish Acad. Sci; 2013. pp. 137-152

[19] Zamski E, Ucko O, Koller D. The mechanism of root contraction in *Gymnarrhena micranatha*, a desert plant. New Phytolpgist. 1983;**95**(1):29-35

[20] Peña-Valdivia CB et al. Anatomical root variations in response to water deficit: Wild and domesticated common bean (*Phaseolus vulgaris* L.). Biological Research. 2010;**43**(4):417-427

[21] North GB, Nobel PS. Radial hydraulic conductivity of individual root tissues of Opuntia ficus- indica (L.) Miller as soil moisture varies. Annals of Botany. 1996;**77**(2):133-142

[22] De Micco V, Aronne G. Morpho-anatomical traits for plant adaptation to drought. In: Aroca R, editor. Plant Responses to Drought Stress, From Morphology to Molecular Features. Berlin, Heidelberg: Springer-Verlag; 2012. pp. 37-61

[23] Konno H et al. Differential changes in cell wall matrix polysaccharides and glycoside-hydrolyzing enzymes in developing wheat seedlings differing in drought tolerance. Journal of Plant Physiology. 2008;**165**(7):745-754

[24] Kerepesi I, Galiba G. Osmotic and salt stress-induced alteration in soluble carbohydrate content in wheat seedlings. Crop Science. 2000;**40**(2):482-487

[25] Miller RH. The prevalence of pores and canals in leaf cuticular membranes. Annals of Botany. 1985;**55**(4):459-471

[26] Miller RH. The morphology and permeability of isolated cuticular membranes of *Hoya carnosa* R. Br. (Asclepiadaceae). Annals of Botany. 1986;**58**(4):407-416

[27] Riederer M, Schreiber L. Waxes: the transport barriers of plant cuticles. In: Hamilton RJ, editor. Waxes: Chemistry, Molecular Biology and Functions. Dundee: The Oily Press; 1995. pp. 131-156

[28] Provost GLe G et al. Soil water stress affects both cuticular wax content and cuticle-related gene expression in young saplings of maritime pine (*Pinus pinaster* Ait). BMC Plant Biology. 2013;**13**:95. DOI: 10.1186/1471-2229-13-95

[29] Moor JP et al. Adaptations of higher plant cell walls to water loss: Drought vs desiccation. Physiologia Plantarum. 2008;**134**(2):237-245

[30] Cameron K, Teece M, Smart LB. Increased accumulation of cuticular wax and expression of lipid transfer protein in response to periodic drying events in leaves of tree tobacco. Plant Physiology. 2006;**140**(1):176-183

[31] Aharoni A et al. The *SHINE* Clade of *AP2* domain transcription factors activates wax biosynthesis, alters cuticle properties, and confers drought tolerance when overexpressed in arabidopsis. The Plant Cell. 2004;**16**(9):2463-2480

[32] Kordyum EL et al. In: Kordyum E, editor. Cellular Mechanisms

of Adaptation of Plants to Adverse Effects of Environmental Factors in Natural Conditions. Kiev: Publ. Nauk. Dumka; 2003. p. 277

[33] Zhang S, Quartararo A, Betz OK, et al. Root vacuolar sequestration and suberization are prominent responses of Pistacia spp. rootstocks during salinity stress. Plant Direct. 2021;**5**(5):e00315. DOI: 10.1002/pld3.315

[34] De Santo AV, Alfani A, De Luca P. Water vapour uptake from the atmosphere by some *Tillandsia* species. Annals of Botany. 1976;**40**(166):391-394

[35] Lange OL, Kilian E, Ziegler H. Water vapor up-take and photosynthesis in lichens: Performance dif-ferences in species with green and blue-green algae as phycobionts. Oecologia. 1986;**71**(1):104-110

[36] Paulino MKSS, de Souza ER, Lins CMT, Dourado PRM, et al. Influence of vesicular trichomes of *Atriplex nummularia* on photosynthesis, osmotic adjustment, cell wall elasticity and enzymatic activity. Plant Physiology and Biochemistry. 2020;**155**(1):177-186. DOI: 10.1016/j.plaphy.2020.07.036

[37] Grammatikopoulos G, Manetas Y. Direct absorp-tion of water by hairy leaves of *Phlomis fruticosa* and its contribution to drought avoidance. Cananadian Journal of Botany. 1994;**72**(12):1805-1811

[38] Schӧnherr J. Calcium chloride penetrates pear leaf cuticles via aqueous pores. Planta. 2000;**212**(1):112-118

[39] Wang X et al. Leaf epidermal water-absorbing scales and their absorption of unsaturated atmospheric water in *Reaumuria soongorica*, a desert plant from the northwest arid region of China. Journal of Arid Environments. 2016;**128**(1):17-29

[40] Perrone R et al. What do leaf anatomy and micro-morphology tell us about the psammophilous Pancratium maritimum L. (Amaryllidaceae) in response to sand dune condi-tions? Flora. 2015;**213**:20-31

[41] Redmann RE. Adaptation of grasses to water stress leaf rolling and stomate distribution. Annals of the Missouri Botanical Garden. 1985;**72**(4):833-842

[42] Ripley EA, Redmann RF. Grassland. In: Mon-teith JL, editor. Vegetation and the Atmosphere. Vol. 2. London: Academic Press; 1976. pp. 349-398

[43] Nedukha OM. Influence of growth conditions on content of monolignols in *Phragmites australis* leaves. The Bulletin of Kharkiv National Agrarial University. Series Biology. 2020;**3**(51):48-57

[44] Hesp PA. Ecological processes and plant adapta-tions on coastal dunes. Journal of Arid Environments. 1991;**21**(2):165-191

[45] Jianu LD, Bercu R, Popoviciu DR. *Silene thymifolia* Sibth. et Sm. (Caryophyllaceae)—A vulner-able species in Romania: Anatomical aspects of vegetative organs. Notulae Scientia Biologicae. 2021;**13**(1):1-7

[46] Leisola M, Pastinen O, Axe DD. Lignin—Designed. BIO-Complexity. 2012;**3**:1-11

[47] Glerum C. Drought ring formation in conifers. Forest Science. 1970;**16**(2):246-248

[48] Lloyd D. Abnormal lignin distribution in wood from severely drought stressed *Pinus radiate* trees.

IAWA Journal /International Association of Wood Anatomists. 2002;**23**(2):161-178

[49] Alvarez S et al. Metabolomic and proteomic changes in the xylem sap of maize under drought. Plant Cell and Environment. 2008;**31**(3):325-340

[50] Lee B-R et al. Peroxidases and lignification in relation to the intensity of water-deficit stress in white clover (*Trifolium repens* L.). Journal of Experimental Botany. 2007;**58**(6):1271-1271

[51] Terzi R et al. Lignification response for rolled leaves of *Ctenanthe setosa* under long-term drought stress. Turkish Journal of Biology. 2013;**37**(5):614-619

[52] Roig-Oliver M, Bresta P, Nadal M, Liakopoulos G, et al. Cell wall composition and thickness affect mesophyll conductance to CO_2 diffusion in *Helianthus annuus* under water deprivation. Journal of Experimental Botany. 2020;**71**(22):7198-7209. DOI: 10.1093/jxb/eraa413

[53] Roig-Oliver M, Nadal M, Flexas JB. *Ginkgo biloba* and *Helianthus annuus* show different strategies to adjust photosynthesis, leaf water relations, and cell wall composition under water deficit stress. Photosynthetica. 2020;**58**(5):1098-1106. DOI: 10.32615/ps.2020.063

[54] Novaković L, Guo T, Bacic A, Sampathkumar A, Johnson K. Hitting the wall-sensing and signaling pathways involved in plant cell wall remodeling in response to abiotic stress. Plants. 2018;7(4):89. DOI: 10.3390/plants7040089

[55] Tenhaken R. Cell wall remodeling under abiotic stress. Frontiers in Plant Science. 2015;**5**:771. DOI: 10.3389/fpls.2014.00771

[56] Roig-Oliver M, Bresta P. Nikolopoulos, Bota J. Dynamic changes in cell wall composition of mature sunflower leaves under distinct water regimes affect photosynthesis. Journal of Experimantal Botany. 2021;**72**(22):7863-7875. DOI: 10.1093/jxb/erab372

[57] Roig-Oliver M, Rayon C, Roulard R, Fournet F, Bota J, Flexas J. Reduced photosynthesis in *Arabidopsis thaliana atpme17.2* and *atpae11.1* mutants is associated to altered cell wall composition. Physiologia Plantarum. 2021;**172**(3):1439-1451. DOI: 10.1111/ppl.13186

[58] Domozych D, Li K, Palacio-Lopez K. The effects of osmotic stress on the cell wall-plasma membrane domains of the unicellular streptophyte, Penium margaritaceum. Protoplasma. 2021;**258**(6):1231-1249. DOI: 10.1007/s00709-021-01644-y

[59] Roig-Oliver M, Douthe C, Bota J, Flexas J. Cell wall thickness and composition are related to photosynthesis in Antarctic mosses. Physiologia Plantarum. 2021;**173**(4):1914-1925. DOI: 10.1111/ppl.13533

[60] Lins U et al. Structure, morphology and composition of silicon biocomposites in the palm tree *Syagrus coronate* (Mart.) Becc. Protoplasma. 2002;**220**(1-2):89-96

[61] Wang L et al. Biosilicified structures for cooling plant leaves: a mechanism of highly efficient mid-infrared thermal emission. Applied Physics Letters. 2005;**87**(19):194105. DOI: 10.1063/1.2126115

[62] Epstein E. Silicon. Annual Review of Plant Physiology and Plant Molecular Biology. 1999;**50**:641-664. DOI: 10.1146/annurev.arplant.50.1.641

[63] Guerriero G, Hausman JF. Legay S (2016) Silicon and the plant extracellular matrix. Frontiers in Plant Science. 2016;7:463. DOI: 10.3389/fpls.2016.00463

[64] Sheng H, Chen S. Plant silicon-cell wall complexes: Identification, model of covalent bond formation and biofunction. Plant Physiology and Biochemistry. 2020. DOI: 10.1016/j.plaphy.2020.07.020

[65] Chen Z, Zhou T, Hu J, Duan H. Quartz crystal microbalance with dissipation monitoring of Dynamic viscoelastic changes of tobacco BY-2 cells under different osmotic conditions. Biosensors. 2021;**11**(5):136. DOI: 10.3390/bios11050136

[66] Hodson MJ et al. Phylogenetic variation in the silicon composition of plants. Annals of Botany. 2005;**96**(6):1027-1046

[67] Ma JF, Takahashi E. Interaction between calcium and silicon in water-cultured rice plants. Plant and Soil. 1993;**148**(1):107-113

[68] Gong HJ et al. Silicon alleviates oxidative damage of wheat plants in pots under drought. Plant Science. 2005;**169**(2):313-321

[69] Ahmad R, Zaheer SH, Ismail S. Role of silicon in salt tolerance of wheat (*Triticum aestivum* L.). Plant Science. 1992;**85**(1):43-50

[70] Ming DF et al. Silicon alleviates PEG-induced water-deficit stress in upland rice seedlings by enhancing osmotic adjustment. Journal of Agronomy and Crop Science. 2012;**198**(1):14-26

[71] Liu P et al. Aquaporin-mediated increase in root hydraulic conductance is involved in silicon-induced improved root water uptake under osmotic stress in *Sorghum bicolor* L. Journal of Experimental Botany. 2014;**65**(17):4747-4756

[72] Ganie SA, Ahammed GJ. Dynamics of cell wall structure and related genomic resources for drought tolerance in rice. Plant Cell Reports. 2021;**40**:437-459. DOI: 10.1007/s00299-020-02649-2

[73] Liu J, Zhang W, Long S, Zhao C. Maintenance of cell wall integrity under high salinity. International Journal of Molecular Sciences. 2021;**22**(6):3260. DOI: 10.3390/ijms22063260

[74] Ljerka K, Lacey S. Plant cuticles shine: advances in wax biosynthesis and export. Current Opinion in Plant Biology. 2009;**12**(6):721-727

[75] Kosma DK et al. The impact of water deficiency on leaf cuticle lipids of Arabidopsis. Plant Physiology. 2009;**151**(4):1918-1929

[76] Amal H et al. Molecular and physiological analysis of drought stress in Arabidopsis reveals early responses leading to acclimation in plant growth. Plant Physiology. 2010;**154**(3):1254-1271

[77] Wu Y, Cosgrove DJ. Adaptation of roots to low water potentials by changes in cell wall extensibility and cell wall proteins. Journal of Experimental Botany. 2000;**51**(350):1543-1553

[78] Hannes V et al. Domains of expansin gene expression define growth regions in the shoot apex of tomato. Plant Molecular Biology. 2003;**53**(3):267-272

[79] Muller B et al. Association of specific expansins with growth in maize leaves is maintained under environmental, genetic, and developmental sources of variation. Plant Physiology. 2007;**143**(1):278-290

[80] Bacon MA. The biochemical control of leaf expansion during drought. Plant Growth Regulation. 1999;**29**(1-2):101-112

[81] Bray EA. Genes commonly regulated by water-deficit stress in *Arabidopsis thaliana*. Journal of Experimental Botany. 2004;**55**(407):2331-2341

[82] Skirycz A et al. Developmental stage specificity and the role of mitochondrial metabolism in the response of Arabidopsis leaves to prolonged mild osmotic stress. Plant Physiology. 2010;**152**(1):226-244

[83] Neumann PM, Azaizeh H, Leon D. Hardening of root cell walls: A growth inhibitory response to salinity stress. Plant, Cell and Environment. 1994;**17**(3):303-309

[84] Cuadrado-Pedetti MB, Rauschert I, Sainz MM, et al. The Arabidopsis TETRATRICOPEPTIDE THIOREDOXIN-LIKE 1 gene is involved in anisotropic root growth during osmotic stress adaptation. Genes. 2021;**12**(2):236. DOI: 10.3390/genes12020236

[85] Chen Z et al. Disruption of the cellulose synthase gene, *AtCesA8/IRX1*, enhances drought and osmotic stress tolerance in *Arabidopsis*. The Plant Journal. 2005;**43**(2):273-283

[86] Rui Y, Dinneny JR. A wall with integrity: surveillance and maintenance of the plant cell wall under stress. New Phytologist. 2020;**225**(4):1428-1439. DOI: 10.1111/nph.16166

[87] Vartapetian BB et al. Functional electron microscopy in studies of plant response and adaptation to anaerobic stress. Annals of Botany. 2003;**91**(2):155-172

[88] Doronin YP. Ocean Physics. Sank-Peterb: Sank-Peterburg Press; 2000. p. 305. UDK535 BBK 22.34.L22 (In Russ.)

[89] Landsberg GS. Optics. Moscow: Physico-Mathematical Literature; 2003. p. 848. ISRN 5-9221-0314-8 (In Russ.)

[90] Voesenek LA et al. How plants cope with complete submergence. New Phytologist. 2006;**170**(2):213-226

[91] Mommer L, Visser EJW. Underwater photosynthesis in flooded terrestrial plants: A matter of leaf plasticity. Annals of Botany. 2005;**96**(4):581-589

[92] Jackson MB, Armstrong W. Formation of aerenchyma and the processes of plant ventilation in relation to soil flooding and submergence. Plant Biology. 1999;**1**(3):274-287

[93] Rijnders JG et al. Ethylene enhances gibberellin levels and petiole sensitivity in flooding–tolerant *Rumex palustris* but not in flooding-intolerant *R. Acetosa*. Planta. 1997;**20**:320-325

[94] Voesenek LAC et al. Plant hormones regulate fast shoot elongation under water: From genes to communities. Ecology. 2004;**85**(1):16-27

[95] Mommer L et al. Submergence–induced leaf acclimation in terrestrial species varying in flooding tolerance. New Phytologist. 2007;**176**(2):337-345

[96] Cho HT, Kende H. Expansins and internodal growth of deepwater rice. Plant Physiology. 1997;**113**(4):1145-1151

[97] Lee Y, Kende H. Expression of α-Expansin and Expansin-Like Genes in Deepwater Rice. Plant Physiology. 2002;**130**(3):1396-1405

[98] Cho HT, Kende H. Expression of expansin genes is correlated with

growth in deepwater rice. The Plant Cell. 1997;**9**(9):1661-1671

[99] Cosgrove DJ. Loosening of plant cell walls by expansins. Nature. 2000;**407**(6802):321-326

[100] Vreeburg RAM et al. Ethylene regulates fast apoplastic acidification and expansin. A transcription during submergence-induced petiole elongation in *Rumex palustris*. The Plant Journal. 2005;**43**(4):597-610

[101] Daisuke S, Tetsushi A, Kazuyuki I. Changes in expansin activity and cell wall susceptibility to expansin action during cessation of internodal elongation in floating rice. Plant Growth Regulation. 2009;**57**(1):79-88

[102] McQueen-Mason SJ, Cosgrove DJ. Expansin mode of action on cell walls. Analysis of wall hydrolysis, stress relaxation, and binding. Plant Physiology. 1995;**107**(1):87-100

[103] Xu K et al. *Sub1A* is an ethylene-response-factor-like gene that confers submergence tolerance to rice. Nature. 2006;**442**:705-708

[104] Fukao T et al. A variable cluster of ethylene response factor-like genes regulates metabolic and developmental acclimation responses to submergence in rice. The Plant Cell. 2006;**18**(8):2021-2034

[105] Saulnier L, Thibault JF. Ferulic acid and diferulic acids as components of sugar-beet pectins and maize bran heteroxylans. Journal of the Science of Food and Agriculture. 1999;**79**(3):396-402

[106] Little ST. Adaptation and acclimation of populations of *Ludwigia repens* to growth in high- and lower CO_2 springs. A Dissertation presented to the graduate school of the University of Florida in partial fulfillment of the requirements for the degree of doctor of philosophy. USA: University Florida; 2003. pp. 1-157

[107] Cleland RE. The capacity for acid-induced wall loosening as afactor in the control of *Avena* coleoptile cell elongation. Journal of Experimental Botany. 1983;**34**(143):676-680

[108] Rayle DL, Cleland RE. The acid growth theory of auxin-induced cell elongation. Plant Physiology. 1992;**99**(4):1271-1274

[109] Cosgrove DJ. Relaxation in a high-stress environmental: the molecular bases of extensible cell wall and cell enlargement. The Plant Cell. 1997;**9**(7):1031-1041

[110] Cosgrove DJ. Wall structure and wall loosening. A look backwards and forwards. Plant Physiology. 2001;**125**(1):131-134

[111] Frost-Christensensen H, Floto F. Resistance to CO_2 diffusion in cuticular membranes of amphibious plants and the implication for CO_2 acquisition. Plant, Cell and Environment. 2007;**30**(1):12-18

[112] Hoffmann-Benning S, Kend H. On the role of abscisic-acid and gibberellin in the regulation of growth in rice. Plant Physiology. 1992;**99**(3):1156-1161

[113] Nedukha OM. Epidermis leaf structural responses of some aquatic plants to constant water environment. Advances of Agricultural Sciences. 2010;**545**:169-178

[114] Nedukha OM. Constant water environment and peculiarity of leaf structure of some water plants. Advances of Agricultural Sciences. 2008;**524**:179-203

[115] Clark AM, Bohnert H. Epidermis–specific transcripts: Nucleotide sequence of a full–length cDNA of $EP1_{12}$ encoding a putative lipid transfer protein. Plant Physiology. 1993;**103**(2):677-678

[116] Liu F et al. Non-specific lipid transfer proteins in plants. Journal of Experimental Botany. 2015;**66**(19):5663-5681

[117] Hauke V, Schreiber L. Ontogenetic and seasonal development of wax composition and cuticular transpiration of ivy (*Hedera helix* L.) sun and shade leaves. Planta. 1998;**207**(1):65-75

[118] Lardizabal KD et al. Purification of a jjoba embryo wax synthase, cloning of its cDNA, and production of high levels of wax in seeds of transgenic *Arabidopsis*. Plant Physiology. 2000;**122**(3):645-656

[119] Brown RM. The biosynthesis of cellulose. Journal of Macromolecular Science. Part A. 1996;**33**(10):1345-1373

[120] Nilsson M. Water-Induced Charge Transport in Microcrystalline Cellulose. Uppsalla: Acta Universitatis Upsaliensis; 2006. pp. 1-54. (Digital Comprehensive Summaries of Uppsala Dissertations from the Faculty of Science and Technology. ISSN 1651-6214 ISBN 91-554-6541-2)

[121] Nedukha OM. Content and types of cellulose in *Sium latifolium* leaves at vegetative growth. Ukrainian Botanical Journal. 2010;**67**(2):199-126

[122] Nedukha OM. Leaf: tissue, cellular and molecular levels. In: Cellular mechanisms of plants adaptation to unfavorable influences of ecological factors in nature conditions. In: Cellilar Mechanisms of Plants Adaptation to Environment in Nature. Kiev: Naukova Dumka; 2003. pp. 177-218

[123] Zenoni S et al. Downregulation of the *Petunia hybrida* α-exansin gene *PhEXP1* reduces the amount of crystalline cellulose in cell walls and leads to phenotypic changes in petal limbs. The Plant Cell. 2004;**16**(2):295-308

[124] Métraux JP, Kende H. The role of ethylene in growth response of submerged deep water rice. Plant Physiology and Biochemistry. 2020;**155**:13-19

[125] Taylor NG. Identification of cellulose synthase *AtCesA7* (IRX3) in vivo phosphorylation sites—A potential role in regulating protein degradation. Plant Molecular Biology. 2007;**64**(1):161-171

[126] Tanaka K et al. Three distinct rice cellulose synthetase catalytic subunit genes required for cellulose synthesis in the secondary wall. Plant Physiology. 2003;**133**(1):73-83

[127] Krzesłowska M. The cell wall in plant cell response to trace metals: polysaccharide remodeling and its role in defense strategy. Acta Physiologiae Plantarum. 2011;**33**(1):35-51

[128] Parre E, Geitmann A. More than a leak sealant. The mechanical properties of callose in pollen tubes. Plant Physiology. 2005;**137**(1):274-286

[129] Clarke AE, Stone BA. Chemistry and biochemistry of β-1,3-glucans. Reviews of Pure and Applied Chemistry. 1963;**13**:134-156

[130] Dmitriev AP. Phytoalexins and Their Role in Plant Resistance. Kiev: NAN of Ukraine; 1999. pp. 1-207

[131] Nedukha OM, Ovrutska II. Effects of water deficit on polysaccaharides content in cell walls of *Alisma plantago-aquatica* L. Reports of National Academy of Sciences of Ukraine. 1998;**N11**:170-173

[132] Nedukha OM. Role of cell walls in plant adaptation to flooding. Journal of Biology and Nature. 2017;**7**(1):38-62

[133] Adler E. Lignin chemistry—Past, present and future. Wood Science and Technology. 1977;**11**:169-218

[134] Shunsaku N et al. Mechanisms for coping with submergence and waterlogging in rice. Rice. 2012;**5** (Article 2):1-14

[135] Ranathunge K et al. Stagnant deoxygenated growth enhances root suberization and lignifications, but differentially affects water and NaCl permeabilities in rice (*Oryza sativa* L.) root. Plant, Cell and Environment. 2011;**34**(8):1223-1240

[136] Tyree MT, Cheung YNS. Resistance to water flow in *Fagus grandifolia* leaves. Canadian Journal of Botany. 1977;**55**(20):2591-2599

[137] Niklas KJ. Plant Biomechanics: An Engineering approach to Plant Form and Function. Chicago: Univ. of Chicago Press; 1992. p. 622

[138] Baucher M et al. Biosynthesis and genetic engineering of lignin. Critical Reviews in Plant Sciences. 1998;**17**(2):125-197

[139] Rastogi S, Dwivedi U. Manipulation of lignin in plants with special reference to O-methyltransferase. Plant Science. 2008;**174**(3):264-277

[140] Moura JCMS et al. Abiotic and biotic stresses and changes in the lignin content and compositon in plants. JIPB—Journal of Integrative Plant Biology. 2010;**52**(4):360-376

[141] Boeger MRT, Poulson ME. Morphological adaptations and photosynthesis rates of amphibious *Veronica anagallis-aquatica* (Scrophulariaceae) under different flow regimes. Aquatic Botany. 2003;**75**(2):123-136

[142] Hano C et al. Differential accumulation of monolinol-derived compounds in elicited flax (*Linum usitatissimum*) cell suspension culture. Planta. 2006;**223**(5):975-989

[143] Pegg J, Edelmann RR, Gladish DK. Immunoprofiling of cell wall carbohydrate modifications during flooding-induced aerenchyma formation in Fabaceae roots. Frontiers in Plant Science. 2020;**10**:1805. DOI: 10.3389/fpls.2019.01805

[144] Pan R, Han H, Medison MB, et al. Aerenchyma formation in the root of leaf-vegetable sweet potato: Programmedcell death initiated by ethylene-mediated H_2O_2 accumulation. Physiologia Plantarum. 2021;**173**:2361-2375. DOI: 10.1111/ppl.1358710.1111/ppl.13587

[145] Weng J-K et al. Emerging strategies of lignin engineering and degradation for cellulosic biofuel production. Current Opinion in Biotechnology. 2008;**19**(2):166-172

Chapter 17

Climate Change and Abiotic Stresses in Plants

Ananya Baidya, Mohammed Anwar Ali and Kousik Atta

Abstract

Any altered physiological conditions which can hamper the growth and development of crop plants that is denoted as stress. The challenges of abiotic stress on plant growth and development are evident among the emerging ecological impacts of climate change. In a compatible environment for one genotype may not be fitable for another. The field of plant abiotic stress encompasses all studies on abiotic factors or stresses from the environment that can impose stress on a variety of species. Abiotic stress induces redox imbalance during reproductive growth. These studies indicate that stress induced over accumulation of ROS leads to pollen abortion and programmed cell death of microspores in developing anthers consequently resulting in male sterility. With these changing climatic conditions climate resilient crops and crop varieties have been recommended as a way for farmers to cope with or adopt to climate change. Integrated physiological and molecular approaches are important for combating multiple abiotic stresses.

Keywords: abiotic stress, climate change, ROS

1. Introduction

Plants encounter changing environments during their entire life cycle, from seedling to reproductive stage, that are often unfavourable to their growth and developmental processes, and they create unique mechanisms to cope with these challenges. There are primarily two types of negative environmental variables. The first are biotic variables, such as disease and herbivore attacks, and the second are abiotic factors, such as drought, heat, cold, nutrient inadequacy, and heavy metal build-up in the soil. Salt, drought, and temperature, for example, have an impact on the geographical distribution of plant species and disturb plant metabolism. As a result, they reduce the quality and quantity of food production in agriculture, lowering the food supply for a growing population, and tolerance mechanisms in plants have been thoroughly explored to overcome these negative impacts. Various environmental variables (biotic and abiotic) activate stress tolerance genes in plants, causing them to become resistant.

2. Heat stress

The average temperature has been determined to be increasing by 0.2°C per year, and it will have to grow by 1.8°C 4°C by the end of the year 2100, making temperature

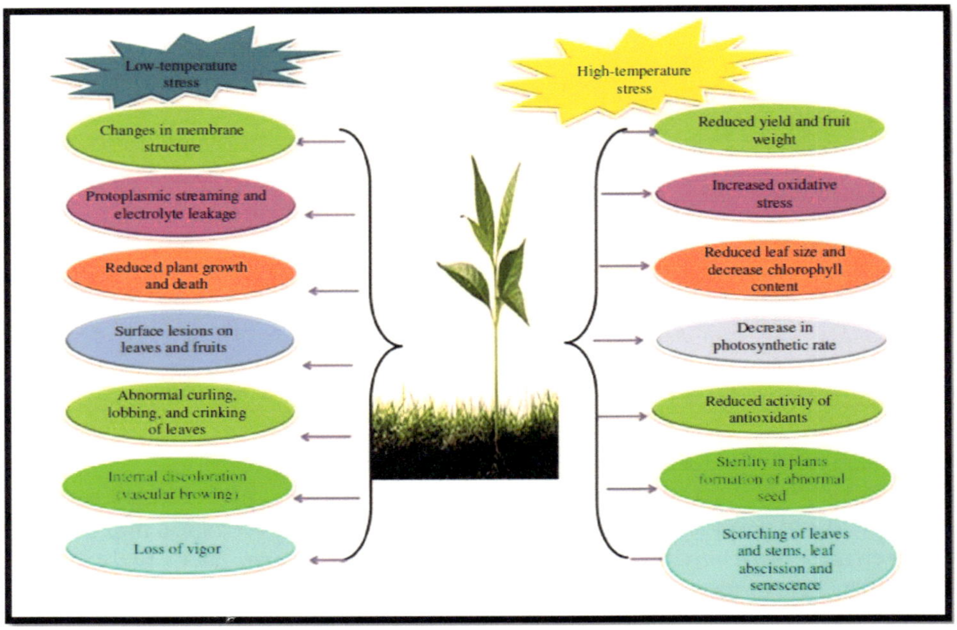

Figure 1.
Effect of low and high-temperature stress in plants. Source: Tiwary et al. [9].

one of the most damaging stresses [1]. Temperature-related climate change is a global concern that has affected plant physiological and biochemical activity, lowering crop output [1, 2]. Plants are subjected to heat stress as a result of rising temperatures, which is dependent on the quality, intensity, and duration of light.

All environmental conditions (biotic and abiotic) contribute to the production of reactive oxygen species (ROS), including heat stress, which damages macromolecules such as DNA, proteins, and lipids [3], and plants are under oxidative stress. Heat stress also changed the expression of genes involved in the creation of osmo-protectants, detoxifying enzymes, transporters, and regulatory proteins [4]. Heat stress, on the other hand, inhibits protein folding, alters membrane (lipid bilayer) fluidity and cytoskeleton arrangement, and has an impact on vegetative and reproductive tissue [5, 6]. A rise in temperature up to a certain point is helpful to plants, since it governs plant circadian rhythms, plant movements (corolla opening/closing), and impacts the geographical dispersion of plants in nature [7]. High temperatures increased the susceptibility of plants to pathogens. When the ambient temperature rose, the infection capability of tobacco mosaic and tomato-spotted wilt viruses increased, causing viral illnesses in tobacco (*Nicotiana tabacum*) and pepper (*Capsicum annuum*), respectively [8]. Wheat genotypes were shown to be more sensitive to *Cochliobolus sativus* (which causes spot blotch) when night time temperatures rose (**Figure 1**) [10].

3. Cold stress

Apart from heat stress, a drop in ambient temperature causes chilling stress in plants, which has a significant effect on cell physiology. Chilling, according to Ruelland et al. [5], promotes cell death by suppressing enzymatic activities, rigidifying biological membranes, stabilising nucleic acids, generating reactive oxygen species (ROS), and impairing photosynthesis. Low temperature causes flowering in plants, which is known as vernalization [11], and upregulates metabolic processes that confer the tolerance strategy of plants, known as the cold-hardening process [12], which results in the accumulation of compatible solutes (sugar), membrane composition changes, and increased synthesis of dehydrin-like proteins [13]. Plants had long- and short-term responses to temperature stress (heat or cold stress). Long-term effects included morphological and phenological adaptations, whereas short-term effects included leaf orientation changes, increased transpiration, and changes in membrane lipid content.

Heat stress causes a reduction in water loss by closing the stomata, as well as increased stomatal densities and larger xylem vessels [14], allowing plants to thrive in these harsh conditions. ROS production is an unavoidable by-product of aerobic activity, and its toxicity is determined by its concentration. It functions as a signalling molecule at low concentrations, but at greater concentrations, it becomes poisonous and causes cell death [15]. Under various stress situations, particularly heat stress, ROS such as H_2O_2, O_2, and $^1O_2$ are produced [16]. Every plant contains an assortment of antioxidant systems to deal with the detrimental effects of ROS. These systems help to lessen the negative effects. Enzymatic antioxidants such as superoxide dismutase (SOD), peroxidase (POD), catalase (CAT), and glutathione S-transferase (GST), as well as non-enzymatic antioxidants such as cysteine, proline, non-protein thiols, and the synthesis of molecular chaperones known as heat-shock proteins (HSPs), are part of the antioxidant system [17]. HSPs are important proteins that are activated by heat stress and target HS-responsive transcription factors, which regulate protein quality by re-energising proteins [18].

4. Salt stress

The impact of salt stress in plants has become an important issue in modern agricultural development, climate change and global food crisis. Survival of plants under adverse environmental conditions relies on integration of stress adaptive metabolic and structural changes into endogenous developmental programs. Almost all crops are sensitive to high salt (NaCl) levels [19], however the degree of sensitivity varies substantially between species and marginally between cultivar types within a species [19]. The plant response to high salt concentrations is complex. The effect of excessive NaCl concentrations on plants results in osmotic stress and creates an ionic imbalance due to accumulation of toxic ions like Na^+ and Cl^-. The plant's developmental growth stage [20] as well as external environmental elements influence the plant's sensitivity to salinity [21]. Salt stress is a major issue for agriculture systems since it reduces crop yield potential [22]. Excess salt in the soil affects plant growth by lowering seed germination, plant height, root length, and fructification [19], and it has also negative impact on mineral homeostasis, in particular Ca^{2+} and K^+. But it also has indirect impacts on the food web segment that depends on that host plant [22]. Salinity induced imbalance of cellular ion homeostasis is coped with regulated ion influx and effect at plasma membrane and vacuoler ion sequestration [23]. All of this occurs as a result of salt-induced oxidative stress [24], ion toxicity, and decreased photosynthetic rate in plants, all of which led to a considerable reduction in overall crop production [25].

Despite this, plants may naturally live and complete their life cycle under extreme salt stress [26]. They have well-developed biological, chemical, and physiological mechanisms [19], which could result in the synthesis of products and the start of processes that improve plant resistance to soluble salts [26]. Depending on the types of modifications they must perform in response to plant salt stress, these systems could be complex or simple [26]. Plant stress sensing and signalling machinery are critical components of their salt stress tolerance network, according to several studies [27]. The most common salt tolerance systems in plants are salt excessively sensitive (SOS) signalling pathways, hyperosmotic sensors, gene regulation in roots, and plant membrane Na^+ and K^+ transporters [27]. Plants' biological salt tolerance adaptations include osmoregulation and hormonal alterations. Other options for improving plant tolerance to salt stress include the use of plant growth-promoting rhizobacteria (PGPR), plant fungal associations, and the application of organic and inorganic amendments [28]. To address this problem, scientists used salt tolerant engineered plants or transgenic salt resistant cultivars, as well as a potential physiological method [27]. However, because the salt tolerance mechanism in plants is genetically complex, it was not very successful [19].

5. Drought stress

Drought is one of the key factors affecting crop output around the world, as crop growth and yield are both influenced by this stress [29]. Drought stress is caused by a lack of rainfall, salt buildup in the soil, significant temperature swings, and excessive light intensity. Due to climate change, climate simulation models that take previous year data and estimate the future have indicated that this stress would become more severe in the near future. Drought stress affects plant growth, water retention, and water efficiency [30, 31], as well as causing changes in physiological, biochemical, morphological, and molecular features [30, 31]. Drought-tolerant/resistant plants have evolved a more efficient drought resilience mechanism to withstand drought stress, however these mechanisms are not well-organised or investigated. Plants, in general, have a mechanism for maintaining cell homeostasis that involves increased water transport into the plant cell. Drought resistance is a cellular defence mechanism that allows cells to survive long periods of drought [32]. Plants, in addition to drought tolerance, tend to undergo a number of metabolic changes in response to drought stress, including decreased ribulose bisphosphate (RuBP) and adenosine triphosphate (ATP) levels, as well as reduced Rubisco activity. Plants reduce substomatal CO_2 conductivity and close stomata to avoid water loss up to their maximum capacity during drought. Water stress reduces the light saturation rate, decarboxylation velocity, ribulose 1,5-bisphosphate regeneration ability, photosystem II (PS-II) efficiency, and stomatal conductance in plants [33]. Drought has a negative impact on amino acids in plants, such as asparagine and glutamic acid, although plants can respond by boosting amino acids and soluble levels to temporarily relieve stress and manage osmotic potential [34].

6. Conclusions

Daily environmental fluctuations can have dramatic effects on plant vegetative growth at multiple levels, resulting in molecular, cellular, physiological and morphological changes. Environmental stress factors such as drought, elevated temperature, salinity and rising CO_2 affect to sustainable agriculture. Plants are even more sensitive to environmental changes during reproductive stages. Changing climate condition imposes different abiotic stresses to plant growth and development. This chapter will provide how different stress condition affects the plant growth and development and how they acclimatised with changing environment.

With this conclusion, there are some future scopes of research. Changing climate and Yield reduction in the late sown crop may be minimised with appropriate hormonal and nutritional interventions. Hormonal boosting may invigorate the plants under stress [35]. The stimulatory effect of hormones does not sustain for a long period when applied externally. Hormone based stimulatory physiology may be enhanced internally through nutritional treatment with boron and zinc which either enhances auxin stimulation or auxin biosynthesis. Smooth and healthy reproductive development is a precondition for the realisation of the yield potential of the crop.

Author details

Ananya Baidya[1*], Mohammed Anwar Ali[2] and Kousik Atta[1]

1 Plant Physiology, Bidhan Chandra Krishi Viswavidyalaya, India

2 Crop Physiology, Agricultural College Bapatla, Angrau, India

*Address all correspondence to: theananya4@gmail.com

References

[1] Hasanuzzaman M, Nahar K, Md MA, Chowdhury RR, Fujita M. Physiological, biochemical, and molecular mechanisms of heat stress tolerance in plants. International Journal of Molecular Sciences. 2013;**14**:9643-9684

[2] Hasanuzzaman M, Hossain MA, da Silva JAT, Fujita M. Plant responses and tolerance to abiotic oxidative stress: Antioxidant defenses is a key factor. In: Bandi V, Shanker AK, Shanker C, Mandapaka M, editors. Crop Stress and Its Management: Perspectives and Strategies. Berlin, Germany: Springer; 2012. pp. 261-316

[3] Singh R, Singh S, Parihar P, Mishra RK, Tripathi DK, Singh VP, et al. Reactive oxygen species (ROS): Beneficial companions of plants' developmental processes. Frontiers in Plant Science. 2016;**7**:1299

[4] Krasensky J, Jonak C. Drought, salt and temperature stress-induced metabolic rearrangements and regulatory networks. Journal of Experimental Botany. 2012;**63**:1593-1608

[5] Ruelland E, Vaultier MN, Zachowski A, Hurry V. Cold signalling and cold acclimation in plants. Advances in Botanical Research. 2009;**49**:35-150

[6] Zinn KE, Tunc-Ozdemir M, Harper JF. Temperature stress and plant sexual reproduction: Uncovering the weakest links. Journal of Experimental Botany. 2010;**61**:1959-1968

[7] Thines B, Harmon FG. Ambient temperature response establishes ELF3 as a required component of the core Arabidopsis circadian clock. Proceedings. National Academy of Sciences. United States of America. 2010;**107**:3257-3262

[8] Moury B, Selassie KG, Marchoux G, Daubeze A, Palloix A. High temperature effects on hypersensitive resistance to Tomato spotted wilt to spovirus (TSWV) in pepper (*Capsicum chinense* Jacq.). European Journal of Plant Pathology. 1998;**104**:489-498

[9] Tiwari S, Patel A, Singh M, Prasad SM. Regulation of temperature stress in plants. Plant life under changing environment. 2020

[10] Sharma RC, Duveiller E, Ortiz-Ferrara G. Progress and challenge towards reducing wheat spot blotch threat in the eastern Gangetic Plains of South Asia: Is climate change already taking its toll? Field Crops Research. 2007;**103**:109-118

[11] Kim DH, Doyle MR, Sung S, Amasino RM. Vernalization: Winter and the timing of flowering in plants. Annual Review of Cell and Developmental Biology. 2009;**25**:277-299

[12] Thomashow MF. Plant cold acclimation, freezing tolerance genes and regulatory mechanisms. Annual Review of Plant Physiology and Plant Molecular Biology. 1999;**50**:571-599

[13] Korn M, Gartner T, Erban A, Kopka J, Selbig J, Hincha DK. Predicting Arabidopsis freezing tolerance and heterosis in freezing tolerance from metabolite composition. Molecular Plant. 2010;**3**:224-235

[14] Srivastava S, Pathak AD, Gupta PS, Shrivastava AK, Srivastava AK. Hydrogen peroxide-scavenging enzymes impart tolerance to high temperature induced oxidative stress in sugarcane. Journal of Environmental Biology. 2012;**33**:657-661

[15] Baxter A, Mittler R, Suzuki N. ROS as key players in plant stress signalling. Journal of Experimental Botany. 2014;**65**:1229-1240

[16] Suzuki N, Koussevitzky S, Mittler R, Miller G. ROS and redox signalling in the response of plants to abiotic stress. Plant, Cell & Environment. 2012;**35**:259-270

[17] Gill SS, Tuteja N. Reactive oxygen species and antioxidant machinery in abiotic stress tolerance in crop plants. Plant Physiology and Biochemistry. 2010;**48**:909-930

[18] Qu AL, Ding YF, Jiang Q, Zhu C. Molecular mechanisms of the plant heat stress response. Biochemical and Biophysical Research Communications. 2013;**432**:203-207

[19] Liang W, Ma X, Wan P, Liu L. Plant salt-tolerance mechanism: A review. Biochemical and Biophysical Research Communications. 2018;**495**:286-291

[20] Phogat V, Pitt T, Cox JW, Simunek J, Skewes MA. Soil water and salinity dynamics under sprinkler irrigated almond exposed to a varied salinity stress at different growth stages. Agricultural Water Management. 2018;**201**:70-82

[21] Di Gioia F, Rosskopf EN, Leonardi C, Giuffrida F. Effects of application timing of saline irrigation water on broccoli production and quality. Agricultural Water Management. 2018;**203**:97-104

[22] Harmon JP, Daigh ALM. Attempting to predict the plant-mediated trophic effects of soil salinity: A mechanistic approach to supplementing insufficient information. Food Website. 2017;**13**:67-79

[23] Hasegawa PM, Bressan RA, Zhu JK, Bohnert HJ. Plant cellular and molecular responses to high salinity. Annuals of Review Plant Physics. 2000;**51**:463-499

[24] Horie T, Kaneko T, Sugimoto G, Sasano S, Panda SK, Shibasaka M, et al. Mechanisms of water transport mediated by PIP aquaporins and their regulation via phosphorylation events under salinity stress in barley roots. Plant & Cell Physiology. 2011;**52**:663-675

[25] Rengasamy P. Soil processes affecting crop production in salt-affected soils. Functional Plant Biology. 2010;**37**:613

[26] Parida AK, Das AB. Salt tolerance and salinity effects on plants: A review. Ecotoxicology and Environmental Safety. 2005;**60**:324-349

[27] Deinlein U, Stephan AB, Horie T, Luo W, Xu G, Schroeder JI. Plant salt-tolerance mechanisms. Trends in Plant Science. 2014;**19**:371-379

[28] Wang SJ, Chen Q, Li Y, Zhuo YQ, Xu LZ. Research on saline-alkali soil amelioration with FGD gypsum. Resources, Conservation and Recycling. 2017;**121**:82-92

[29] Atta K, Pal AK, Jana K. Effects of salinity, drought and heavy metal stress during seed germination stage in ricebean [*Vigna umbellata* (Thunb.) Ohwi and Ohashi]. Plant Physiology Reports. 2021;**26**(1):109-115

[30] Atta K, Chettri P, Pal AK. Physiological and biochemical changes under salinity and drought stress in Ricebean [*Vigna umbellata* (Thunb.) Ohwi and Ohashi] seedlings. International Journal of Environment and Climate Change. 2020;**10**(8):58-64. DOI: 10.9734/ijecc/2020/v10i830218

[31] Fathi A, Tari DB. Effect of drought stress and its mechanism in plants.

International Journal of Life Science. 2016;**10**:1

[32] Basu S, Ramegowda V, Kumar A, Pereira A. Plant adaptation to drought stress. F1000Research. 2016;**5**:1554

[33] Flexas J, Bota J, Loreto F, Cornic G, Sharkey TD. Diffusive and metabolic limitations to photosynthesis under drought and salinity in C_3 plants. Plant Biology. 2004;**6**:269-279

[34] Pinheiro C, Chaves MM. Photosynthesis and drought: Can we make metabolic connections from available data? Journal of Experimental Botany. 2010;**62**:869-882

[35] Waraich EA, Ahmad R, Ashraf MY, Saifullah, and Ahmad, M. Improving agricultural water use efficiency by nutrient management in crop plants. Acta Agriculturae Scandinavica. 2011;**61**:291-304